2006 第一届结构工程新进展国际论坛文集
Proceedings of the First International Forum on Advances in Structural Engineering—2006

新型结构材料与体系
Emerging Structural Materials and Systems

中国建筑工业出版社
CHINA ARCHITECTURE & BUILDING PRESS

图书在版编目（CIP）数据

新型结构材料与体系/第一届结构工程新进展国际论坛文集编委会编．—北京：中国建筑工业出版社，2006
（2006第一届结构工程新进展国际论坛文集）
ISBN 978-7-112-08690-0

Ⅰ．新… Ⅱ．第… Ⅲ．结构材料—国际学术会议—文集 Ⅳ．TB3-53

中国版本图书馆CIP数据核字（2006）第126821号

责任编辑：赵梦梅 刘婷婷
责任设计：肖广慧
责任校对：汤小平

2006第一届结构工程新进展国际论坛文集
Proceedings of the First International Forum on Advances in Structural Engineering—2006
新型结构材料与体系
Emerging Structural Materials and Systems
*
中国建筑工业出版社出版、发行（北京西郊百万庄）
新 华 书 店 经 销
北京蓝海印刷有限公司印刷
*
开本：787×1092毫米 1/16 印张：25¼ 字数：615千字
2006年11月第一版 2007年5月第二次印刷
印数：1501—3000册 定价：**58.00**元
ISBN 978-7-112-08690-0
（15354）

版权所有 翻印必究
如有印装质量问题，可寄本社退换
（邮政编码100037）
本社网址：http://www.cabp.com.cn
网上书店：http://www.china-building.com.cn

前　言 | Preface

当前，中国正在从事着几乎是世界上最大规模的工程建设，而这一建设高潮还会随着我国全面建设小康社会进程的推进而持续相当一段时间。持续的、大规模的工程建设，为结构工程技术的发展、学科的建设、人才的培养、队伍的壮健，特别是创新能力的提升，提供了历史性的机遇。中国土木工程（包括结构工程）事业取得的巨大成就引起了世人的关注。可以预期，在新的世纪里，结构工程事业必将得到更加迅猛的发展。

正是基于这种时代背景，在建设部的支持下，由中国建筑工业出版社、同济大学《建筑钢结构进展》编辑部、香港理工大学《结构工程进展》编委会联合主办、清华大学土木工程系承办的首届"结构工程新进展论坛"在北京举行。

本届论坛的主题为"新型结构材料与体系"。我们荣幸地邀请到 11 位特邀报告人，他们的报告主题涵盖了现代结构计算、分析、设计理论和施工技术、新型结构材料及应用、新型组合结构和混合结构设计理论与技术等领域。

我们感谢论坛特邀报告人，他们不仅在大会上做了精彩的大会发言，而且还奉献了精心准备的章节（论文），使得本书顺利出版。

感谢参加本次论坛的所有代表，正是大家的热心参与和帮助，才使得本次论坛能如期顺利举行。

感谢建设部、中国建筑工业出版社、同济大学《建筑钢结构进展》编辑部、香港理工大学《结构工程进展》编委会的支持。

最后，我们感谢论坛承办方清华大学土木工程系给予的帮助和支持。

目 录 | Contents

前言
Preface

第一章　Chapter 1
结构工程研究的若干新进展　　　　　　　　　　　　　袁　驷　韩林海　滕锦光　－2－
Some Recent Advances in Structural Engineering Research

第二章　Chapter 2
索穹顶结构的新形式及分析与施工中的新方法　　　　　　　　董石麟　袁行飞　－10－
New Forms and Methods for Analysis and Construction of Cable Domes

第三章　Chapter 3
索和膜结构在我国的应用、发展及存在问题　　　　　　　　　　　　张其林　－48－
Application, Development and Expectation of Cable and Membrane Structures in China

第四章　Chapter 4
新型结构材料的发展与应用　　　　　　　　　　　　　　　　　　　孙　伟　－80－
The Development and Application of New Structural Meterials

第五章　Chapter 5
The Development Trend of Advanced Building Materials　　　Z. J. Li　H. S. Chen　－120－
（新型建筑材料的发展趋势）

第六章　Chapter 6
高强高性能工程结构材料与现代工程结构及其设计理论的发展　　叶列平　陆新征 等　-208-
High Strength/Performance Structural Materials and the Developments of
Modern Engineering Structures and the Design Theory

第七章　Chapter 7
现代钢管混凝土结构研究的若干关键问题　　韩林海　陶　忠　-250-
Several Key Issues on Modern Concrete-Filled Steel Tubular Structures

第八章　Chapter 8
High Performance Steels and Their Use in Steel and Steel-Concrete Structures　　B. Uy　-296-
（高性能钢材及其在钢结构和钢—混凝土组合结构中的应用）

第九章　Chapter 9
ISIS Technologies for Civil Engineering Smart Infrastructure　　A. Mufti　C. Klowak　-314-
（ISIS 加拿大创新/可持续发展的土木工程技术）

第十章　Chapter 10
Structural Applications of FRP Composites in Construction　　J. G. TENG　-354-
（FRP 在结构工程中的应用）

首届"结构工程新进展国际论坛"简介　　-393-

首届论坛特邀报告人简介　　-396-

第一章 | Chapter 1

结构工程研究的若干新进展

袁　驷　清华大学土木工程系
韩林海　清华大学土木工程系
滕锦光　香港理工大学结构与土木工程系

摘　要：本章简要论述了结构工程学科在若干领域中的一些新进展，并对发展趋势和前景作了展望。
关键词：结构工程，进展，技术，展望

SOME RECENT ADVANCES IN STRUCTURAL ENGINEERING RESEARCH

S. yuan[1], L. H. Han[1] and J. G. Teng[2]

1. Department of Civil Engineering, Tsinghua University
2. Department of Civil and Structural Engineering, The Hong Kong Polytechnic University

Abstract: This paper provides a brief summary of some recent advances in structural engineering research. The future prospect and trend of the area are also discussed.
Keywords: Structural engineering, Research, Advances, Technology, Prospect

1. 引言

当前，我国正在从事着几乎是世界上最大规模的工程建设，而这一建设高潮还会随着我国全面建设小康社会进程的推进而持续相当一段时间。持续的、大规模的工程建设，为结构工程技术的发展、学科的建设、人才的培养、队伍的壮健，特别是创新能力的提升，提供了历史性的机遇。

在建设部的支持下，由中国建筑工业出版社、同济大学《建筑钢结构进展》编辑部、香港理工大学《结构工程进展》编委会联合主办、清华大学土木工程系承办的首届"结构工程新进展论坛"在北京举行。

为将论坛办成工程领域高水准的品牌论坛，论坛组织者为"结构工程新进展论坛"确立了以下几个方面的特色：

（1）论坛采取特邀报告人专题演讲的形式，其人选在由业内专家所组成的学术委员会

推荐的基础上产生。这一坚持学术立场、相对独立的人选推荐机制，促使专业人员对于分享特邀报告人在专业上的最新成就，将抱有极大的兴趣以及期待，从而扩大论坛影响力，在整体上保证了论坛的高水准。

（2）论坛在结构工程领域中遴选出四个明确的主题，以每年一个主题的形式轮流出现。这四个主题分别为：新型结构材料与体系；钢结构；混凝土结构；结构防灾、监测与控制。相对明确而集中的主题既可以突出体现近期该领域的重大科研成果，又能以四年为一个周期，很好地反映研究成果的更新及交替，与本论坛的主旨达到高度契合。

（3）每年一卷的"论坛文集"是本论坛区别于其他学术会议的根本，也是"论坛"的意义所在。文集不同于以往会议将大量投稿论文结集出版的情况，而是由每位特邀报告人对演讲内容进行深度展开或延伸，充分表达其学术成果，再加以整理出版。"论坛文集"不仅是论坛的重要标志，更重要的是，它将成为结构工程研究领域的风向标，能够深入、及时反映该领域最新的研究成果，为广大科研人员及研究生提供一个既有前瞻性又有可行性的、能引导研究方向的重要文献资料。同时，"论坛文集"也为工程师应用新技术提供宝贵的信息。"论坛文集"以丛书的形式连续出版，也将成为结构工程科研历程的全记录，对整个研究工作的沿革将起着承上启下的作用。

本届论坛的主题为"新型结构材料与体系"。应论坛主办者的邀请，本文作者代表承办方和组织者，围绕论坛的主题，合作贡献一篇综述性的发言报告。

有关结构工程最新进展的全面论述，超出了作者的知识和能力范围，且已有相应的文献可以参阅。如，有关我国2020年的工程科学技术的发展研究，可见工程技术发展研究综合专题组（2004），刘西拉等（2005）；有关土木工程学科及其各个领域的研究现状、进展和展望的全面论述，可见国家自然科学基金委员会工程与材料科学部学科发展战略研究报告（2006～2010年）（国家自然科学基金委员会工程与材料科学部，2006）。此外，本论坛的每位特邀报告人都会对相应的专门领域的进展给出较全面透彻的述评。

有鉴于此，作者根据"论坛"的特点，结合作者熟悉的领域，围绕本届论坛的主题，对结构材料和结构技术以及结构分析计算理论等方面的一些新进展作一简要论述，并对相关的结构工程技术的发展趋势作了展望。

2. 结构工程技术的一些新进展

2.1 新型结构材料应用空前活跃

结构材料是结构工程的基础，结构材料的发展和进步往往会使结构工程发生质的变化（刘西拉等，2005）。人类一直在寻求能大量使用的、高效能的新型建筑结构材料。随着可持续发展意识的不断普及，符合环保要求的新型建筑结构材料也日显重要。

目前，建筑结构工程中应用最为广泛的建筑材料仍然是工业革命后发展起来的混凝土和钢材。近年来，混凝土和钢材也正在逐渐向高性能化的方向发展，例如高强、高性能混凝土的应用正在逐渐增多（Li 和 Chen，2006；孙伟，2006），耐火、耐候钢的研发成功以及在一些实际工程中的应用。随着高性能混凝土和高性能钢材的出现，传统的结构设计理论和设计方法也需要不断完善和发展，以适应结构材料发展的要求。

FRP（Fiber Reinforced Polymer）复合材料是近年来在结构工程中开始广泛应用的一种新型结构材料（Teng等，2002；Teng，2006a，2006b；滕锦光和叶列平，2006）。FRP的应用面非常广，在结构加固中的应用已相当普及，并取得了良好的经济效果。在新建结构中，应用也正在逐渐增多。此外，不锈钢、铝合金等材料在建筑结构中的应用也有不少报道。

2.2 新型结构技术得到不断发展

组合结构和混合结构是目前结构工程领域研究和应用的热点话题之一。它们是工程技术进步和现代施工技术向工业化生产发展的必然产物。

组合结构一般是指由两种或两种以上材料在构件层次的组合。如常见的钢—混凝土组合板、组合梁，型钢混凝土（SRC），钢管混凝土和FRP（Fiber Reinforced Polymer）约束混凝土结构等。

组合结构的特点在于如何优化地组合不同材料，通过组成材料之间的相互作用，充分发挥组成材料的优点，尽可能避免或减少其弱点所带来的不利效应；而且，通过不同材料的组合，使施工过程比钢筋混凝土结构（广义地说，也是一种组合结构）更为便捷。此外，组成组合结构不同材料之间的相互贡献、协同互补和共同工作的优势，还使其具有较好的耐火性能及火灾后可修复性（李国强等，2006）。例如，钢管混凝土利用钢管和混凝土两种材料在受力过程中的相互作用，即钢管对其核心混凝土的约束作用，使混凝土处于复杂应力状态之下，从而使混凝土的强度得以提高，延性得到改善。同时，由于混凝土的存在，可以延缓或避免钢管过早地发生局部屈曲，从而可以保证其材料性能的充分发挥。此外，在钢管混凝土的施工过程中，钢管还可以作为浇筑其核心混凝土的模板，与钢筋混凝土相比，可节省模板费用，加快施工速度。总之，通过钢管和混凝土组合而成为钢管混凝土，不仅可以弥补两种材料各自的缺点，而且能够充分发挥二者的优点，这也正是钢管混凝土组合结构的优势所在（韩林海和陶忠，2006）。

混合结构一般是指由不同材料的结构构件组合而成的结构、结构体系，如常见的钢—混凝土混合结构剪力墙、钢或钢管混凝土框架—钢筋混凝土核心筒结构等。也就是说，混合结构是对结构构件在结构、结构体系层次进行组合。

混合结构工作的特点在于如何很好地发挥其组成结构之间的组合优势，达到取长补短和共同工作的优势（徐培福等，2005），不仅使结构构件的特性达到较为充分的发挥，同时也使结构体系具有优越的整体力学特性。

这些新型结构构件和结构体系的出现为现代结构工程技术的发展奠定了良好的基础。

2.3 结构工程科学研究手段不断提高

深入开展工程结构工作机理的研究，既是结构技术发展的需要，也是结构工程科学向更高层次发展的重要基础。

近年来结构理论分析和实验技术的发展，为深入开展结构工程科学的研究工作创造了条件。主要体现在：实验技术不断更新和完善，如利用互联网的远程试验技术（肖岩等，2006），采用光纤等先进传感器的结构实地实时监测（欧进萍，2006；Teng等，2003），利用激光技术（Teng 等，2001）和数字摄影技术（Maas 和 Hampel 2006）对结构变形的精确测量等；实验设备在不断更新，使更多的科研机构有能力进行大型复杂的实验研究，例如足尺实验、地震台、精细模型实验、火灾实验等。实验技术和设备的发展，不仅进一步提高人们对结构破坏全过程的认识，也使实验研究更接近于实际结构的工作状态。

结构分析的手段也在不断提高，例如大型有限元分析软件的不断完善，使一些复杂的结构、结构体系的非线性静、动力力学性能的精细分析成为可能。此外，结构计算与结构试验交互仿真理论与技术也在近期取得了实质性的进展。

科学研究手段的提高，促进了结构工程设计理论以及结构技术向更高层次的发展。

2.4　工程全"生命周期"结构设计理论及第三代结构技术

所谓工程的全"生命周期"是包括工程建造、使用和老化的全过程（刘西拉等，2005）。在这一过程中，工程结构的风险来源不完全相同。施工建造阶段的风险来自于对未完成结构和它的支撑系统缺乏可靠的分析，以及对人为错误的失控；老化阶段的风险是来自结构或材料功能在长期自然环境和使用环境中的逐渐退化或劣化。

以往的工程设计通常仅考虑在使用阶段工程的安全性，现在还需考虑安全以外的内容，如结构的使用功能能否得到保证、耐久性以及结构体系在极端条件下的抗倒塌性能等问题。因此，比较科学的做法是考虑全"生命周期"的综合决策。只有综合考虑这些因素，才可能做出更为合理的决策。

近年来结构技术创新的一个重要特征是吸收和融合其他学科的先进技术。这些技术包括材料、信息与通信、计算机仿真、传感器等。高性能结构材料的应用，设计理论和结构形式的革新，结构功能由被动承载向智能响应的转变，正在孕育着新的一代结构技术——第三代结构技术（Teng 等，2003）。

2.5　结构分析计算理论不断取得新进展

现代结构分析计算已不再简单地满足于能够对大型工程问题进行一般的分析计算，而是追求求解的质量以及对疑难问题的求解效力，故有两个主要的发展方向：一是加大求解问题的"量"，诸如各类大型结构分析软件在不断地增加求解功能，扩大求解问题的类型和覆盖面；二是提高求解问题的"质"，即从理论上突破，从而从本质上改变求解的质量。后一方向是学术研究的重点，下面简述几点进展。

随着近年来计算技术的发展，特别是常微分方程（ODE）求解器的出现，使得半解析的有限元线法（FEMOL）得以问世（Yuan，1993），将有限元法（FEM）提升为面向"常微"的半解析方法。FEMOL 对 FEM 的改进是优点，同时也是缺点；它提高了结线方向的求解精度，但仍摆脱不了 FEM 离散带来的离散方向的误差。近年来，受结构力学中矩阵位移法的启发，一个自然合理的 FEM 超收敛计算的单元能量投影（EEP）法（袁驷等，2004）得以提出，它不仅使一维 FEM 的超收敛计算结果（m 次单元）可以达到最佳的 2m 阶精度，而且顺利地应用于半解析的 FEMOL 中，一举克服了 FEMOL 的解答"一手硬、一手软"的弱点，使解答的整体质量有本质性的提高和改善。

结构分析计算理论中另一个新进展是杆系结构动力特性和弹性稳定分析的精确方法和算法。虽然基于动力刚度法的 Wittrick-Williams 算法能够给出精确的结构频率和失稳荷载，但是近30年来始终未能很好地解决振型计算的问题。新近提出的导护型 Newton 法可以同时计算精确的振型和频率，而且效率很高，比商业软件 ANSYS 有过之无不及（Yuan 等，2003；袁驷等，2005）。

3. 结语

国家自然科学基金委员会最近出版发行了结构工程学科学科发展战略研究报告（2006~2010 年）（国家自然科学基金委员会工程与材料科学部，2006）。该报告较为全面地阐述和展望了结构工程科学发展的过去、现在和未来。

本文围绕本次论坛所重点关注的问题，结合作者所熟悉的领域，简要论述了结构工程学科的一些新进展，远非全面完整，意在抛砖引玉。

可以预期，在新的世纪里，结构工程新材料、结构分析理论和结构工程新技术必将得到更加迅猛的发展，使第三代的结构技术得以全面建立并广泛应用。

参考文献

[1] 工程技术发展研究综合专题组. 2020 年的中国工程科学技术发展研究（周光召主编）. 北京：中国科学技术出版社，2004，上卷，474-558

[2] 国家自然科学基金委员会工程与材料科学部. 学科发展战略研究报告（2006 年~2010 年）. 建筑、环境与土木工程Ⅱ（土木工程卷）. 北京：科学出版社，2006

[3] 韩林海，陶忠. 现代钢管混凝土结构研究的若干关键问题. 2006 第一届结构工程新进展国际论坛文集—新型结构材料与体系. 北京：中国建筑工业出版社，2006：250-294

[4] 李国强，韩林海，楼国彪，蒋首超. 钢结构及钢—混凝土组合结构抗火设计，北京：中国建筑工业出版社，2006

[5] Li, Z. J. and Chen, H. S. (2006). The development trend of advanced building materials. 2006 第一届结构工程新进展国际论坛文集—新型结构材料与体系. 北京：中国建筑工业出版社，2006：120-206

[6] 刘西拉，袁驷，宋二祥. 关于我国工程建设技术发展的战略思考. 土木工程学报，2005，38（12）：1-7

[7] Maas, H. G. and Hampel, U. Photogrammetric techniques in civil engineering material testing and structure monitoring, Photogrammetric Engineering and Remote Sensing 2006, 72 (1): 39-45

[8] 欧进萍. 重大工程结构损伤积累、健康监测与安全评定. 国家自然科学基金委员会工程与材料科学部，学科发展战略研究报告（2006 年~2010 年），建筑、环境与土木工程Ⅱ（土木工程卷）. 北京：科学出版社，2006：139-160

[9] 孙伟. 新型结构材料的发展与应用. 2006 第一届结构工程新进展国际论坛文集—新型结构材料与体系. 北京：中国建筑工业出版社，2006：80-118

[10] Teng, J. G. Structural applications of FRP composites in construction. 2006 第一届结构工程新进展国际论坛文集—新型结构材料与体系. 北京：中国建筑工业出版社，2006a：354-392

[11] Teng, J. G. Fibre-reinforced polymer composites in construction: current research and future challenges, Proceedings, Edited by L. H. Han, J. P. Ru and Z. Tao, Ninth International Symposium on Structural Engineering for Young Experts, August 18-21 2006, Fuzhou & Xiamen, China, Science Press, 2006b: 31-

41（Keynote Paper）

[12] Teng, J. G., Chen, J. F., Smith, S. T. and Lam, L. FRP Strengthened RC Structures. John Wiley & Sons Ltd, 2002

[13] Teng, J. G., Ko, J. M., Chan, T. H. T., Ni, Y. Q., Xu, Y. L., Chan, S. L., Chau, K. T. and Yin, J. H. Third-generation structures: intelligent high-performance structures for sustainable urban systems, Proceedings, International Symposium on Diagnosis, Treatment and Regeneration for Sustainable Urban Systems, 13-14 March, Japan, 2003: 41-55（Invited Paper）

[14] 滕锦光，叶列平．新型复合材料与结构．国家自然科学基金委员会工程与材料科学部，学科发展战略研究报告（2006年~2010年），建筑、环境与土木工程II（土木工程卷）．北京：科学出版社，2006：177-192

[15] Teng, J. G., Zhao, Y. and Lam, L. Techniques for buckling experiments on steel silo transition junctions, Thin-Walled Structures, 2001, 39 (8): 685-707

[16] 肖岩，易伟建．结构试验学研究的过去、现在和展望．国家自然科学基金委员会工程与材料科学部，学科发展战略研究报告（2006年~2010年），建筑、环境与土木工程II（土木工程卷）．北京：科学出版社，2006：126-138

[17] 徐培福，傅学怡，王翠坤和肖丛真．复杂高层建筑结构设计．北京：中国建筑工业出版社，2005

[18] Yuan S. The Finite Element Method of Lines, Science Press, Beijing-New York, 1993

[19] 袁驷，王枚．一维有限元后处理超收敛解答计算的EEP法，工程力学，2004，21（2）：1-9

[20] Yuan S., Ye, K. S., Williams, F., Kennedy, D. Recursive second order convergence method for natural frequencies and modes when using dynamic stiffness matrices, Int. J. Numer. Meth. Engng, 2003, 56: 1795-1814

[21] 袁驷，叶康生，Williams, F., Kennedy, D. 杆系结构自由振动精确求解的理论和算法，工程力学（增刊），第十四届全国结构工程学术会议论文集（特邀报告），2005，22：1-6

第二章 | Chapter 2

索穹顶结构的新形式及分析与施工中的新方法

董石麟　袁行飞

浙江大学空间结构研究中心，浙江杭州玉泉，310027　E-mail：yuanxf@zju.edu.cn

摘　要：论文在传统索穹顶结构基础上提出了几种新型索穹顶结构形式，并对分析与施工中的新方法如整体可行预应力确定法、节点平衡法、索杆膜协同分析、施工分析等作了介绍。论文最后还对该类结构的模型试验和刚性屋面应用等作了分析，得出了一些可供工程应用参考的结论。

关键词：索穹顶，新形式，整体可行预应力，节点平衡法，协同分析，施工分析

NEW FORMS AND METHODS FOR ANALYSIS AND CONSTRUCTION OF CABLE DOMES

S. L. Dong　X. F. Yuan

Space Structures Research Center, Zhejiang University, Hangzhou, 310027, P. R. C

Abstract：Cable domes have been employed as lightweight, large span roofs. In this paper, several new forms of cable domes are first presented. Then methods for analysis of cable domes including global prestress determination, nodal equilibrium, cooperative work of cable, strut and membrane, and new construction methods are proposed. Experimental research and application of rigid covering material are finally introduced. Some conclusions are drawn to provide reference for design of cable domes.

Keywords：cable dome, new form, global prestress, nodal equilibrium, cooperative work, construction analysis

1. 传统索穹顶结构发展及特点

索穹顶结构是一种支承于周边受压环梁上的张力集成体系或全张力体系。由于其外形类似于一个穹顶，且主要构件又是钢索，因此被命名为索穹顶（Cable Dome）。

索穹顶结构是由美国工程师盖格尔（Geiger）根据富勒（Fuller）的张拉整体结构思想开发的。早在20世纪40年代，Fuller就认为宇宙的运行是按照张拉整体的原理进行的，由此他设想真正高效的结构体系应该是压力与拉力的自平衡体系。1948年，他的学生——雕塑家Snelson完成了第一个张拉整体模型见图1（a），并以此为基础设计了城市雕塑见

图 1 (b)。这一事件证实了富勒的设想，并被公认为现代张拉整体结构发展的一个起点。富勒由此受到更大的鼓励与启发，于 20 世纪 60 年代初发表了张拉集成体系的概念和初步理论。他在 1962 的专利中较详细地描述了他的结构思想：即在结构中尽可能地减少受压状态而使结构处于连续的张拉状态，从而实现他的"压杆的孤岛存在于拉杆的海洋中"的设想，并第一次提出了 Tensegrity 这一概念。继富勒的"张拉整体结构"专利后，法国的 Emmerich 于 1963 年提出了"构造的自应力索网格"专利，美国的 Snelson 于 1965 年提出了"连续拉、间断压"的专利。这些研究进一步推动了张拉整体结构的发展。

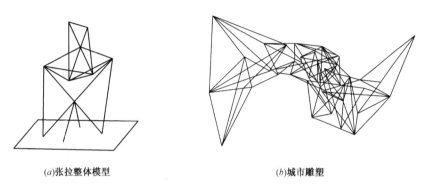

(a)张拉整体模型 (b)城市雕塑

图 1　Snelson 的张拉整体模型和城市雕塑

自从张拉整体概念提出以来，各国学者（Emmerich、Vilnay、Pugh、Motro、Hanaor 等）对各种形式的张拉整体结构进行了研究，但很长时间来这种结构除了艺术雕塑方面的应用和模型实验研究外，没有功能性建筑出现。

1986 年，美国著名工程师盖格尔首次根据富勒的张拉整体结构思想，发明了支承于周边受压环梁上的一种索杆预应力张拉整体穹顶即索穹顶结构，并把它成功地应用于汉城奥运会的体操馆（Gymnastic Arena，圆平面 $D=119.8$m，见图 2）和击剑馆（Fencing Arena，圆平面 $D=89.9$m）。Geiger 设计的索穹顶是由连续的张力索和不连续的受压脆杆构成，荷载从中央的张力环通过一系列辐射状的脊索、环索和中间的斜索传递至周边的压力环。为了减少屋盖表面薄膜材料的费用，在设计时使结构曲面最小；为了增加整体刚度，在脊索

(a)外景　　　　　　　　　　　　　　(b)内景

图 2　汉城体操馆

之间增加了谷索。Geiger 还发现这类结构的重量随跨度的增加并不显著增加，且造价增加也很少，这使得索穹顶结构具有很好的经济性。继汉城体操馆和击剑馆之后，盖格尔和他的公司又相继建成了美国伊利诺伊州大学的红鸟体育馆（Redbird Arena，椭圆平面 76.8m×91.4m）和佛罗里达州的太阳海岸穹顶（Sun Coast Dome，圆平面 D=210m），这些工程的建成充分显示了其广阔的应用前景。

由美国工程师 M. P. Levy 和 T. F. Jing 设计的佐治亚穹顶（Georgia Dome）是 1996 年亚特兰大奥运会主要比赛场馆的屋盖结构（见图 3），它是在 Geiger 穹顶基础上进一步发展后实现的。这个被命名为双曲抛物型"Hyper Tensegrity Dome"的索穹顶为 790ft×630ft（240.79m×192.02m）的椭圆形平面，由联方型索网、三道环索、桅杆及中央桁架组成。整个结构只有 156 个节点，分别在 78 根桅杆的两端。屋盖周边由四个弧段组成，端部弧段及中部弧段的半径不等，结构的节点均采用焊接节点。据有关报道这个结构的耗钢量不到 $30kg/m^2$。继亚特兰大体育馆后，他们又成功设计了圣彼得堡的雷声穹顶（Thunder Dome，D=210m）等体育馆。此外，另一类穹顶的建成更进一步显示了这一结构形式的应用前景，它就是 M. P. Levy 和 T. F. Jing 在沙特阿拉伯利雅德大学体育馆中实现的可开启索穹顶。

(a)外景　　　　　　　　　　　　　　　(b)内景

图 3　亚特兰大佐治亚穹顶

传统的索穹顶结构形式主要有肋环形和葵花形两种，分别由 Geiger 和 Levy 设计并应用到工程中，因此这两种形式又分别被命名为 Geiger 型和 Levy 型。

Geiger 型索穹顶的代表工程为图 2 所示的汉城体操馆穹顶，由图 4 所示结构形式简图可见 Geiger 型索穹顶是由中心受拉环、径向布置的脊索、斜索、压杆和环索组成，并支承于周边受压环梁上。由于它的几何形状类似于平面桁架系结构，而桁架系结构平面外刚度较小，所以在不对称荷载作用下容易出现失稳。

Levy 型索穹顶的代表工程为图 3 所示的佐治亚穹顶，其结构形式简图见图 5。它将辐射状布置的脊索改为葵花形（三角化型）布置，使屋面膜单元呈菱形的双曲抛物面形状，较好地解决了 Geiger 型穹顶存在的索网平面内刚度不足容易失稳的问题。

综上所述，索穹顶是一种受力合理、结构效率高的结构体系，它由连续的拉索和不连续的压杆组成，完全体现了 Fuller 关于"压杆的孤岛存在于拉杆的海洋中"的思想。其主要特点如下：

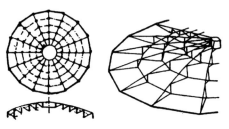
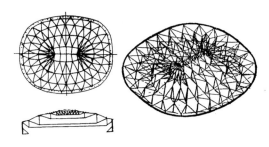

图4 Geiger 设计的汉城体操馆穹顶　　　　图5 Levy 设计的佐治亚穹顶

（1）全张力状态。张拉整体索穹顶结构由连续的拉索和不连续的压杆组成，连续的拉索构成了张力的海洋，使整个结构处于连续的张力状态，即全张力态。

（2）预应力提供刚度。索穹顶结构中的索在未施加预应力前是几乎没有自然刚度的，它的刚度完全由预应力提供。索穹顶结构的刚度与预应力的分布和大小有密切关系。

（3）力学性能与形状有关。索穹顶结构的工作机理和力学性能依赖于其自身的拓扑形状。只有合理的结构形态，才能有良好的工作性能。

（4）力学性能与施工方法有关。索穹顶结构的力学性能很大程度上取决于预应力状态，而预应力的形成又与施工过程有直接关系，所以选择合理、有效的施工方法是实现结构良好力学性能的保证。

（5）自平衡体系。无论在成形态还是受荷态，它都是压力和拉力的有效自平衡体系，在周边支承柱处可不产生水平反力。

2. 新型索穹顶结构的形式及特点

在综合考虑结构构造、几何拓扑和受力机理的基础上提出了几种新型索穹顶结构形式：Kiewitt 型穹顶（图6）、混合型穹顶（图7，8）和鸟巢形穹顶（图9）。其中混合 I 型（图7）为肋环形和葵花形的重叠式组合，混合 II 型（图8）为 Kiewitt 型和葵花形的内外式组合。

与传统索穹顶结构相比，新型索穹顶结构具有如下特点：

（1）脊索布置新颖，网格划分均匀；

（2）刚度分布均匀，降低预应力水平；

（3）节点构造简单，施工操作方便；

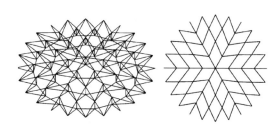
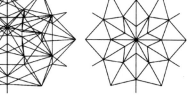

图6 Kiewitt 型穹顶　　　　　　　图7 混合 I 型穹顶
　　　　　　　　　　　　　　（肋环形和葵花形的重叠式组合）

(4) 使柔性薄膜和刚性屋面的铺设更为简便可行；

(5) 鸟巢型穹顶的脊索沿内环切向布置，连接两边界的脊索贯通，可省去内上环索。

这些新型索穹顶形式的提出大大丰富了现有索穹顶结构的形式，使这一结构更具生命力。

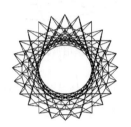
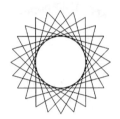

图 8　混合 II 型穹顶　　　　　　　　图 9　鸟巢形穹顶
（Kiewitt 型和葵花形的内外式组合）

3. 索穹顶结构整体可行预应力分析法

整体可行预应力分析法是充分考虑索穹顶结构的对称性，利用平衡矩阵奇异值分解法提出的一种适用于求解各种索穹顶结构的整体自应力模态的方法，由于该方法求解过程中二次用到了奇异值分解，故又命名为二次奇异值法。

对于给定的空间铰接结构体系，设杆件数为 b，非约束节点数为 N，排除约束节点中某些自由度不被约束的情况，则非约束位移数（自由度）为 $3N$。该结构体系的平衡方程如下：

$$At = f \tag{1}$$

其中 A 为 $3N \times b$ 矩阵，称为平衡矩阵；t 为 b 维杆件内力矢量；f 为 $3N$ 维节点力矢量。设 A 的秩为 r，可得自应力模态数 $s = b - r$，独立机构位移数 $m = 3N - r$。

对平衡矩阵 A 进行按比例选列主元高斯消元法消元或奇异值分解可得 s 种自应力模态。一般的预应力状态是各单位自应力模态的线性组合，记为：

$$T\alpha = T_1\alpha_1 + T_2\alpha_2 + L + T_s\alpha_s \tag{2}$$

式中 α 为自应力模态组合因子，可取任意实数。

索穹顶结构是索杆组合的空间预应力体系，按上述步骤可求得独立机构位移模态和单位自应力模态，不同的是其中的索是一种单向约束构件，只能承受拉力。杆虽为双向约束构件，但由于张拉整体结构特有的"压杆的孤岛存在于拉杆的海洋中"的构造思想，只能承受压力。这种杆受压、索受拉的预应力状态称为可行预应力状态。考虑到索穹顶结构的对称性，特提出结构整体可行预应力状态，该状态除了满足杆受压、索受拉条件外，还具有同类（组）杆件初始内力相等和整体自应力平衡等特点，这种预应力状态能使索穹顶结构最终达到理想设计状态。

索穹顶结构是一种杆件拓扑关系较有规律的对称结构体系，因此结构中的索和杆内力分布具有一定的规律性，具体来说即对一实际的索穹顶结构，位于等同地位（位置）的杆件属于同一类（组）杆件，其初始内力值也应该是相同的。如图 10 所示结构，尽管总杆件数 b

为49，但相应的杆件类只有7类，分别为①第一道上斜索、②第二道上斜索、③第一道下斜索、④第一道竖杆、⑤第一道环索、⑥第二道下斜索和⑦中心竖杆，因此结构对应的初始预应力值也只有不同的7组。

对索穹顶结构，先从一般预应力状态 $X = T_1\alpha_1 + T_2\alpha_2 + L + T_s\alpha_s$ 出发，找到一组 α，使同组杆件预应力值相同，设该预应力为 X，有：

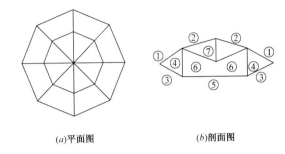

(a)平面图　　(b)剖面图

图10　肋环形索穹顶

$$T_1\alpha_1 + T_2\alpha_2 + L + T_s\alpha_s = X \quad (3)$$

对于具有 n 组杆件数的结构，X 可记为：

$$X = \{x_1 \quad x_1 \quad x_1 \quad L \quad x_i \quad x_i \quad x_i \quad L \quad x_n \quad L \quad x_n\}^T$$

为更好地用矩阵表示，整理（3）式如下：

$$T_1\alpha_1 + T_2\alpha_2 + L + T_s\alpha_s - X = 0 \quad (4)$$

简记为：

$$\tilde{T}\tilde{\alpha} = 0 \quad (5)$$

式中，$\tilde{T} = [T_1 \quad T_2 \quad L \quad T_s \quad -e_1 \quad -e_2 \quad L \quad -e_n]$，$T_i$ 为单位独立自应力模态；基向量 e_i 由相应第 i 类杆件轴力为 -1（索力为 $+1$）、其余杆件轴力为 0 组成，即 $e_i = \{0 \quad L \quad 0 \quad 1 \quad 1 \quad L \quad 0 \quad 0\}^T$，未知数为 $\tilde{\alpha} = [\alpha_1 \quad \alpha_2 \quad L \quad \alpha_s \quad x_1 \quad x_2 \quad L \quad x_n]^T$。对 $\tilde{T}$ 进行奇异值分解如下：

$$\tilde{T} = UDV^T \quad (6)$$

设矩阵 $\tilde{T}$ 的秩为 r'，则整体自应力模态数 $\bar{s} = s + n - r'$，V 中第 $r'+1$ 列至第 $s+n$ 列为 $\tilde{\alpha}$ 的解，即 $\tilde{\alpha} = [v_{r'+1} \quad L \quad v_{s+n}]$，由 $\tilde{\alpha}$ 中第 $s+1$ 行到第 $s+n$ 行可得 n 组杆件对应的预应力值。

对肋环形索穹顶和葵花形索穹顶，$\tilde{T}$ 为 $b \times (s+n)$ 维矩阵，其秩 r 为 $(s+n-1)$，可得 $\bar{s} = s + n - r' = 1$，即为一种预应力分布，该分布同时满足杆受压、索受拉条件，所以是一种整体可行预应力分布。对其他类型如 Kiewitt 型等索穹顶结构，满足同组杆件预应力值相同的解大于 1，设分别为 X_1, X_2, L, X_w，$w > 1$。此时可再根据杆受压、索受拉条件对求得的若干组预应力向量进行组合 $X_1\beta_1 + X_2\beta_2 + L + X_w\beta_w$，从而得到整体可行预应力分布。

值得特别指出的是在用整体可行预应力一般概念进行预应力设计时，杆件的正确分组是能否求得满足整体平衡的预应力分布的关键。若杆件分组与实际受力情况不符，则按该分组计算得到的预应力不能使结构各节点受力平衡。

［算例1］如图10所示一Geiger型索穹顶，设置一道环索，杆件数 $b = 49$，其中压杆数 9，拉杆数 40，总节点数 $N = 26$，其中非约束节点数 18，非约束自由度数为 54，杆件类型数 $n = 7$，求得 $r(A) = 43$，独立自应力模态数 $s = 6$，独立的位移机构模态数 $m = 11$。考虑对称性后，利用奇异值分解可得到 $r'(\tilde{T}) = 12$，故整体自应力模态数 $\bar{s} = 6 + 7 - 12 = 1$，考察各元素 x_i 均满足索受拉杆受压，故该模态同时为整体可行预应力模态，该模态分布如表1所示：

Geiger 型索穹顶整体自应力模态分布　　　　表1

组数编号	1	2	3	4	5	6	7
整体自应力模态	1.0	0.624	6.519	−0.456	8.497	0.289	−0.757

［算例2］如图11所示一 Levy 型索穹顶，设置三道环索，杆件数 $b=161$，其中压杆数25，拉杆数136，总节点数 $N=58$，其中非约束节点数50，非约束自由度数为150，杆件类型数 $n=15$，求得 $r(A)=150$，独立自应力模态数 $s=11$，独立的位移机构模态数 $m=0$。考虑对称性后，利用奇异值分解可得到 $r'(\overline{T})=25$，独立的整体自应力模态数 $\bar{s}=11+15-25=1$，考察 $\{v_1\}$ 中各元素 x_i 均大于零，故该整体自应力模态同时为整体可行预应力模态，该模态分布如表2所示：

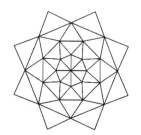
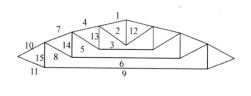

图11　三道环索 Levy 型索穹顶（不设内环）

Levy 型索穹顶整体自应力模态分布　　　　表2

组数编号	1	2	3	4	5	6	7	8
整体自应力模态	0.0076	0.0076	0.0199	0.0103	0.0103	0.0504	0.0351	0.0351
组数编号	9	10	11	12	13	14	15	
整体自应力模态	0.1506	0.1424	0.1424	−0.0029	−0.0026	−0.0124	−0.0593	

［算例3］如图12所示一 Kiewitt 型索穹顶，设置两道环索，杆件数 $b=145$，其中压杆数

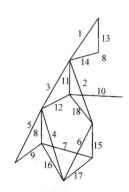

(a)三维图　　　　(b)杆件分组编号

图12　kiewitt 型索穹顶

19，拉杆数126，总节点数56，其中非约束节点数38，非约束自由度数为114，杆件组数 $n=18$，分组编号见图33-26b。求得 $r(A)=114$，独立自应力模态数 $s=31$，独立的位移机构模态数 $m=0$。考虑对称性后，利用奇异值分解可得 $r'(\overline{T})=45$，$\overline{s}=18+31-45=4$，即考虑对称性后可得到4组整体自应力模态，各组整体自应力模态值大小分别见表3。

4组整体自应力模态值　　　　　　　　　　　　　　表3

自应力模态＼杆件组号	1	2	3	4	5	6
第1组	0.00537	0.13391	-0.15171	-0.02416	-0.06294	0.12782
第2组	0.10320	-0.05343	0.18287	0.07584	0.09113	-0.02037
第3组	-0.02727	-0.01827	-0.01012	0.14128	-0.21793	-0.01434
第4组	0.00129	0.00163	-0.00042	-0.12822	0.16055	0.00203
自应力模态＼杆件组号	7	8	9	10	11	12
第1组	0.13804	0.02837	0.04655	0.04028	0.01633	0.06943
第2组	0.17280	0.03562	0.05695	0.03260	0.02412	-0.00840
第3组	0.00423	-0.01171	0.15806	-0.02191	-0.00995	-0.03145
第4组	0.00525	-0.01473	0.19870	0.00101	0.00055	0.00087
自应力模态＼杆件组号	13	14	15	16	17	18
第1组	0.00271	0.00082	0.02731	0.02391	0.04342	-0.02003
第2组	0.05220	0.01580	0.03419	0.03099	0.05435	0.03711
第3组	-0.01380	-0.00418	0.00084	-0.12622	0.00133	0.00568
第4组	0.00065	0.00020	0.00104	-0.15870	0.00165	0.00021

4. 索穹顶结构预应力模态的节点平衡法

4.1 肋环形索穹顶结构初始预应力分布的快速计算法

考虑到肋环形索穹顶为一轴对称结构，它的计算模型可取一榀平面径向桁架。针对不设内拉环和设有内拉环两种情况，分别有计算简图13和图14：

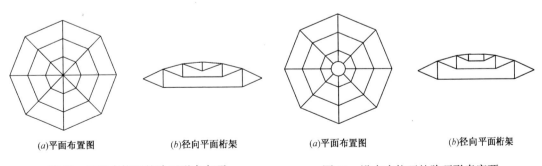

(a)平面布置图　　(b)径向平面桁架　　　　(a)平面布置图　　(b)径向平面桁架

图13　不设内拉环的肋环形索穹顶　　　图14　设有内拉环的肋环形索穹顶

其中图13（b）径向平面桁架中的中心竖线为等效竖杆，等效竖杆内力 $V_{0,equ}$ 与结构中心竖杆实际内力 V_0 的关系为

$$V_{0,equ} = \frac{2}{n}V_0 \qquad (7)$$

图13（b）和14（b）径向平面桁架中的水平线为等效环索，等效环索内力 $H_{i,equ}$ 与结构环索实际内力 H_i 的关系由图15可得

$$H_{i,equ} = 2H_i\cos\phi_n = 2H_i\cos\left(\frac{\pi}{2} - \frac{\pi}{n}\right) = 2H_i\sin\frac{\pi}{n} \qquad (8)$$

式（7）、（8）中 n 为结构平面环向等分数。

分别以图13（b）和图14（b）所示简化平面桁架为基础，对各节点建立平衡关系，可推导各类杆件内力计算公式如下。

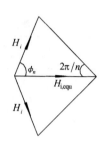

图15 环索内力示意图

（1）不设内拉环的情况

由平面桁架的对称性并引入边界约束条件（包括对称面的对称条件），可进一步简化为图16所示的半榀平面桁架，由机构分析可知该结构为一次超静定结构。由图17所示各类杆件内力示意图，可得以中心竖杆内力 V_0 为基准的各脊索、压杆、斜索和环索内力计算公式：

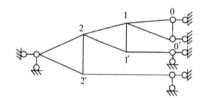

图16 简化半榀平面桁架

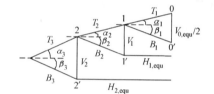

图17 各类杆件内力示意图

当 $i = 1$ 时，$T_1 = -\dfrac{1}{n\sin\alpha_1}V_0$，$B_1 = \dfrac{1}{n\sin\beta_1}V_0$ \qquad (9)

当 $i \geq 2$ 时，

$$\left.\begin{aligned}
T_i &= \frac{(\cot\alpha_1 + \cot\beta_1)(1 + \tan\alpha_2\cot\beta_2) L (1 + \tan\alpha_{i-1}\cot\beta_{i-1})}{n\cos\alpha_i}(-V_0) \\
B_i &= T_i\sin\alpha_i/\sin\beta_i \\
V_{i-1} &= -T_i\sin\alpha_i \\
H_{i-1} &= -\frac{\cot\beta_i}{2\sin\dfrac{\pi}{n}}V_{i-1}
\end{aligned}\right\} \qquad (10)$$

（2）设有内拉环的情况

对设有内拉环（图18）的索穹顶，仍以竖杆内力 V_0 为基准，可得各脊索、压杆、斜索和环索的一般性内力计算公式：

当 $i = 1$ 时，

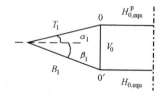

图18 内环处节点内力示意图

$$T_1 = -\frac{1}{\sin\alpha_1}V_0, \quad B_1 = -\frac{1}{\sin\beta_1}V_0$$
$$H_0^p = -\frac{\cot\alpha_1}{2\sin\frac{\pi}{n}}V_0, \quad H_0 = -\frac{\cot\beta_1}{2\sin\frac{\pi}{n}}V_0 \tag{11}$$

当 $i \geq 2$ 时，

$$T_i = \frac{(\cot\alpha_1 + \cot\beta_1)(1+\tan\alpha_2\cot\beta_2)L(1+\tan\alpha_{i-1}\cot\beta_{i-1})}{\cos\alpha_i}(-V_0)$$
$$B_i = T_i\sin\alpha_i / \sin\beta_i$$
$$V_{i-1} = -T_i\sin\alpha_i$$
$$H_{i-1} = -\frac{\cot\beta_i}{2\sin\frac{\pi}{n}}V_{i-1} \tag{12}$$

4.2 葵花形索穹顶结构初始预应力分布的快速计算法

对于圆形平面的葵花形索穹顶，若环向分为 n 等分，其 $1/n$ 不设内环的索穹顶示意图见图 19（a）。与肋环形索穹顶相类同，只要分析研究一肢半榀桁架便可。如图 19（b）所示，这是一次超静定结构，其中节点 0、1、3、5 在一个对称平面内，节点 2、4 在相邻的另一个对称平面内。

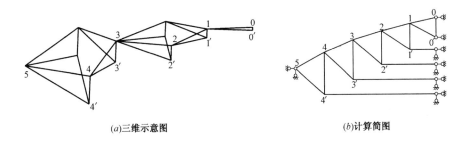

（a）三维示意图　　　　　　　　　　（b）计算简图

图 19　$1/n$ 不设内环的葵花形索穹顶示意图和计算简图

（1）不设内拉环的情况

图 20 为一不设内环的葵花形索穹顶。为方便各类杆件内力三向分解，引入变量 f_i、

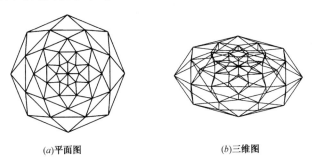

（a）平面图　　　　　　　　　　（b）三维图

图 20　不设内环的葵花形索穹顶

$d_{i,i+1}$、α_i、β_i、$\varphi_{i,i+1}$ 和 $\varphi_{i+1,i}$，详见图 21。其中 $\varphi_{i,i+1}$ 代表由节点 i、$i+1$ 组成的杆件与通过节点 i 的径向轴线的夹角；$\varphi_{i+1,i}$ 代表由节点 $i+1$、i 组成的杆件与通过节点 $i+1$ 的径向轴线的夹角；α_i 代表脊索与水平面夹角；β_i 代表斜索与水平面夹角。

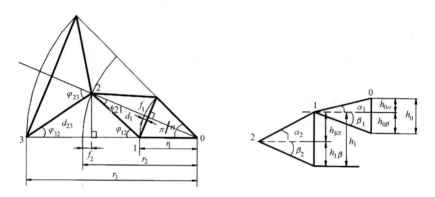

图 21 不设内环的葵花形穹顶平面及脊索、斜索与水平面夹角示意图

由图 21 所示几何关系可得

$$\varphi_{i,i+1} = \tan^{-1}\left(\frac{r_{i+1}\sin\frac{\pi}{n}}{r_{i+1}\cos\frac{\pi}{n} - r_i}\right), \quad \varphi_{i+1,i} = \tan^{-1}\left(\frac{r_i\sin\frac{\pi}{n}}{r_{i+1} - r_i\cos\frac{\pi}{n}}\right) \tag{13}$$

当 $i = 1$ 时，$\alpha_1 = \tan\left(\dfrac{h_{0\alpha}}{r_1}\right)$，$\beta_1 = \tan\left(\dfrac{h_{0\beta}}{r_1}\right)$ \hfill (14)

当 $i \geq 2$ 时，

$$\alpha_i = \tan\left(\frac{h_{i-1,\alpha}}{\sqrt{\left(r_i\sin\frac{\pi}{n}\right)^2 + \left(r_i\cos\frac{\pi}{n} - r_{i-1}\right)^2}}\right), \beta_i = \tan\left(\frac{h_{i-1,\beta}}{\sqrt{\left(r_i\sin\frac{\pi}{n}\right)^2 + \left(r_i\cos\frac{\pi}{n} - r_{i-1}\right)^2}}\right) \tag{15}$$

以中心竖杆的实际内力 V_0 为基准，对各节点建立平衡关系，可得各脊索、竖杆、斜索和环索的一般性内力计算公式：

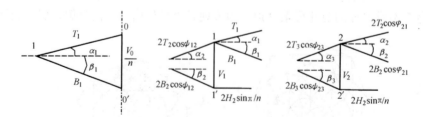

图 22 不设内环的节点内力示意图

当 $i = 1$ 时，$T_1 = -\dfrac{1}{n\sin\alpha_1}V_0$，$B_1 = -\dfrac{1}{n\sin\beta_1}V_0$ \hfill (16)

当 $i=2$ 时,

$$\left.\begin{array}{l} T_2 = \dfrac{(\cot\alpha_1 + \cos\beta_1)}{2n\cos\alpha_2\cos\varphi_{12}}(-V_0), \quad B_2 = T_2\sin\alpha_2/\sin\beta_2 \\ V_1 = -2T_2\sin\alpha_2, \quad H_1 = \dfrac{\cos\beta_2\cos\varphi_{1,2}}{\sin\dfrac{\pi}{n}}B_2 \end{array}\right\} \quad (17)$$

当 $i \geqslant 3$ 时,

$$\left.\begin{array}{l} T_i = \dfrac{(\cot\alpha_1+\cot\beta_1)(1+\tan\alpha_2\cot\beta_2)L(1+\tan\alpha_{i-1}\cot\beta_{i-1})\cos\varphi_{21}\cos\varphi_{32}L\cos\varphi_{i-1,i-2}}{2n\cos\alpha_i\cos\varphi_{12}\cos\varphi_{23}L\cos\varphi_{i-1,i}}(-V_0) \\ B_i = T_i\sin\alpha_i/\sin\beta_i \\ V_{i-1} = -2T_i\sin\alpha_i \\ H_{i-1} = \dfrac{\cos\beta_i\cos\varphi_{i-1,i}}{\sin\dfrac{\pi}{n}}B_i \end{array}\right\} \quad (18)$$

(2) 设有内拉环的情况

图 23 为一设有内环的葵花形索穹顶。同样,为方便各类杆件内力三向分解,引入变量 f_i,$d_{i,i+1}$,α_i,β_i,$\varphi_{i,i+1}$ 和 $\varphi_{i+1,i}$,详见图 24,各变量含义同前。

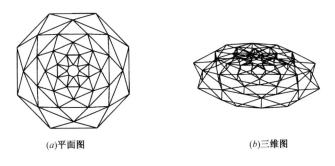

(a)平面图　　　(b)三维图

图 23　设有内环的葵花形索穹顶

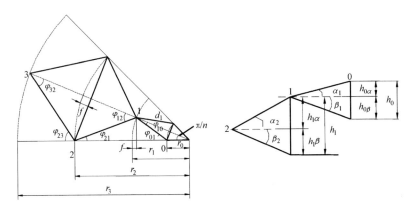

图 24　设有内环的葵花形穹顶平面及脊索、斜索与水平面夹角示意图

由图 24 所示几何关系可得

当 $i \geq 0$ 时，$\varphi_{i,i+1} = \tan^{-1}\left(\dfrac{r_{i+1}\sin\dfrac{\pi}{n}}{r_{i+1}\cos\dfrac{\pi}{n} - r_i}\right)$, $\varphi_{i+1,i} = \tan^{-1}\left(\dfrac{r_i \sin\dfrac{\pi}{n}}{r_{i+1} - r_i\cos\dfrac{\pi}{n}}\right)$ (19)

当 $i \geq 0$ 时，
$$\alpha_i = \tan^{-1}\dfrac{h_{i-1,\alpha}}{\sqrt{\left(r_i - r_{i-1}\cos\dfrac{\pi}{n}\right)^2 + \left(r_{i-1}\sin\dfrac{\pi}{n}\right)^2}}$$

$$\beta_i = \tan^{-1}\dfrac{h_{i-1,\beta}}{\sqrt{\left(r_i - r_{i-1}\cos\dfrac{\pi}{n}\right)^2 + \left(r_{i-1}\sin\dfrac{\pi}{n}\right)^2}} \quad (20)$$

仍以内环的竖杆内力 V_0 为基准进行推导，可得各脊索、竖杆、斜索和环索的一般性内力计算公式：

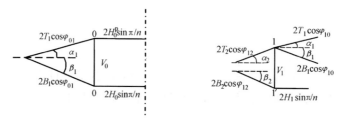

图 25 设有内环的节点内力示意图

当 $i = 1$ 时，
$$\left.\begin{array}{l} T_1 = -\dfrac{1}{2\sin\alpha_1} V_0, \quad B_1 = -\dfrac{1}{2\sin\beta_1} V_0 \\ H_0^p = -\dfrac{\cot\alpha_1 \cos\varphi_{01}}{2\sin\dfrac{\pi}{n}} V_0, \quad H_0 = -\dfrac{\cot\beta_1 \cos\varphi_{01}}{2\sin\dfrac{\pi}{n}} V_0 \end{array}\right\} \quad (21)$$

当 $i = 2$ 时，
$$\left.\begin{array}{l} T_2 = \dfrac{(\cot\alpha_1 + \cot\beta_1)\cos\varphi_{10}}{2\cos\alpha_2 \cos\varphi_{12}}(-V_0), \quad B_2 = T_2\sin\alpha_2/\sin\beta_2 \\ V_1 = -2T_2\sin\alpha_2, \quad H_1 = \dfrac{\cos\beta_2 \cos\varphi_{1,2}}{\sin\dfrac{\pi}{n}} B_2 \end{array}\right\} \quad (22)$$

当 $i \geq 3$ 时，
$$\left.\begin{array}{l} T_i = \dfrac{(\cot\alpha_1 + \cot\beta_1)(1+\tan\alpha_2\cot\beta_2)L(1+\tan\alpha_{i-1}\cot\beta_{i-1})\cos\varphi_{10}\cos\varphi_{21}\cos\varphi_{32}L\cos\varphi_{i-1,i-2}}{2\cos\alpha_i \cos\varphi_{12}\cos\varphi_{23}L\cos\varphi_{i-1,i}}(-V_0) \\ B_i = T_i\sin\alpha_i/\sin\beta_i \\ V_{i-1} = -2T_i\sin\alpha_i \\ H_{i-1} = \dfrac{\cos\beta_i \cos\varphi_{i-1,i}}{\sin\dfrac{\pi}{n}} B_i \end{array}\right\}$$

(23)

5. 考虑索杆膜协同工作的索穹顶结构找形分析

索穹顶结构由下部的索杆部分和上部的膜部分共同组成。目前对该类结构受力性能的研究一般仅考虑索杆部分承受荷载，而忽略薄膜部分对整个结构承载能力的影响。这并不能真实反映索穹顶结构的实际工作状况，并使计算结果偏于不安全。本节对考虑膜材与索杆协同工作的索穹顶结构的找形方法进行了探讨。

国内外有关索膜结构找形分析的研究众多。用于薄膜结构初始形态分析的主要方法有力密度法、动力松弛法和非线性有限元法等。这三种方法在弹性力学的本质上是一致的，且各具优缺点。目前应用较广的方法是非线性有限元法，这是由于非线性有限元法具有较高的计算精度，并且随着计算机运行速度的大幅度提高，该方法的计算速度也在一定程度上得到了提高。本节膜部分的形态分析采用非线性有限元法中的支座移动法；索杆部分的形态分析为一个找力过程，采用基于上节提出的整体可行预应力分析法。

索穹顶结构的找形与一般索膜结构的找形相比，既有共性又有其独特之处，一般膜结构和索杆结构的找形方法同样适用于索穹顶的找形。但是一般索膜结构找形时，其边界控制点为固定点，而索穹顶索杆部分与膜部分相交处为脊索，它是柔性构件，并非固定约束，这就存在着两部分的协调问题。索穹顶结构一般的协同找形方法是先将膜和索杆相交的边界处固定并分别对膜和索杆找形，然后将两者共同边界处约束去除再进行协同分析。此种找形方法的缺点是随着膜部分预应力的增加，索杆部分的位移将增大，所找形状并非初始形状，同时索杆内力也较难控制。为了克服上述缺点，保证索穹顶结构索杆与膜相交处控制点的位置变化不大，现提出索穹顶结构找形的逐次逼近法，通过索杆部分的多次找力来找到满意的形状。逐次逼近法的基本过程为：

（1）建立建筑师所要求结构初始形状的平面投影，包括膜部分、索杆部分、边索及谷索并对其进行单元划分；

（2）根据索穹顶结构索杆部分的初始形状，找到其初始平衡预应力分布并将预应力赋予索杆部分（1 次找力）；

（3）以脊索和边、谷索为边界，运用支座移动法对膜部分进行找形（此时假设建筑控制点固定不动），得到建筑控制点（选取脊索与压杆相交点）处反力（1 次找形）；

（4）求建筑控制点处反力作用下索杆结构的内力，并对步骤（2）所求得的索杆内力进行修正（2 次找力）；

（5）重复步骤（3）～（4），进行多次找形找力计算，直到满足精度要求或反力不能再减小（此时结构体系的平衡矩阵与反力向量为非相容方程）；

（6）去除结构体系中的多余约束，恢复结构的约束条件，对结构整体进行平衡迭代并得到结构的最后几何形状。

［算例 4］一帐篷形结构，曲面为旋转的悬链面。膜的预应力为 16kN/m，膜厚 1mm，弹性模量 $E = 10$MPa，泊松比 0.4。膜平面内圆半径 $a = 10$m，外圆半径 $a = 50$m，帐篷高 22.924m，其曲面方程为：$z = -a[\ln(\sqrt{x^2 + y^2 - a^2} + \sqrt{x^2 + y^2}) - \ln a] + h$。表 4 给出了代表节点（见图 26）$Z$ 方向的精确值与本文计算值的比较。从表 4 数据可以看出本文计算值

与精确值吻合较好。

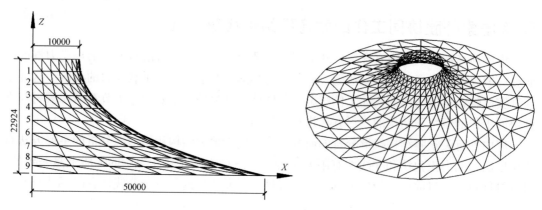

图 26　算例 4 计算简图及最终成形图

算例 4 节点坐标本文值与精确值的比较（m）　　　　　　　　　表 4

节点号	1	2	3	4	5	6	7	8	9
精确值	20.5	18.0	15.6	13.1	10.6	8.03	5.50	3.34	1.52
本文值	20.4	18.0	15.6	13.1	10.5	8.02	5.50	3.34	1.53
误差（%）	0.19	0.19	0.19	0.17	0.15	0.10	0.04	-0.0	-0.1

［算例 5］计算模型选自文献［1］，结构直径 24m，高 6.5m，模型尺寸和找形初态索杆布置如图 3，实际模型在图 27（a）基础上覆盖膜材。找形的初始信息为：膜材厚度取 1mm，初始预应力为 2.0MPa，膜弹性模量 $E=100$MPa，泊松比为 $\mu=0.3$；谷索和边索预应力设为 20kN，截面面积取 $2cm^2$；索弹性模量 $E=180$GPa，截面面积取 $8cm^2$；杆弹性模量 $E=210$GPa，截面面积 $100cm^2$。结构索杆部分初始预应力取值见表 5。

计算模型各杆件初始预应力及找形过程内力分布（kN）　　　　　　　　　表 5

内力\单元	1	2	3	4	5	6	7	8	9	10	11
找力 1	300	137	65	300	137	65	260	130	-14	-45	-48
情况 1	290	122	50	308	142	65	266	133	-15	-48	-58
找力 2	281	110	27	314	145	77	272	137	-15	-48	-57
情况 2	281	110	26	314	145	77	272	137	-15	-48	-57

为了便于比较，分别对两种情况进行了计算，第一种情况是一般的协同找形方法（找形过程见 1 节）；第二种情况为利用本文提到的逐次逼近法进行计算（此时建筑控制点选 $A\sim F$ 点，由于 $D\sim F$ 点反力很小，故表 7 中未列出其反力）。计算结果见表 6～表 8（表中的找力 1 表示对索杆进行 1 次找力，找形 1 表示对膜进行 1 次找形，其余类推）。由表 7～

表9可以看出：

（1）结构找形完成后其节点位移很小但是杆件的内力进行了重分布，且最里圈脊索内力改变的百分比最大，这主要是由于谷索的拉力对索杆部分的中心节点产生了一个向下的压力造成的；

（2）第一种情况控制点位移较大，有0.2064m，第二种情况控制点位置基本不变；

（3）第一种情况谷索内力由原来的20kN降为16kN，索杆内力改变由索杆、膜部分及谷索变位和内力改变共同完成，索杆内力较难控制。第二种情况谷索内力基本保持不变，控制点位移很小，索杆内力基本由节点反力和初始预应力共同决定，因此较好控制；

（4）在计算过程中，第一种情况在靠近平衡态时收敛较慢，第二种情况收敛迅速，算例2杆件内力只需修正1次即进行2次找力，就基本保持恒定。

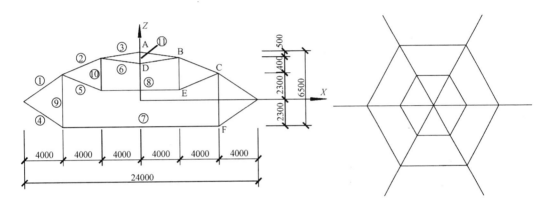

图27（a） 索杆部分找形初态投影图

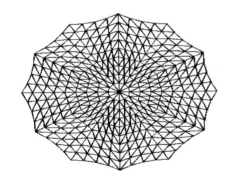

图27（b） 膜部分找形终态投影图

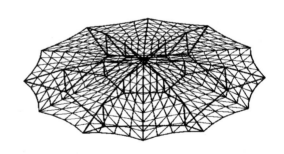

图27（c） 索穹顶整体结构找形终态图

计算模型各杆件初始预应力及找形过程内力分布（kN）　　　表6

单元	1	2	3	4	5	6	7	8	9	10	11
找力1	30	13	65	30	13	65	26	13	−14	−45	−48
情况1	29	12	50	30	14	65	26	13	−15	−48	−58
找力2	28	11	26	31	14	77	27	13	−15	−48	−57
情况2	28	11	26	31	14	77	27	13	−15	−48	−57

注：表中脊索内力为各单元平均值。

膜找形过程中各控制点处反力值（N） 表7

找形	1次找形完成时反力						2次找形完成时反力					
节点	A		B		C		A		B		C	
方向	X	Z	X	Z	X	Z	X	Z	X	Z	X	Z
反力	0.	37	1.	5.	2.	4.	0.	0.	-0	-0	-0	-0

节点位移变化对比表 表8

节点	初 态		情 况 1		情 况 2	
	X坐标	Z坐标	X向位移	Z向位移	X向位移	Z向位移
A	0.000	4.200	0.0000	-0.2064	0.0000	-0.0002
B	4.000	3.700	-0.0107	-0.1091	0.0000	0.0001
C	8.000	2.300	-0.0111	-0.0497	0.0000	0.0000

算例分析表明：

（1）逐次逼近法概念清晰，迭代收敛迅速，找形结果精确，能解决索穹顶等具有自应力模态结构的精确找形问题；

（2）索杆部分的内力分布较好控制，可以在修正杆件内力时对其进行重新调整，使结构整体受力合理；

（3）索穹顶结构的边、谷索及上部膜部分的内力分布都对其整体受力性能有较大影响，很有必要对其进行整体协同分析。

6. 索穹顶结构的施工分析

索穹顶的施工成形分析一直是国内外学者研究的热点和难点，一方面成形过程分析与结构的实际张拉过程密切相关，而现有的这方面资料很少，仅有几篇也只是对几个实际工程施工过程的简单描述；另一方面，体系在施工过程中伴随着索杆的大变形和大转角，存在刚体位移，刚度矩阵奇异，无法采用一般的非线性有限单元法进行计算。考虑到任何结构在一定的荷载（包括自重）作用下都会通过形状和内力的调整达到一个平衡态，可以忽略杆系在各阶段的移动和变位过程，只需关注在各阶段施工完成后体系所处的平衡态，即此时各杆件内力和节点坐标，由此索穹顶结构的施工成形分析转化为求解结构在各施工阶段平衡态的问题，该问题属于形态分析范畴。目前其主要方法有：力密度法，非线性有限元法和动力松弛法等。考虑到索穹顶结构由连续拉的索和间断压的杆组成，索穹顶结构的施工方法可以分别为张拉环索、下斜索和脊索或调节压杆长度。现有的资料通常采用由外及里逐圈张拉下斜索的施工方法。

本节采用动力松弛法、控制构件原长进行平衡态的找形分析，并尝试对由外及里逐圈张拉下斜索、只张拉外圈下斜索、由外及里逐圈调节压杆长度等多种施工方案进行探讨。研究表明，各施工方案都能成形且获得满意的结果，基于动力松弛法、控制构件原长的找形方法是有效和正确的。

这里需要补充说明的是在确定施工方案时，认为结构的初始预应力分布已确定，故索和杆的原长已知。根据施工过程索是否直接张拉可将索分为主动索和被动索，其中主动索是指直接张拉的索，其下料长度除原长外需另加一段牵引长度，牵引长度根据实际施工情况确定。被动索是指不直接张拉的索，其下料长度同原长。

6.1 分析思路

索穹顶结构的施工过程实质上是将已知原长的构件按照一定的拓扑关系组装到位的过程，并且这个安装过程实际是按阶段分批安装的，存在明显的阶段性，所以其设计必须跟踪实际的施工过程，两者密不可分，具体的设计和施工过程基本如下：

（1）根据平衡矩阵理论确定体系各单元在设计构形的自应力分布，同时根据荷载大小确定预应力水平系数即整体自应力模态组合系数从而确定初始预应力；对于多整体自应力模态体系需根据某优化目标确定最终的预应力分布；

（2）选择初始截面，考虑结构自重修正初始预应力分布，由杆件初内力确定各索杆原长；

（3）确定施工方案，从而确定施工的各个阶段；

（4）运用动力松弛法对每个阶段进行找形分析。首先假定结构初始形状，在初始形状下求解各单元的内力和节点力，然后判断节点力和外荷载是否满足节点平衡。如果平衡，那么给定的初始形状就是结构的平衡形状；如果不平衡，那么需要修改结构形状，然后根据变形协调方程重新求解在修改后形状下的单元内力和节点力，重新判断是否满足节点平衡，按照上述步骤迭代，直至体系达到平衡，由此确定各阶段各单元内力及各节点位置；

（5）张拉成形，铺设膜材，施工结束。

索穹顶在张拉成形前，部分索处于松弛状态，部分张拉索内力也很小，所以不能采用直线单元，而必须考虑自重引起的索垂度的影响，采用抛物线单元来模拟索单元，其曲线方程为下式：

$$z = \frac{qr}{2hl}x(l-x) + \frac{c}{l}x \tag{24}$$

式中：l 为悬索两端节点的水平距离；c 为悬索两端节点的竖向高差；r 为悬索两端节点的弦向距离；h 为悬索张力 t 的水平分量，q 为单位长度重量。根据相容条件建立已知原长的构件变形后节点坐标与构件内力之间的关系[14]：

$$\frac{l^4+c^2l^2}{r^3EA}t^3 + \frac{8s_0l^3-8l^4-4c^2l^2+c^4}{8lr^2}t^2 + \frac{w^2l^2}{12EAr}t + \frac{w^2c^2}{16l} - \frac{w^2l}{24} = 0 \tag{25}$$

式中：t 为索张力，$w = qr$ 为整段索重量，s_0 为索原长，E，A 分别为索材弹性模量和截面面积。

当索穹顶张拉成形后，索内力往往很大，此时可以忽略垂度影响，认为索单元为直线单元，根据直线单元小变形理论求得原长 s_0 和索内力 t 关系，即：

$$s_0 = \frac{EAr}{t+EA} \tag{26}$$

由平衡矩阵理论得到的初始预应力分布是没有考虑自重作用时的初内力分布，为消除结构自重引起的对结构初内力的影响，一般可以采用以下几种方法：

（1）将由自重等效成的节点荷载向量施加在体系上，将最终的内力分布作为设计形状修正后的内力分布，但此时结构在自重作用下发生变形，变形后的位形已经不是设计要求的位形；

（2）利用杆件原长不变原则进行不断迭代，求得考虑自重影响后的理想设计形状及预应力分布；本文直接将自重等效节点荷载作为外力作用在节点上，由多次迭代不断逼近理想设计形状和真实预应力分布。

6.2 动力松弛法

动力松弛法是根据达朗贝尔原理建立运动控制方程，其基本思想是将结构离散为空间节点位置上具有一定虚拟质量的质点，在设定的非平衡构形下，这些离散的质点将产生沿不平衡力方向的运动，从宏观上使结构的总体不平衡力趋于减小，当体系的动能达到极大值时，所有的速度分量设定为零，在当前不平衡力作用下重新开始运动，如此反复直至结构的动能趋近于零，体系达到静力平衡点。根据达朗贝尔原理，以中心有限差分形式表示，t 时刻节点 i 在 x 方向的运动可表示为

$$R_{ix}^t - \frac{C_{ix}(v_{ix}^{t+\Delta t/2} + v_{ix}^{t-\Delta t/2})}{2} = \frac{M_{ix}(v_{ix}^{t+\Delta t/2} - v_{ix}^{t-\Delta t/2})}{\Delta t} \tag{27}$$

式中：R_{ix}^t 是节点不平衡力，C_{ix} 是阻尼系数，v_{ix} 是 x 方向的速度，Δt 是微小时间增量，M_{ix} 是虚拟质量以用来优化收敛速度。当采用动态阻尼时，$C_{ix}=0$，此时方程（4）可简写为：

$$v_{ix}^{t+\Delta t/2} = v_{ix}^{t-\Delta t/2} + \frac{R_{ix}^t \Delta t}{M_{ix}} \tag{28}$$

则 $t+\Delta t$ 时刻 i 节点的 x 坐标改变为

$$x_i^{t+\Delta t} = x_i^t + \Delta t \cdot v_{ix}^{t+\Delta t/2} \tag{29}$$

同理可得到 i 节点 y, z 方向的速度分量和当前坐标（$t+\Delta t$ 时刻）以及结构所有其他节点的相应值。然后根据式（25）求得当前坐标时各构件的内力值和节点不平衡力，不断重复这个迭代过程，直到节点不平衡力和体系动能足够小为止。

现给出动力松弛法求解各施工阶段平衡态迭代步骤：

（1）假定一初始构形，设定体系动能 W_0，按式（25）求解在此构形下各构件的内力；

（2）求解节点不平衡力 R，由式（28），（29）求解节点速度 v_1，x_1 及体系动能 W_1；

（3）若最大节点不平衡力 R_{max} 和体系动能 W_1 同时小于某极小值，那么此初始构形就是理想构形；否则转（4）；

（4）如 $W_1 > W_0$，则 $W_0 = W_1$，$v_0 = v_1$，$x_0 = x_1$；如 $W_1 < W_0$，则按式（31）、式（32）重新计算 v_1, x_1 及体系动能 W_1，然后赋值 $W_0 = W_1$，$v_0 = v_1$，$x_0 = x_1$；

（5）重复（1）～（4），直到满足精度要求。

6.3 算例分析

如图 28（a）所示一 Geiger 型索穹顶结构（设内拉环），跨度为 100m，设有两道环索，根据对称性和几何拓扑关系，体系杆件类型分为 13 种，几何参数和杆件分组编号见图 28（b），杆件截面见表 9。钢材密度为 $q=7850$ kg/m³，索弹性模量 $E=170$ GPa，杆的弹性模量 $E=210$ GPa。

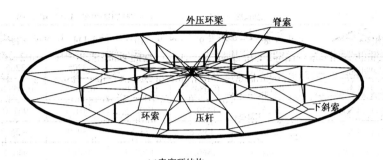

(a) 索穹顶结构

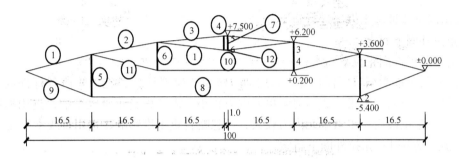

(b) 几何尺寸及杆件分组编号

图 28 Geiger 型索穹顶结构

依据文献分析方法，可求得体系的整体自应力模态 $\bar{s}=1$。根据预应力优化理论，求得同时满足可行性条件、应力条件、变形条件及以最低预应力水平系数为优化目标的预应力水平系数 $\beta=16847913$；根据本文理论将结构自重等效为节点荷载进行修正得到修正后的预应力分布，结果列于表 9。可以发现，自重引起的索杆张力结构内力变化相对于整体的预应力来说很小，但自重在施工成形阶段的影响却是至关重要的，所以在模拟施工阶段时必须考虑自重影响。

截面参数及初始预应力分布　　　　　　　　　表9

单元组号	1	2	3	4	5	6	7
截面积（mm²）	3205	3205	3205	3205	25819	25819	25819
整体自应力模态	0.12384	0.06941	0.03897	0.07506	−0.02640	−0.01080	−0.00306
理想预应力（kN）	2086.4	1169.4	656.6	1264.5	−444.8	−182.0	−51.6
自重修正值（kN）	−125.2	−162.5	−147.3	−283.7	−21.6	3.0	6.5
修正后预应力（kN）	1961.2	1006.9	509.3	980.8	−466.4	−179.0	−45.1
原长（m）	16.828	16.673	16.536	0.258	9.001	6.000	3.000

续表

单元组号	8	9	10	11	12	13
截面积(mm^2)	7996	3391	4250	3205	3205	3205
整体自应力模态	0.15582	0.08487	0.10129	0.05353	0.05740	0.02987
理想预应力（kN）	2625.3	1429.9	1706.5	901.9	967.0	503.2
自重修正值（kN）	256.6	139.8	73.8	39.0	-26.3	-13.7
修正后预应力（kN）	2881.9	1569.6	1780.3	940.9	940.7	489.5
原长（m）	17.304	17.314	8.778	16.818	0.258	16.572

拟采用以下几种施工方法进行施工：

1）张拉外圈下斜索；

2）由外及里逐圈张拉下斜索；

3）由外及里逐圈调节压杆长度。采用动力松弛法模拟计算各施工阶段各杆件内力及各节点位置，其中未完全张拉成形前索单元内力较小，需考虑自重垂度影响，选用抛物线索单元。施工张拉最后一步索单元内力较大，可忽略自重垂度影响，选用直线索单元。

（1）施工方法1：张拉外圈下斜索

此施工方法中，外圈下斜索为主动索，其余索为被动索。具体施工步骤如下：

第一步：计算单元原长，按原长组装被动索和压杆，连接主动索。主动索下料长度为原长加牵引长度5m；

第二步：逐步逐根张拉外圈下斜索到位。

计算结果见表10，表11，张拉过程杆件位置见图29（a）。

方法1 各施工阶段平衡态单元内力（kN）　　　　　表10

单元组号	1	2	3	4	5	6	7
第一步	385.3	124.9	20.2	39.0	22.4	-41.4	-6.9
第二步	1961.2	1006.9	509.3	980.8	-466.4	-179.0	-45.1

单元组号	8	9	10	11	12	13
第一步	10.2	1.8	486.4	257.2	199.2	103.8
第二步	2881.9	1569.6	1780.3	940.9	940.7	489.5

方法1 各施工阶段平衡态的节点坐标（m）　　　　　表11

节点号	1		2		3		4		5		6	
步骤	x	z	x	z	x	z	x	z	x	z	x	z
第一步	33.443	-3.072	31.967	-11.951	16.968	-0.495	16.970	-6.495	0.499	0.637	0.499	-2.363
第二步	33.500	3.600	33.500	-5.400	17.000	6.200	17.000	0.200	0.500	7.500	0.500	4.500

（2）施工方法2：由外及里逐圈张拉下斜索

此施工方法中，下斜索为主动索，其余索为被动索。具体施工步骤如下：

第一步：计算单元原长，按原长组装被动索和压杆，连接主动索。主动索下料长度为原长加牵引长度4m；

第二步：逐步逐根张拉外圈下斜索到位；

第三步：逐步逐根张拉中圈下斜索到位；

第四步：逐步逐根张拉里圈下斜索到位。

计算结果见表12～13，张拉过程杆件位置见图29（b）。

方法2 各施工阶段平衡态单元内力（kN） 表12

单元组号	1	2	3	4	5	6	7	8	9	10	11	12	13
第一步	299.2	289.4	285.3	550.7	17.2	9.5	5.2	16.5	8.8	5.6	2.6	4.0	2.2
第二步	312.5	305.5	301.4	582.0	-122.4	9.9	6.0	891.5	485.5	4.8	2.6	4.1	2.2
第三步	1577.4	747.2	736.5	1421.8	-391.2	-147.4	4.3	2458.6	1339.4	1552.5	819.8	4.0	2.2
第四步	1961.2	1006.9	509.3	980.8	-466.4	-179.0	-45.1	2881.9	1569.6	1780.3	940.9	940.7	489.5

方法2 各施工阶段平衡态的节点坐标（m） 表13

节点号 步骤	1		2		3		4		5		6	
	x	z	x	z	x	z	x	z	x	z	x	z
第一步	33.599	-3.807	32.860	-12.777	17.032	-5.749	16.772	-11.744	0.500	-6.361	0.499	-9.361
第二步	33.610	3.855	33.451	-5.145	17.032	2.003	16.714	-3.989	0.500	1.371	0.499	-1.629
第三步	33.523	3.647	33.490	-5.353	17.056	6.402	16.995	0.403	0.500	6.174	0.499	3.174
第四步	33.500	3.600	33.500	-5.400	17.000	6.200	17.000	0.200	0.500	7.500	0.500	4.500

（3）施工方法3：由外及里逐圈改变压杆长度

具体施工步骤如下：

第一步：计算单元原长，按原长组装全部索，连接压杆。压杆长度由外及里分别由原长缩短3m，2m，1m；

第二步：逐步逐根调节外圈压杆到位；

第三步：逐步逐根调节中圈压杆到位；

第四步：逐步逐根调节里圈压杆到位。

计算结果见表14～15，张拉过程杆件位置见图29（c）。

方法3 各施工阶段平衡态单元内力（kN） 表14

单元组号	1	2	3	4	5	6	7	8	9	10	11	12	13
第一步	133.3	35.4	5.6	10.6	-46.5	-17.5	-3.2	415.6	223.0	189.1	99.6	54.1	28.9
第二步	366.1	111.3	20.0	38.7	-125.8	-28.4	-4.5	914.9	498.1	477.8	250.2	176.2	91.5
第三步	1726.5	852.8	469.7	906.6	-417.7	-158.6	-24.6	2608.2	1420.8	1631.3	862.3	721.1	374.4
第四步	1961.2	1006.9	509.3	980.8	-466.4	-179.0	-45.1	2881.9	1569.6	1780.3	940.9	940.7	489.5

方法3 各施工阶段平衡态的节点坐标（m）　　　　　表15

节点号 步骤	1		2		3		4		5		6	
	x	z	x	z	x	z	x	z	x	z	x	z
第一步	33.194	0.919	33.439	-5.077	16.536	0.487	16.963	-3.491	0.499	-1.646	0.499	-3.646
第二步	33.608	3.851	33.452	-5.148	17.000	5.349	16.970	1.349	0.499	5.796	0.499	3.796
第三步	33.514	3.628	33.494	-5.372	17.034	6.321	16.997	0.321	0.500	7.037	0.500	5.037
第四步	33.500	3.600	33.500	-5.400	17.000	6.200	17.000	0.200	0.500	7.500	0.500	4.500

三种施工方法比较结果表明：

1）张拉成形后，不管单元设计内力还是形状都能达到设计要求，表明设计方法和程序的正确性；

2）三种不同施工方法的比较表明各施工方法虽然张拉单元不一样，施工成形顺序也不一样，但只要各构件原长一定，那么结构最终都能达到理想的设计形状，同时获得设计初内力分布；

3）现有资料显示现有工程较多采用第2种施工方法，即由外及里逐圈张拉下斜索的施工方法。本文认为，此方法较第1种方法即只张拉外圈下斜索方法有更多的可调空间，因为杆件制作、施工过程中不可避免地存在施工误差，使实际结构与理想状态存在偏差，为使结构能张拉成形，必须进行局部内力和节点位置的调整，而由外及里逐圈张拉下斜索的施工方法更容易调节和控制。第1种施工方法由于现场只需要张拉数量较少索，所以施工比较方便。由外及里逐圈改变压杆长度的第3种方法同样也能成形，但对压杆制作和相关构造有较高要求；

4）预应力在成形过程中是逐步形成的，在未张拉成形以前，预应力水平均未达到设计水平，只有最后张拉完成，才最终生成索要求的预应力，达到设计形状。

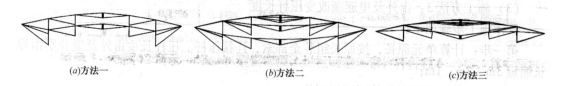

(a)方法一　　　　　(b)方法二　　　　　(c)方法三

图29　三种施工方法各阶段杆件变位示意图

7. 模型试验

为验证理论分析的正确性，设计加工一直径为5m的Kiewitt型索穹顶结构模型进行试验研究。通过试验，可以进一步了解把握该体系的力学性能和成形机理，为实际工程应用提供依据。

7.1 模型设计

本模型主要由自平衡支承平台、桅杆和拉索等3部分组成，其中拉索包括脊索、斜索和环索，结构模型见图30。

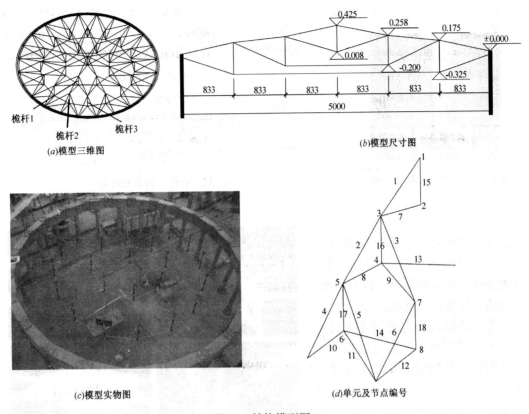

图30 结构模型图

支承平台和桅杆、拉索构成整个结构的自平衡体系。本试验模型支承平台由24件基本组装单元拼装而成以来模拟张力结构周边刚性支座。每个组装单元主要由环梁和立柱组成，环梁采用250mm×180mm×12mm×10mm焊接工字形梁，立柱采用ϕ114×6无缝钢管，环梁和立柱通过焊接于柱顶的盖板用4根ϕ16普通螺栓连接于环梁下翼缘。为解决由于地面高低不平整而使部分立柱悬空，特地将立柱设计成高度可调的，即在靠近立柱底部加一刻有正、反螺纹的可调套筒，如图31所示。

桅杆作为索穹顶结构中承受压力构件必须满足强度和稳定要求。本试验模型中桅杆有四种类型，即中心桅杆，桅杆1、2和3，结构中所处位置见图30（a），中心桅杆采用ϕ20×3无缝钢管，其余桅杆采用ϕ15×3无缝钢管，所有桅杆长度都是可调的，即中间采用有正、反螺纹的套筒连接。

内、外两圈环索采用ϕ5高强钢丝，脊索和斜索采用ϕ3高强钢丝。索与索头的连接采用挤压式直接锚固。索头设计成U字形，中间开槽7mm以用来插桅杆端部节点叶片，然后用ϕ6销钉销住。环索是闭合的，索两端索头通过正、反螺纹套筒连接，同时通过套筒可以调节索闭合长度以满足索的微调效果。如图32所示。

本试验的节点设计主要有两种：索与桅杆的连接、外圈索与环梁的连接。

图 31　支承平台基本组装单元

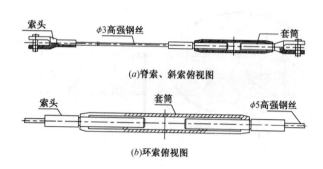

图 32　脊索、斜索和环索

索杆的连接通过设计成 U 字形的索头、中间开槽插桅杆端部节点叶片、然后用 φ6 销钉销住来完成，这种连接方式安装简单、传力明确。部分索、杆连接节点见图 33。

索穹顶结构内力通过外圈脊索和斜索传到环梁上从而使包括环梁在内的整个结构组成有效的整体自平衡体系。试验采用 A、B 两种螺杆来连接外圈索和环梁，螺杆一端连接索头，另一端穿过环梁腹板 φ16 开孔用螺帽锚住。其中螺杆 A 端部伸出两片叶片以用来连接外圈脊索 5、6（斜索 11、12）索头，每片叶片上留有 φ6 孔，销钉穿过索头和螺杆上的孔销住。螺杆 B 端部伸出一片叶片同样留有 φ6 孔，以穿销钉来连接脊索 4（斜索 10）索头，索在结构中位置如图 30（d）所示，螺杆连接节点如图 34 所示。

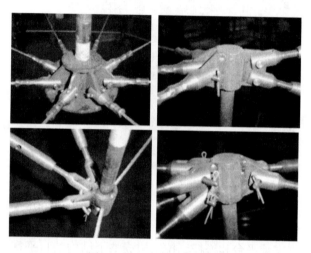

图 33　索、杆连接节点图

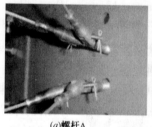

(a) 螺杆 A

(b) 螺杆 B

图 34　螺杆

试验用电阻应变法来测量杆件内力，采用 TS3860 静态电阻应变仪，如图 35 所示。电阻应变片的型号为 BX120-5AA，灵敏度系数为 2.08%±1%。为了消除试验过程中偏心受力和温度引起的误差，在所有测点的正反两面各贴一片应变片，将两片应变片串联后与温度补偿片接成半桥电路进行测试。节点位移采用百分表测试。

本试验选定两组对称面杆件进行测试，对称面位置如图 36 所示。每组对称面有 18 类杆件，其中脊索（JC）编号 1~6，斜索（XC）编号 7~12，环索（HC）编号 13、14，桅杆

（P）编号 15~18，除环索和中心桅杆外其余各类杆件均有两个测点，试验共有33个测点，测试结果取两组平均值。节点编号由里及外依次为1~8。测点编号具体见图30（d）。

图35 电阻应变仪

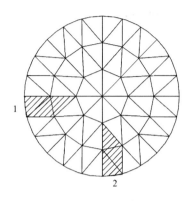

图36 两组测点布置

考虑到各种构件加工尺寸、材料存在的差异，同时为了考察应变片粘贴的有效性，因此在内力测试前需对所有贴应变片的构件进行标定，即测定构件的应变（ε）—内力（F）关系，部分单元标定结果见图37。

可以发现，拉索和桅杆在标定荷载范围内其应变—内力关系基本上呈线性关系，同时也反映了构件截面加工尺寸和材料存在较大的差异和离散性，不能简单测个别几根索或桅杆的应变—内力曲线来模拟其他索、杆的应变—内力关系。

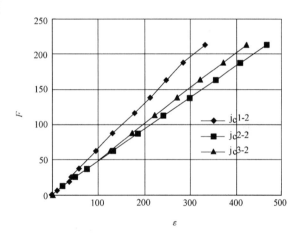

图37 脊索1、2、3标定曲线

7.2 试验

从理论上讲，Kiewitt型索穹顶结构按特定原长组装张拉到位，那么结构内预应力就是设计初始预应力。表16给出了试验初始预应力实测值（F_{im}）与理论值（F_{ic}）之间的比较。可以发现，初始预应力分布测试结果（平均值F_{ip}）和理论计算基本吻合，表明控制杆件原长张拉成形后的、由多组整体自应力模态组合得到的初始预应力分布基本上能达到理论设计值；但个别索的初内力相差还是比较大，如斜索4（XC4）达到了36.76%，桅杆1、2和3的误差也都超过了10%。误差原因分析详见7.3节。

模型各测点初始预应力 表16

构件编号	构件	F_{im}	F_{ip}	F_{ic}	误差%	构件编号	构件	F_{im}	F_{ip}	F_{ic}	误差%
1	JC1-1 JC1-2	2.37 2.18	2.275	2.30	-1.09	10	XC4-1 XC4-2	0.67 0.19	0.43	0.68	-36.76
2	JC2-1 JC2-2	1.80 1.64	1.72	1.74	-1.15	11	XC5-1 XC5-2	0.47 0.68	0.575	0.73	-21.23
3	JC3-1 JC3-2	1.38 1.31	1.345	1.37	-1.82	12	XC6-1 XC6-2	1.05 0.92	0.985	0.97	1.55
4	JC4-1 JC4-2	0.80 0.70	0.75	0.61	22.95	13	HC1	1.25	1.25	1.14	9.65
5	JC5-1 JC5-2	1.18 1.07	1.125	1.14	-1.32	14	HC2	3.71	3.71	4.39	-15.49
6	JC6-1 JC6-2	1.50 1.70	1.6	1.64	-2.44	15	P中	-3.76	-3.76	-3.61	4.16
7	XC1-1 XC1-2	1.43 1.66	1.545	1.58	-2.22	16	P1-1 P1-2	-0.46 -0.50	-0.48	-0.41	17.07
8	XC2-1 XC2-2	0.58 0.55	0.565	0.61	-7.38	17	P2-1 P2-2	-0.77 -0.64	-0.705	-0.61	15.57
9	XC3-1 XC3-2	0.29 0.20	0.245	0.23	6.52	18	P3-1 P3-2	-0.63 -0.68	-0.655	-0.58	12.93

本试验静载试验包括3种荷载工况：

1）工况1：1/8跨节点均布荷载；

2）工况2：半跨均布节点荷载；

3）工况3：满跨均布节点荷载。各种荷载工况所加节点荷载均为6.2 kg，且均为一次加载。加载采用手工加质量块来模拟节点荷载，加载时用钢丝制成加载环，将加载环一端固定于桅杆节点上，然后把质量块置于环上即可完成加载。各种荷载工况的内力实测值（F_m）与内力计算值（F_c）及位移实测值（U_m）与位移计算值（U_c）见表17、18。

由表17可以发现，模型在各种荷载工况作用下内力实测值与理论值基本上是吻合的，但个别杆件存在较大的误差，如脊索4（JC4）误差超过了30%，桅杆误差也超过了15%；需要指出的是各种荷载工况下存在的误差相当一部分是由于初始预应力分布的不均匀引起的。表17数据表明，模型在各种荷载工况下的节点位移试验值与理论值吻合得较好。多种荷载工况试验结果表明，理论计算是正确的，试验设计也是合理有效的。但有些方面有待进一步探讨和改进，如试验室的加载系统需要改善，电阻应变仪的漂移造成读数误差如何解决等。

各荷载工况构件内力分布   表17

构件编号	构件	荷载工况1			荷载工况2			荷载工况3		
		F_m	F_c	误差/%	F_m	F_c	误差/%	F_m	F_c	误差/%
1	JC1	2.327	2.241	3.84	2.236	2.182	2.45	2.098	2.116	-0.81
2	JC2	1.785	1.718	3.89	1.735	1.697	2.20	1.594	1.613	-1.18
3	JC3	1.332	1.319	0.98	1.312	1.301	0.89	1.263	1.270	-0.60
4	JC4	0.809	0.603	34.01	0.816	0.626	30.29	0.780	0.568	37.52
5	JC5	1.165	1.119	4.12	1.172	1.125	4.19	1.068	1.099	-2.85
6	JC6	1.425	1.588	-10.25	1.425	1.596	-10.68	1.534	1.559	-1.56
7	XC1	1.385	1.528	-9.33	1.392	1.538	-9.48	1.528	1.471	3.87
8	XC2	0.576	0.601	-4.13	0.639	0.657	-2.68	0.556	0.645	-13.68
9	XC3	0.303	0.236	28.47	0.278	0.266	4.36	0.247	0.258	-4.07
10	XC4	0.703	0.716	-1.81	0.732	0.756	-3.22	0.757	0.754	0.36
11	XC5	0.475	0.736	-35.43	0.552	0.823	-32.90	0.593	0.803	-26.17
12	XC6	1.080	1.035	4.32	1.100	1.056	4.19	0.986	1.066	-7.46
13	HC1	1.309	1.281	2.19	1.364	1.262	8.06	1.340	1.233	8.66
14	HC2	3.947	4.697	-15.98	4.224	4.914	-14.04	4.132	4.835	-14.55
15	P中	-3.670	-3.583	2.44	-3.600	-3.486	3.27	-3.425	-3.381	1.30
16	P1	-0.496	-0.459	8.06	-0.508	-0.464	9.47	-0.533	-0.451	18.15
17	P2	-0.820	-0.649	26.47	-0.881	-0.680	29.53	-0.773	-0.666	16.21
18	P3	-0.676	-0.623	8.49	-0.704	-0.652	8.03	-0.741	-0.640	15.83

各荷载工况各节点位移   表18

	荷载工况1			荷载工况2			荷载工况3		
	U_m	U_c	误差/%	U_m	U_c	误差/%	U_m	U_c	误差/%
2	0.63	0.64	-1.56	1.55	1.41	9.93	2.32	2.25	3.11
4	1.81	1.63	11.04	2.67	2.88	-7.29	2.09	2.07	-3.87
6	1.65	1.61	2.48	2.26	2.23	1.35	1.43	1.35	5.92
8	1.87	1.73	8.09	2.31	2.22	4.05	1.47	1.42	3.52

7.3 误差分析与结论

本试验结果表明测试值与理论计算值基本上能吻合，但存在一定的误差，实验误差的产生一般可以归结为三个方面，即系统误差、偶然误差和过失误差，本试验误差主要由以下几方面引起：

1）模型的误差：包括构件长度的误差，连接节点制作误差，周边环梁加工安装误差等。由于本试验初始预应力值和各荷载工况节点荷载都较小，所以杆件长度变化量很小，一般约几毫米。然而实际模型加工安装时很难保证其精度，误差可能也达到这个量级，即模型加工时的误差都直接影响了试验精度；

2）测试仪器引起的误差：整个试验过程一般持续半天时间，在这段时间内，静态应变仪应变值存在一定的漂移，而构件应变值本身就很小，仪器的漂移对试验结果有相当的影响。各种荷载工况作用时的节点位移采用百分表读数，此时会由于百分表摆放不正及人为读数误差引起一定的误差；

3）杆件最终的内力是通过应变—内力关系曲线反算得到，所以杆件标定时的误差及曲线斜率的取值误差都将引起测试结果的精度。

试验结果表明：

1）试验结果和理论分析基本上能吻合，表明本模型设计和节点构造是合理有效的，理论计算方法是正确可行的，试验基本上能达到预期效果；

2）张拉各构件到设计长度时，多整体自应力模态组合后的体系初始预应力分布就是设计初始预应力值；

3）试验中存在一定的问题，有待进一步改善，如测试仪器和方法的改进，索具的加工需要专门厂家的制作以保证精度和强度，如何引进参数（如缺陷）考虑系列误差导致的结果偏差等。

8. 刚性屋面索穹顶结构方案设计与分析

国外已建索穹顶工程的屋面覆盖材料比较单一，几乎全部采用柔性织物膜，而以刚性材料作为屋面的工程很少有报道。尽管集覆盖和承重于一身的膜材使索穹顶结构更具轻质感和流动感，但由此带来的加工制作以及施工维护等费用大大增加了索穹顶结构的造价。与以膜材为屋面材料的索穹顶结构相比，刚性屋面索穹顶结构则具有以下几个显著特点：可以采用通用的刚性材料如压型钢板、铝板等作屋面材料；刚性屋面可以增加屋面的整体刚度，提高结构抵抗不对称荷载的能力；另外与膜材的材料成本、加工制作、铺设和围护等费用相比，刚性屋面的索穹顶结构造价低，施工简便，具有良好的技术经济指标。可以预见，刚性屋面的索穹顶结构是有良好发展前景的。

结合某一工程要求，提出了刚性屋面索穹顶结构的设计方案，对其结构布置、索杆初始内力分布、荷载作用下结构内力和变位计算以及结构用钢指标等作了介绍，并对边界支承结构的刚度以及屋面支撑系统对索穹顶结构初始预应力分布及外荷载作用下受力性能的影响进行了研究，为该类工程的设计和应用提供了理论依据和技术保证。

8.1 刚性屋面索穹顶结构设计方案

本工程拟采用索穹顶结构和周边立体桁架相结合的结构体系。其中索穹顶为肋环形布置，由中心受拉环、径向布置的脊索、斜索、压杆和环索组成。索穹顶支承于周边立体桁架环梁，环梁支承于拱形立体桁架柱。

屋面为刚性屋面系统。径向脊索节点间设置方钢管主檩条，主檩间设置环向次檩条，上铺金属屋面板。

结构三维图、平面图和立面图如下：

第二章 索穹顶结构的新形式及分析与施工中的新方法　　39

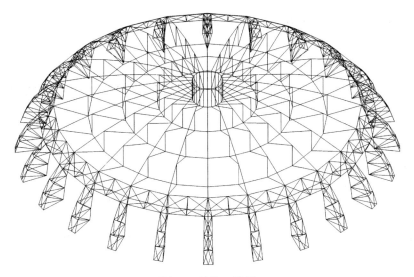

图 38　结构三维图

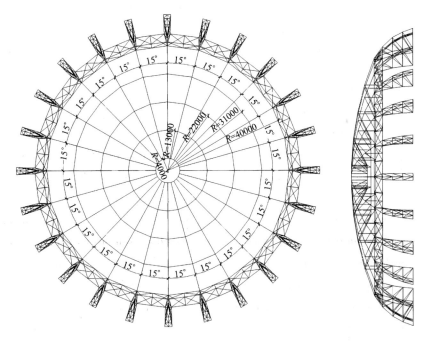

图 39　结构平面图和立面图

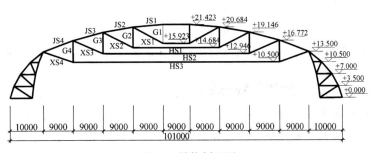

图 40　结构剖面图

索穹顶结构杆件编号见图40，杆件截面配置见表19：

索穹顶结构杆件截面配置 表19

编号	截面	编号	截面	编号	截面
JS1	37ϕ7	XS2	37ϕ7	HS3	199ϕ7
JS2	37ϕ7	XS3	37ϕ7	G1	ϕ114×4.0
JS3	37ϕ7	XS4	91ϕ7	G2	ϕ114×4.0
JS4	91ϕ7	HS1	37ϕ7	G3	ϕ114×4.0
XS1	37ϕ7	HS2	91ϕ7	G4	ϕ180×4.0

索穹顶结构由索和杆组成，索在未施加预应力前几乎没有自然刚度，所以必须施加预应力。索穹顶结构的刚度与预应力的分布和大小有密切关系。根据结构几何形状，可惟一确定肋环形索穹顶结构的单位初始预应力分布。根据结构荷载情况以及结构使用阶段索不退出工作这一原则，可确定结构所需预应力水平为 $\beta = 26.13$，此时结构初始内力分布见表20。

索穹顶结构单位预应力和初始预应力分布（kN） 表20

杆件编号	JS1	JS2	JS3	JS4	XS1	XS2	XS3	XS4
单位预应力	12.22	14.28	19.57	32.63	2.14	5.41	12.76	35.26
初始预应力	319.38	373.03	511.34	852.53	55.87	141.45	333.36	921.23
杆件编号	HS1	HS2	HS3	G1	G2	G3	G4	
单位预应力	18.58	44.98	128.12	-1.00	-2.40	-4.99	-11.15	
初始预应力	485.46	1175.20	3347.80	-26.13	-62.83	-130.40	-291.30	

荷载取值：

1）永久荷载

    a. 结构自重 0.20 kN/m^2；

    b. 屋面板及檩条自重 0.30 kN/m^2。

2）可变荷载

    a. 活荷载 标准值 0.30 kN/m^2；

    b. 雪荷载 基本雪压 0.40 kN/m^2；

    c. 风荷载 基本风压 0.45 kN/m^2。

采用 MSTCAD 和 ANSYS 软件进行结构建模和计算分析，其中拉索为索单元，压杆为杆单元。计算结果显示结构应力和变位均满足要求。其中结构最大变位为 0.23m，小于跨度的 1/400，表明结构整体刚度较好。

索穹顶部分用钢量按 80m 直径计算约为 18kg/m^2，按 100m 直径计算约为 12kg/m^2。钢结构支承部分（包括立体桁架环梁和柱）用钢量约为 50kg/m^2。支承部分也可采用钢筋混凝土结构。上述用钢指标说明结构具有较好的经济性。

8.2 支承结构对索穹顶初内力分布及外荷载作用下结构性能的影响

索穹顶必须支承在周边受压环梁上才能工作。不同刚度的支承结构将影响索穹顶的初内力分布和其在外荷载作用下的受力性能。为定量地研究支承结构对索穹顶的影响，本小节对支承于不同刚度支承结构的索穹顶进行了整体分析计算。

以本文提出的刚性屋面索穹顶结构方案为例，支承结构包括周边立体桁架环梁和拱形立体桁架柱。分别对支承结构和包括支承结构和索穹顶在内的整体结构施加径向单位力，得水平变位为 Δ_1 和 Δ_{1+0}，则支承结构水平刚度为 $K_1 = \dfrac{1}{\Delta_1}$，整体结构水平刚度为 $K_{1+0} = \dfrac{1}{\Delta_{1+0}}$，索穹顶水平刚度为两者之差，即 $K_0 = K_{1+0} - K_1 = \dfrac{\Delta_1 - \Delta_{1+0}}{\Delta_1 \Delta_{1+0}}$。定义索穹顶与支承结构水平刚度之比为索穹顶相对水平刚度 λ，$\lambda = \dfrac{K_0}{K_1} = \dfrac{\Delta_1 - \Delta_{1+0}}{\Delta_{1+0}} = \dfrac{\Delta_1}{\Delta_{1+0}} - 1$。当支承结构水平刚度 K_1 无穷大时，有 $\lambda = 0$，此时可认为索穹顶边界节点为理想不动铰支座。对具有不同相对水平刚度 λ 的索穹顶进行分析，得表22所示初内力分布。偏差为理想不动铰支座索穹顶初内力和当前水平刚度索穹顶的初内力之差与理想不动铰支座索穹顶初内力之比。表23为不同相对水平刚度的索穹顶结构在均布荷载 $q = 1.0 \text{kN/m}^2$ 作用下节点最大变位。

不同相对水平刚度的索穹顶结构初始内力分布（kN） 表21

杆件编号		JS1	JS2	JS3	JS4	XS1	XS2	XS3	XS4
$\lambda = 0$	初内力	319.38	373.03	511.34	852.53	55.87	141.45	333.36	921.23
$\lambda = 2\%$	初内力	312.97	365.51	501.05	835.47	54.72	138.62	326.71	903.22
	偏差	-2.05	-2.06	-2.05	-2.04	-2.10	-2.04	-2.04	-1.99
$\lambda = 5\%$	初内力	303.84	354.81	486.40	811.10	53.08	134.58	317.24	877.43
	偏差	-5.11	-5.14	-5.13	-5.11	-5.26	-5.10	-5.08	-4.99
$\lambda = 8\%$	初内力	295.21	344.69	472.56	788.10	51.53	130.78	308.30	853.06
	偏差	-8.19	-8.22	-8.21	-8.18	-8.42	-8.16	-8.13	-7.99
杆件编号		HS1	HS2	HS3	G1	G2	G3	G4	
$\lambda = 0$	初内力	485.46	1175.2	3347.8	-26.13	-62.83	-130.4	-291.3	
$\lambda = 2\%$	初内力	475.74	1151.8	3282.4	-25.59	-61.57	-127.8	-285.6	
	偏差	-2.04	-2.03	-1.99	-2.11	-2.05	-2.03	-2.00	
$\lambda = 5\%$	初内力	461.91	1118.3	3188.6	-24.83	-59.78	-124.1	-277.3	
	偏差	-5.10	-5.09	-4.99	-5.24	-5.10	-5.08	-5.05	
$\lambda = 8\%$	初内力	448.82	1086.8	3100.0	-24.10	-58.09	-120.6	-269.6	
	偏差	-8.16	-8.13	-7.99	-8.42	-8.16	-8.13	-8.05	

不同相对水平刚度的索穹顶结构节点最大变位（mm）　　表22

相对水平刚度		$\lambda=0$	$\lambda=2\%$	$\lambda=5\%$	$\lambda=8\%$
初内力	u_x	0	1.64	3.99	6.20
	U_y	0	1.64	3.99	6.20
	U_z	0	1.13	2.74	4.26
均布荷载	u_x	17.59	18.16	19.28	20.79
	U_y	17.59	18.16	19.28	20.79
	U_z	153.54	154.48	155.21	155.90

由表21、22可知，当索穹顶相对刚度$\lambda\leq5\%$时，结构最大初内力影响约5%，最大变位3.99mm，仅为跨度的1/20000，此时索穹顶边界可按刚性边界计算（边界节点为理想不动铰支座）。当索穹顶相对刚度$\lambda>5\%$时，索穹顶边界按弹性边界计算，此时必须考虑支承结构刚度对索穹顶结构受力性能的影响。

8.3 屋面支撑对索穹顶结构性能的影响

肋环形索穹顶结构由于其脊索放射状布置，几何形状类似于平面桁架，平面外刚度较小，在不对称荷载作用下容易出现失稳。布置支撑系统可以改善结构的受力性能，并能提高结构整体稳定性。

为研究屋面支撑对索穹顶结构性能的影响，特考虑如下几种屋面支撑进行分析计算：（1）不布置屋面支撑；（2）仅径向布置屋面支承；（3）仅环向布置屋面支撑；（4）环向和径向同时布置屋面支撑。屋面支撑布置位置类似斜索，详见图41。支撑体系无初内力，仅在外荷载作用下参加工作，且受压时退出工作。

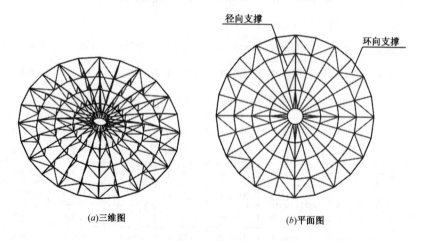

(a)三维图　　(b)平面图

图41　屋面支撑布置示意图

图 42 为不设屋面支撑的索穹顶结构前三阶振型图。表 23 和表 24 分别为不同屋面支撑的索穹顶结构节点最大变位和结构振动频率。上述结果显示：按斜索方向加设外圈环向支撑可以有效减小变位，提高结构的振动频率；但径向支撑使结构的刚度分布趋于不均匀，并不一定能提高结构的基频，对控制结构变位的作用也较小。

不同屋面支撑的索穹顶结构节点最大变位（mm） 表 23

屋面支撑		不布置	径向布置	环向布置	径向、环向同时布置
均布荷载	u_x	17.59	18.76	17.68	17.68
	U_y	17.59	18.76	17.68	17.68
	U_z	153.54	151.91	148.73	148.73

不同屋面支撑的索穹顶结构振动频率 表 24

屋面支撑		不布置	径向布置	环向布置	径向、环向同时布置
频率	一阶	1.09	0.83	1.33	1.34
	二阶	1.34	0.92	1.49	1.51
	三阶	1.60	0.92	1.64	1.82

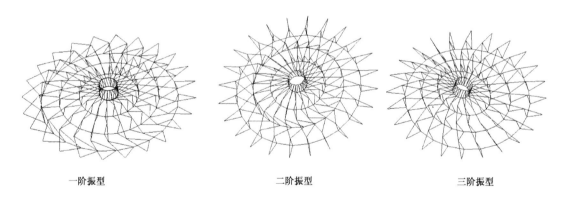

一阶振型　　　　　　　二阶振型　　　　　　　三阶振型

图 42　不设屋面支撑的索穹顶结构前三阶振型图

对布置在索穹顶脊索面的支撑进行了分析，结果显示脊索在外荷载作用下内力减小，脊索面的支撑由于不受拉力而不参与共同工作。

9. 结论

本文对某一工程刚性屋面索穹顶结构设计方案进行了计算分析，验证了该方案具有良好的受力性能和经济指标。对不同刚度边界支承结构对索穹顶结构初始预应力分布及外荷载作用下受力性能的影响进行了研究，表明索穹顶结构相对水平刚度 $\lambda \leqslant 5\%$，可按刚性边界计算，否则必须考虑边界刚度对结构初始内力和受力性能的影响，即引入边界弹簧刚度或对索穹顶和边界结构进行整体分析。对屋面支撑对索穹顶结构性能的影响进行了研究，表明布置在索穹顶脊索面上的支撑不起作用，按斜索方向布置环向支撑可以有效提高结构刚度，但径向支撑的布置使结构刚度分布趋于不均匀，因此，不一定能提高结构基频。

参考文献

［1］Barnes M R. Form finding and analysis of Tensile Memberanes ［J］. Engineering tension structures by dynamic relaxation ［J］

［2］FULLER R B. Tensile-integrity structures ［P］. U. S：Patent 3063521, 1962. U

［3］Geiger D H. The design and construction of two cable domes for the Korean Olympics ［A］. Shells, Membranes and Space Frame, Proceedings of IASS-ASCE. International Symposium ［C］. Osaka, Japan. Elsevier, 1986, 2：265-272

［4］LEVY MP. The Georgia dome and beyond achieving light weight-long span structures ［A］. Proceedings of IASS-ASCE. International Symposium ［C］. Atlanta, USA. ASCE：1994, 560-562

［5］Haug E. and Powell G. H. Finite element analysis of nonlinear membrane structures. IASS Pacific Symp. Part II on Tension Structures and Space Frames. Tokyo and Kyoto (1972)

［6］Kai-Uwe Bletzinger, Ekkehard Ramm. A general finite element approach to the form finding of tensile structure by the updated reference strategy ［J］. International Journal of Space Structures, 1999, 14 (2)：131-145

［7］Maurin B, Motro R. The Surface Stress Density Method as a Form-finding Tool for Tensile Memberanes ［J］. Engineering Struc-tures, 1998, 20 (8)：712-719

［8］Pellegrino S and Calladine CR. Matrix Analysis of Statically and Kinematically Indeterminate Frameworks ［J］. International Journal of Solids Structures, 1986, 22 (4)：409-428

［9］Pellegrino S. Structural Computation with the Singular Value Decomposition of Equilibrium Matrix ［J］. International Journal of Solids and Structures, 1993, 30 (21)：3025-3035

［10］Sheck H. J. The force density method for form finding and computation of general networks. Comp. Meth. Appl. Mech. Eng. 3 (1974) 115-134

［11］Vassart N, Motro. Multiparametered Form-finding Method：Application to Tensegrity Systems ［J］. International Journal of Space Structures, 1999, 14 (2)：147-154

［12］王志明，宋启根. 张力膜结构的找形分析 ［J］. 工程力学, 2002, 19 (1)：52-56

［13］冯虹，钱素萍，袁勇. 索膜结构分析理论研究综述与展望. 同济大学学报, 自然科学版, 2002, 30 (9)：1033-1037

［14］罗尧治. 索杆张力结构的数值分析理论研究. 浙江大学博士学位论文, 2000年

［15］袁行飞. 索穹顶结构的理论分析和实验研究. 浙江大学博士学位论文, 2000年

［16］詹伟东. 葵花形索穹顶结构的理论分析和实验研究. 浙江大学博士学位论文, 2004年

［17］胡宁. 索杆膜空间结构协同分析理论及风振响应研究. 浙江大学博士学位论文, 2004年

［18］陈联盟. Kiewitt型索穹顶结构的理论分析和试验研究. 浙江大学博士学位论文, 2005年

［19］郑君华. 矩形平面索穹顶结构的理论分析和实验研究. 浙江大学博士学位论文, 2006年

［20］袁行飞，董石麟. 索穹顶结构整体可行预应力概念及其应用 ［J］. 土木工程学报, 2001, 34 (2)：33-37, 61

[21] 董石麟，袁行飞．肋环形索穹顶初始预应力分布的快速计算法［J］．空间结构，2003，9（2）：3-8，19

[22] 董石麟，袁行飞．葵花形索穹顶初始预应力分布的快速计算法［J］．建筑结构学报，2004，25（6）：9-14

[23] 袁行飞，董石麟．索穹顶结构的新形式及其初始预应力确定［J］．工程力学，2005，22（2）：22-26

[24] 陈联盟，袁行飞，董石麟．Kiewitt型索穹顶结构自应力模态分析及优化设计［J］．浙江大学学报（工学版），2006，40（1）：73-77

[25] 郑君华，董石麟，詹伟东．葵花形索穹顶结构的多种施工张拉方法及试验研究．建筑结构学报，27（1）：112-117．2006

[26] 刘锡良．现代空间结构．天津：天津大学出版社，2003

[27] 刘锡良，夏定武．索穹顶与张力集成穹顶．空间结构，1997，3（2）：10-17

[28] 陈志华，王小盾，刘锡良．张拉整体结构的力密度法找形分析．建筑结构学报，1999，20（5）：29-35

[29] 唐建民，钱若军，蔡新．索穹顶结构非线性有限元分析．空间结构，1996（1）：12-17

[30] 张华，单建．张拉膜结构的动力松弛法研究．应用力学学报，2002，19（1）：84-86

[31] 沈世钊，徐崇宝．悬索结构设计．北京：中国建筑工业出版社．1977

[32] 姜群峰．松弛索杆体系的形态分析和索杆张力结构的施工成形研究．浙江大学硕士学位论文，2004

[33] 周岱．斜拉网格结构的非线性静力、动力和地震响应分析．浙江大学博士学位论文．1997

[34] 网架结构设计与施工规程（JGJ7-91）［S］．北京：中国建筑工业出版社，1991

[35] 汪树玉，杨德铨，刘国华等．优化原理、方法与工程应用［M］．杭州：浙江大学出版社，1991

第三章 | Chapter 3

索和膜结构在我国的应用、发展及存在问题

张其林

同济大学土木工程学院

上海市四平路1239号，200092，E-mail：zhangqilin@mail.tongji.edu.cn

摘　要：近十年来，新型结构体系在我国得到了非常广泛的应用。本文介绍了新材料和新体系的发展概况和在我国的应用和研究情况。针对其中发展迅速应用极广的建筑索结构和建筑膜结构，本文围绕设计计算理论、施工过程跟踪和工作应力检测等几方面，详细介绍了已有的研究成果和应用情况，提出了应进一步研究的相关问题。

关键词：索和膜结构，设计计算理论，施工过程跟踪，应力检测，研究进展

APPLICATION, DEVELOPMENT AND EXPECTATION OF CABLE AND MEMBRANE STRUCTURES IN CHINA

Q. L. Zhang

College of Civil Engineering, Tongji University, Shanghai, 200092, P. R. C

Abstract: In recent ten years new structure systems have been obtained widely applications in China. The development of new material and new systems in structure engineering is briefly reviewed, and its application and research progress in China are introduced in this paper. As new structure systems, cable and membrane structures have developed very quickly and obtained very wide applications in recent years in China. The research achievement and its applications in aspects of design and computation, construction process simulation, and working stress checking of cable and membrane structures are explained in details in this paper. Further work in this field is pointed out.

Keywords: Cable and membrane structures, design and computation, construction process simulation, working stress checking, research progress

1. 引言

　　土木工程的原始材料是土、木、石，质量大是其最主要的特点。这决定了原始土木结构中的基本单元或基本构件必然是以受压及受弯为主的柱、拱、壳、梁和板，基本体系必

然是桁架、框架、拱架等[1]-[3]。

轻质高强金属材料的出现及应用为土木结构体系的革命提供了契机[4]。一方面，轻质高强材料使得结构构件可以做得越来越细长，这促进了结构稳定分析和设计理论的发展，促进了组合构件和高效截面的发展和应用；另一方面，轻质高强材料又使得体系中受拉的基本构件数量增多、作用更重要，这促进了张拉体系和杂交体系的发展和应用[5]。

材料领域的成果促成了现代玻璃、合金和高分子材料的发展，这为人类实现对结构物的各类现代建筑功能（透光、防腐、质感等）提供了硬件方面的可能性，由此促进了新材料结构体系的应用和发展。机电和控制学科的成果也为人类实现结构物的各类现代使用功能和结构功能（可开启、可展开、可折叠等，隔震与隔振、主动控制、被动控制等）提供了技术条件，由此促成了新功能结构体系的应用和发展[5][6][7]。

20世纪30年代起，工业革命的成果开始充分应用于土木工程领域。其特点是钢筋混凝土截面、钢与钢筋混凝土组合构件的广泛应用和研究。钢筋混凝土截面是一种钢材与混凝土材料的组合截面，它充分利用了钢筋的高强度抗拉性能和混凝土材料良好的抗压性能。钢与钢筋混凝土组合梁及组合板则更进一步利用了钢材和混凝土的各自性能[8][9]。

20世纪60年代，西方发达国家进入土木建设的高潮期。与此同时，高强钢材、玻璃、合金、高分子等材料也已进入实用阶段。各类新型结构体系的发展始于20世纪60年代，并一直持续至今[10]-[16]。

我国大规模的土木建设高潮始于20世纪90年代。近十几年来我国新型结构体系的应用和发展是始于20世纪60年代的国际新型结构体系发展的继续，我国在这一领域的应用和研究在本质上是一个向国际先进水平模仿、学习和追赶的过程，在局部方面有我们自己的特点也有所突破，在总体上我们越来越接近国际先进水平[1]。

新型结构可分为新型材料结构和新型体系结构两大类。

1.1 新型材料结构

1.1.1 传统材料的改良

混凝土作为第一大建筑材料、钢材作为第二大建筑材料的高强度化研究已取得了较大的进展，并开始广泛应用于建筑结构中。

1.1.2 玻璃材料

玻璃是第三大建筑材料，但近十年来，利用玻璃的结构功能将玻璃作为主体结构材料的设计和应用研究已得到了越来越多的重视，并已建成了若干具有示范意义的玻璃结构体系，图1给出了近年国外建造的部分玻璃结构。

1.1.3 合金材料

建筑配件很早就采用不锈钢、铝合金、铝镁锰合金等材料。近十年来，合金材料作为结构主体构件材料的研究和应用也取得了较大进展，图2为部分实例。

图 1 玻璃结构体系

图 2 铝合金结构体系

1.1.4 高分子材料

碳纤维材料已广泛应用于结构加固，目前作为高强度拉索材料的研制也取得了一定进展。膜材料从聚脂纤维的 PVC、玻璃纤维的 PTFE 到无纤维的 ETFE，其研究和应用在国内取得了令人瞩目的成果。图 3 为国内兴建的具有代表性的这三类膜材料建筑。

图 3 膜结构应用实例

1.2 新型体系结构

1.2.1 高效结构

通过截面形式、结构布置和构件组合尽可能挖掘一种材料的潜能或充分发挥各种材料的优势。例如，通过改变薄钢板的形状可以大大提高同等质量钢板组成构件的极限承载力，并设计建造由薄钢板组成的轻型钢结构建筑（图 4）；根据钢材的冷拔效应可以制造高强度钢索，采用索网结构布置可以使绝大部分构件受拉，从而充分发挥拉索的抗拉强度，用很少的材料跨越很大的跨度；或通过拉索和刚性梁的组合充分利用拉索的抗拉强度和刚性梁的截面抗弯刚度，并通过拉索为刚性梁提供支承点，从而充分发挥各类构件的受力性能（图 5）。

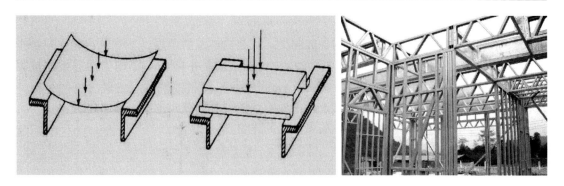

图 4　利用截面形式设计的轻型钢结构建筑

图 5　利用拉索的抗拉性能通过结构布置设计的高效索结构体系

1.2.2　杂交体系

采用不同的典型体系杂交组成单个建筑物可以生成新的集成结构体系。杂交体系可以通过传统技术和现代技术的对比，形成强烈的建筑视觉效果；或通过不同现代体系的组合来充分反映当代科技的发展水平。图 6 左图的体系中，采用铝框玻璃和拉索加劲的现代结构来覆盖传统砖石房屋间的街道，实现了与巴黎卢浮宫广场中建造玻璃金字塔类似的建筑效果；而图 6 右图的中国航海博物馆采用了直接汇交钢管桁架、索网和点支式玻璃幕墙、钢筋混凝土框架体系，充分反应了当代建筑结构设计计算理论和加工安装技术的发展水平。

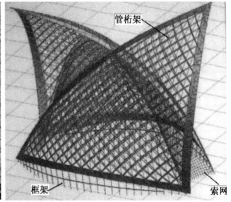

图 6　杂交体系

1.2.3 新功能结构

新功能包括新建筑功能和新结构功能两类。从结构体系的角度，新建筑功能主要是指能满足屋盖可开启和可展开等要求，而新结构功能主要是指能满足结构主动或被动抗震（振）等要求[7]。前者的结构中广泛采用了机械传动和控制等装置，主要为空间结构体系；后者广泛采用了阻尼器、传感器、运动激励和控制等装置，主要为高层建筑。

图 7 为近年来国内建造完成的屋盖可开启的建筑。

(a) 上海旗忠网球馆

(b) 昆明花博会舞台

(c) 南通体育场

图 7 可开启结构体系

纵上所述，建筑结构循材料的发展大致经历了由土木石→钢与钢筋混凝土→索与膜的路径，可由图 8（a）表示。而建筑结构循体系的发展大致经历了以受压为主→受拉为主→充气→组合（拉压组合）→多功能的路径，可由图 8（b）表示。

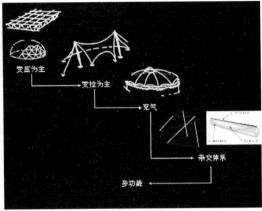

(a)循材料的发展　　　　　　　　　　　　(b)循体系的发展

图8　结构体系发展历程的示意图

近十年来，索和膜结构在我国的发展十分迅速、应用极其广泛，国内许多研究者在索和膜结构的计算方法方面进行了很多研究工作。但是，也正是这几年，商用软件在国内得以大量普及和推广，以至造成了这样一种现象：一个没有索和膜结构专业知识的工程师似乎都会以为自己可以借助商用软件完成这类结构的计算和分析。本文就近几年来计算和设计方法方面的研究成果作了比较详细的介绍，同时指出了商用软件在这类结构计算分析方面的局限性。对于索和膜结构的安全应用，计算和设计只是完成了第一个阶段的工作，实现施工张拉是第二个阶段的工作，建成及使用过程中结构安全性检测是第三个阶段的工作。后两个阶段工作的难度和重要性绝对不亚于第一阶段的工作，遗憾的是，即使在已经出现了如此多的工程破坏事故以后，关于后两个阶段的研究工作迄今尚未得到研究者的足够重视。本文就作者在这两方面的研究工作，作了比较全面的介绍，希望抛砖引玉，吸引更多的研究者投入其中，以完善索和膜结构的理论和技术体系。

2. 索结构计算设计理论的研究进展

2.1　材料和节点[17]

索由索体、护层和锚具组成，如图9所示。

图9　索的组成

2.1.1 索体

索体分为钢丝束索体和钢丝绳索体两种。其中钢丝束索体（spiral strand）由单股高强度钢丝或钢绞线扭绞而成，扭绞方式为平行和半平行，如图10所示；钢丝绳索体（wire rope）由钢丝束围绕绳芯扭绞而成，绳芯可采用纤维芯或钢芯，如图11所示。

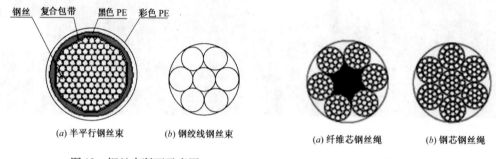

图10 钢丝束断面示意图　　　　　图11 钢丝绳断面示意图

钢丝束索体用高强度钢丝的直径一般为5mm和7mm，钢丝绳索体用高强度钢丝直径小于5mm。建筑结构中的钢丝绳索体，除膜结构中部分对拉索柔软性有要求的绳索可采用纤维钢丝绳外，其他应采用钢芯钢丝绳。

2.1.2 护层

设置护层是为了满足索体的防腐要求。

索体的防腐有简单防护和多层防护两种。简单防护是指对高强钢丝和钢绞线镀锌、锌铝、防腐漆、环氧喷涂，或对光索体包裹防护套；多层防护是指对高强钢丝和钢绞线经防腐处理后再对索体包裹防护套或润滑材料加防护套。

室内非腐蚀环境中的索体可采用简单防护处理，其他情况的索体应采用多层防护处理，具体要求宜根据不同工程不同索材在设计中注明。

2.1.3 锚具

按构造方式，锚具可分为冷铸锚、热铸锚、压接锚、夹片锚和墩头锚等。冷铸锚（Cast socket with epoxy and steel ball）的锚杯内采用冷铸带钢丸有机结合剂以固定索体。热铸锚（Alloy-filled cast socket）的锚杯内采用低熔点合金以填充固定索体。压接锚（Pressed ferrule）采用低合金、合金或其他高强钢材做成索套，在高压下与索体挤压成形固定。夹片锚（Clamping-yoke socket）采用由锚环或锚板和夹片组成的锚头固定索体的钢丝。墩头锚（Stamped-end socket）使索体中的钢丝穿过对应的锚板孔眼后进行墩头予以固定。

建筑结构中的索锚具一般采用压接锚、热铸锚或冷铸锚。

按连接方式锚具可分为销轴式连接和螺纹式连接；按调节形式可分为固定式连接和可调式连接，对于可调式连接可采用螺栓调整、套筒调整；按组成形式可分为叉耳式、单耳式。如图12和图13所示。

冷铸锚锚杯的坯件宜选用锻件，热铸锚锚杯的坯件可选用锻件或铸件。锻件材料应为优质碳素结构钢或合金结构钢。销轴、螺杆的坯件应选用锻件，其材料宜选用优质碳素结构钢或合金结构钢。压接锚和墩头锚的锚具组件宜采用低合金结构钢或合金结构钢。冷铸锚的铸体材料主要为环氧树脂和钢丸，热铸锚的铸体材料为锌铜合金。

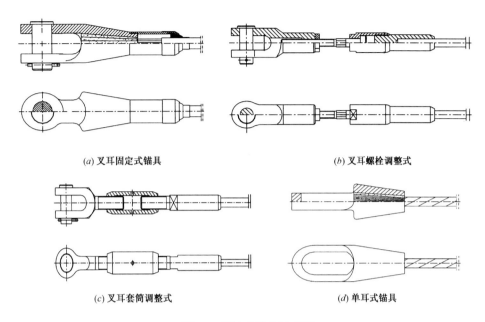

(a) 叉耳固定式锚具　　　　　　　(b) 叉耳螺栓调整式

(c) 叉耳套筒调整式　　　　　　　(d) 单耳式锚具

图 12　销轴式锚具示意图

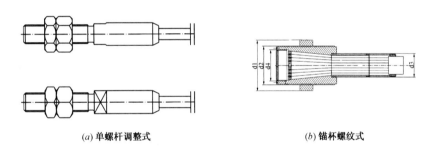

(a) 单螺杆调整式　　　　　　　(b) 锚杯螺纹式

图 13　螺纹式锚具示意图

锚具的强度应符合钢索破断后而锚具和连接件均不能破断的准则，必要时应通过试验来确定。

2.1.4　索的弹性模量和强度

高强钢丝和钢绞线的弹性模量在 1.9×10^5 MPa 的数量级上，在拉伸范围内一般可保持常数。索体的弹性模量一般小于钢丝和钢绞线自身的弹性模量，降低幅度取决于索体的绞合方式。对于建筑结构中使用的张紧索，要求钢丝束索体的弹性模量不应小于 1.9×10^5 MPa；钢丝绳索体的弹性模量，单股不应小于 1.4×10^5 MPa，多股不应小于 1.1×10^5 MPa。

拉索抗拉强度按下式设计或验算：

$$N_{\max}/A \leqslant f \tag{1}$$

式中　N_{max}——拉索的最大内力设计值；
　　　A——拉索截面积；
　　　f——拉索的强度设计值，$f = f_k / \gamma_R$；
　　　f_k——拉索的破断应力，为破断力 P 除以截面积 A；
　　　γ_R——拉索的抗力分项系数，取 2.0。

2.1.5　索的连接节点

（1）节点构造

常用的索与索、索与刚性构件的连接节点构造见图 14 所示。

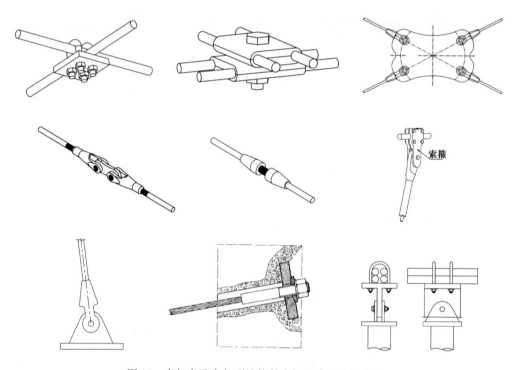

图 14　索与索及索与刚性构件之间的常用连接节点

（2）节点强度

索与索之间相连的连接件计算按与索等强原则考虑。索与其他刚性构件相连的连接件计算按实际内力考虑。

2.2　索结构体系

建筑索结构体系可以由某一类基本形式构成，也可以由几类基本形式组成，所以结构形式十分丰富。索结构的基本形式可总结和归纳为以下几种。

2.2.1　索桁架

上下弦均采用拉索，竖杆采用刚性撑杆。当承受向下荷载时，下弦索为受力索而上弦索为稳定索，反之亦然。

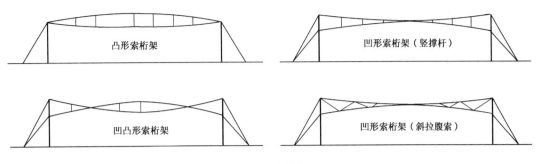

图 15　索桁架

2.2.2　索网

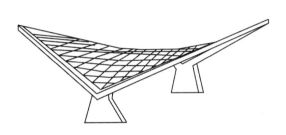
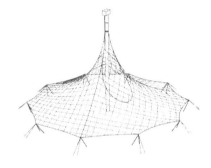

图 16　索网

索网是采用两向拉索在一个曲面上编织而成的，马鞍形和桅杆顶撑形是索网的两种基本形式。

2.2.3　索网格

图 17　索网格

两向相交布置的索桁架构成了索网格结构。

2.2.4　索穹顶

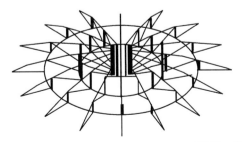

图 18　索穹顶

索穹顶由上弦索、斜索、环索和刚性撑杆及内外刚环圈组成,是一种全张力体系。

2.2.5 张弦梁和悬挂梁

张弦梁由刚性上弦梁、下弦拉索和刚性撑杆组成。这类结构在承受向下荷载时具有卓越的工作性能,但在承受向上荷载时必须注意下弦拉索不会因受拉而退出工作。

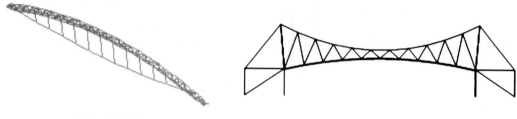

图 19 张弦梁　　　　　　　　图 20 悬挂索

悬挂索中受拉索位于上弦位置,刚性梁位于下弦位置,与张弦梁恰好相反,腹杆可是刚性杆或柔性索。这类体系类似于悬索桥结构。

2.2.6 斜拉索

斜拉索结构中桅杆支撑斜拉索为刚性屋盖提供了支撑吊点。这类体系类似于斜拉桥梁结构。

2.2.7 弦支穹顶

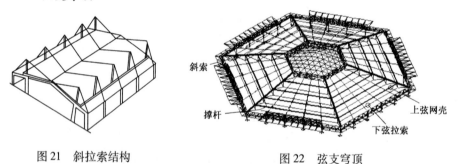

图 21 斜拉索结构　　　　　　图 22 弦支穹顶

弦支穹顶的结构布置类似于索穹顶,但上弦采用刚性构件代替拉索,所以结构体系相当于上弦的单层网壳受其下的索体系加劲支承。

2.2.8 体内预张力

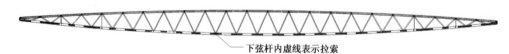

图 23 体内预应力索结构

体内预张力索结构的主要刚度由截面提供,拉索仅改善了结构初始状态的内力分布。

2.3 设计计算的基本概念和术语

在设计计算索结构时必须考虑结构的刚度特性和施工过程中结构的变形特性。所以,必须首先按照结构的刚度特性和施工特性将索结构体系进行分类。此外，索结构经历了加工放样、施工张拉成形、承受荷载变形的阶段，结构计算和设计也应分别考虑到这些阶段。

2.3.1 主动索和被动索

主动索：被张拉并控制预张力值的索段；

被动索：因主动索张拉致结构变形后才受拉的索段。

2.3.2 零状态、初始状态和工作状态

零状态：加工放样时各构件和结构的几何状态；

初始状态：主动索张拉到控制值时的结构几何和内力分布状态；

工作状态：在外部荷载作用下结构自初始状态起变形受力后的状态。

图 24 为索结构三个状态相互关系的示意。

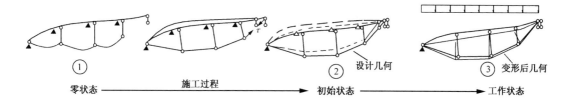

图 24　索结构的三个状态

2.3.3 刚性结构、柔性结构和半刚性结构

刚性结构：结构体系各方向的刚度主要由构件截面刚度提供；

柔性结构：结构体系各方向的刚度主要由索张力提供；

半刚性结构：结构体系某些方向的刚度主要由构件截面刚度提供，其他方向刚度主要由索张力提供。

图 25 为结构节点的预张力刚度和刚度示意。

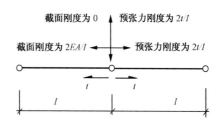

图 25　索结构节点的刚度

2.3.4 索杆体系和索梁体系

变形协调结构：张拉索时各构件同时变形和受力，也称索梁体系；

变形不协调结构：张拉索时各构件不同时变形和受力或有被动索退出工作，也称索杆体系。

图 26 为索梁体系和索杆体系示意。

 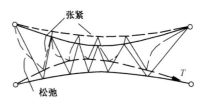

图 26　索梁体系和索杆体系

2.4 索结构初始状态的计算分析

索结构初始状态的确定和分析是索结构设计的关键。一旦初始状态分析得以完成，工作状态的计算本质上是针对具有自相平衡初内力的结构的计算，借用一般商用软件即可完成。

索结构初始状态计算可分为三种情况：控制几何确定内力；控制内力确定几何；同时确定内力和几何。

2.4.1 控制几何确定内力

（1）索杆体系[18][19]

对于索杆体系，因为无须考虑施工过程的变形协调条件而仅需满足平衡条件，可以直接建立索杆体系每个节点的平衡条件，引入边界条件后进行矩阵分析以确定结构在给定几何下的一组内力分布。可以采用最小二乘法确定与理想内力分布最为接近的一组内力。

平衡方程的一般表达式：

$$[A]_{N \times M} \{S\}_M = \{F\}_N \tag{2}$$

上式中 N 为索杆部分节点非约束自由度总数。$[A]_{N \times M}$ 为 $N \times M$ 阶系数矩阵。$\{S\}_M = [s_1 s_2 \Lambda s_M]^T$ 为 M 阶索杆单元预应力分布矢量。$\{F\}_N = [f_1 f_2 \Lambda f_N]^T$ 为 N 阶自重荷载矢量。

采用高斯消去法对公式（2）进行矩阵变换运算，可得：

$$\begin{bmatrix} [A_{r,r}] & [A_{r,k}] & [A_{r,l}] \\ [O_{N-r,r}] & [O_{N-r,k}] & [O_{N-r,l}] \end{bmatrix} \begin{Bmatrix} \{S_r\} \\ \{S_k\} \\ \{S_l\} \end{Bmatrix} = \begin{Bmatrix} \{F_r\} \\ \{F_{N-r}\} \end{Bmatrix} \tag{3}$$

上式中，r 为矩阵 $[A]_{N \times M}$ 的秩。$\{S_l\}$ 为对 l 根主动索所预先给定的控制预张力向量。$\{S_k\}$ 为求解 $\{S\}_M$ 所需确定的 k 根索杆的未知预应力向量，$k = M - r - l$。

一般索结构的预张力分布表达式为：

$$\{S\}_M = \begin{Bmatrix} \{S_r\} \\ \{S_k\} \\ \{S_l\} \end{Bmatrix} = \begin{Bmatrix} [A_{r,r}]^{-1}(\{F_r\} - [A_{r,l}]\{S_l\}) \\ \{O_k\} \\ \{S_l\} \end{Bmatrix} + \begin{bmatrix} -[A_{r,k}] \\ [I_{k,k}] \\ [O_{l,k}] \end{bmatrix} \{S_k\}$$

或

$$\{S\}_M = \{C_0\} + \sum_{i=1}^{k} S_{ki} \{C_i\} \tag{4}$$

上式中，$\lfloor I_{k,k} \rfloor$ 为 $k \times k$ 阶单位对角阵。$\{O_k\}$，$\lfloor O_{l,k} \rfloor$ 分别为 k 阶零矢量和 $l \times k$ 阶零矩阵，S_{ki} 表示索段未知预应力向量 $\{S_k\}$ 中第 i 个元素。$\{C_0\}$ 和 $\{C_i\}$ 为已知系数向量。

由式（4）可见，与给定几何相对应的预张力分布可以有无穷多个线性组合可能。假定期望的预张力分布为 $\{S_0\}$，自平衡的预张力应该为与 $\{S_0\}$ 最为接近的一组数值。根据最小方差原则可以确定这样的一组初始状态预张力。

预张力方差为：

$$M_s = \|\{S\}_M - \{S_0\}\| = (\{S\}_M - \{S_0\})^T (\{S\}_M - \{S_0\}) \tag{5}$$

根据 M_s 最小的条件推导后可得：

$$[BB_s]\{S_k\} = \{B_s\}$$
$$\{S_k\} = [BB_s]^{-1}\{B_s\} \tag{6}$$

上式中，$[BB_s] = \begin{bmatrix} [C_1] \\ \Lambda \\ [C_i] \\ \Lambda \\ [C_k] \end{bmatrix} [\{C_1\}\Lambda\{C_j\}\Lambda[C_k]]$，$\{B_s\} = \{S_0\} - \begin{bmatrix} [C_1] \\ \Lambda \\ [C_i] \\ \Lambda \\ [C_k] \end{bmatrix} \{C_0\}$。

由式（6）确定了向量 $\{S_k\}$ 后回代入式（4）可得到整个结构中的索段预张力分布 $\{S\}_M$。

索桁架、索网格、索穹顶结构常采用这一方法确定初始状态。

（2）索梁体系

对于索梁体系，必须考虑施工过程的结构变形协调条件，即自零状态至初始状态，结构构件随着主动索的张拉同时变形受力。所以，结构的内力分布除了满足平衡条件外还需满足变形协调条件。

对于刚性结构，自零状态至初始状态结构的变形很小、可以忽略不计，可以直接对主动索采用撤杆加力的方法求解结构体系与主动索控制力相对应的预张力分布。

对于半刚性结构和柔性结构，必须采用逆迭代法求解结构满足给定几何的初始状态内力分布，求解后还可同时得到结构的零状态几何。

张弦梁或弦支穹顶常采用这一方法确定初始状态。

2.4.2 控制内力确定几何

对于索网结构，有时要求索网内力或力密度均匀，并在此条件下确定索网的曲面几何。这时，采用有限单元法或力密度法可以确定给定内力下的索网几何[5]。

2.4.3 同时确定几何和内力

当建筑设计对结构几何控制不很严格的情况下，可以同时假定初始状态几何和内力分布，当内力分布在假定的几何条件下不平衡时，迭代计算得到满足平衡条件的内力分布及其对应的几何。如果这样得到的几何与预期相差较大时，可经数次试算以确定合适的结构几何和对应的平衡内力。采用商用软件可以较容易地完成这类问题的计算。

2.5 索结构工作状态分析中的若干问题[17]

2.5.1 关于预张力值的分项系数

按照《建筑结构荷载规范》GB 50009—2001，按恒载取预张力的分项系数。即，对结构不利时，初始状态内力的分项系数均取 1.2；对结构有利时取 1.0。

2.5.2 关于结构分析方法

索结构的工作状态原则上必须采用几何非线性方法进行分析。进行地震分析时除了按照初始状态刚度进行反应谱分析外，对于复杂结构还应进行非线性时程计算。

2.5.3 是否容许索退出工作

在永久荷载效应控制的组合作用下，不能容许索退出工作；在短期荷载效应控制的组合作用下，容许索退出工作，但索的退出不应导致结构破坏。

3. 膜结构计算设计理论的研究进展[20][21]

3.1 材料

膜材料由基材和涂层组成，或由无基材的高分子高强度材料构成。

基材可分为玻璃纤维、聚脂类、聚乙烯醇类、聚酰胺类纤维等。涂层可分为聚四氟乙烯（PTFE）、聚偏氟乙烯（PVDF）、聚氯乙烯（PVC）、氟化树脂等。常用的PVC膜材是指聚脂纤维基材加PVC涂层材料，常用的PTFE膜材是指玻璃纤维基材加PTFE涂层材料。一般而言，PVC膜材的强度和耐久性都要较PTFE的差。图27为由基材和面层组成的膜的构成、受力及应力应变关系示意。

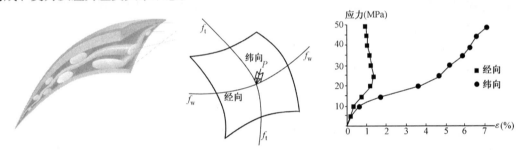

图27 膜材构成及应力应变关系示意

膜材是一种各向异性材料，应力应变曲线呈现明显的非线性特征，在经纬向不同比例荷载作用下有一定差异。一般取单向加载条件下在正常应力水平范围内的应力应变曲线斜率的平均值作为材料的计算弹性模量。

3.2 膜结构体系

膜结构由膜面和支承体系组成。膜结构体系可由某一基本形式构成，也可由几类基本形式组成。膜结构的基本形式有以下几种。

3.2.1 张拉膜结构

张拉膜结构中，支承体系与膜面一起构成结构，没有膜面的共同作用支承体系不能单独工作。这里，膜面包括膜及其与膜相连的加劲索（脊索、谷索、边索等）。图28为一张拉膜结构实例。

图28 张拉膜结构

图29 框支膜结构

3.2.2 框支膜结构

框支膜结构中支承体系作为主结构可以独自工作,膜面只是维护结构,不参与支承体系共同作用。图 29 为框支膜结构实例。

3.2.3 气承式膜结构和气囊式膜结构

气承式膜结构在膜结构中充气形成结构内外的气压差以支承膜面。气囊式膜结构是在封闭的膜内充气以支承膜面,如图 30 所示。

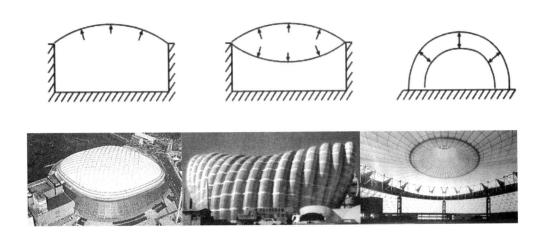

图 30 气承式膜结构和气囊式膜结构

3.2.4 充气加劲膜结构

图 31 为充气加劲膜作为受弯桁架梁的工作原理示意。

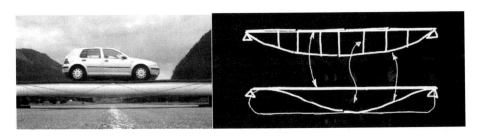

图 31 充气加劲膜结构

这里,斜环索相当于桁架梁受拉下弦,不可压缩空气相当于刚性撑杆,膜管上部的刚性构件相当于桁架梁中的受压上弦。

3.2.5 可展和开合膜结构

可展膜结构是指视需要可以将膜面展开和收起的膜结构。开合膜结构是指可以通过移动开启和闭合的膜结构,如图 32 所示。

(a)可展膜结构　　　　　　　　　　(b)开合膜结构

图 32　可展膜结构和开合膜结构

3.3　膜结构的设计计算理论

与索结构类似,膜结构的计算也可分为零状态、初始状态和工作状态三个阶段。其中,膜结构的初始状态确定一般称为找形分析;膜结构的零状态确定一般称为裁剪分析;膜结构的工作状态分析是指在外部荷载作用下以初始状态几何和应力分布为初始条件的非线性计算和分析。

3.3.1　膜结构的找形分析

膜结构的找形分析实际上包括两个类型:膜面找形;膜面和支承体系的协同作用。

（1）膜面找形[22][23]

膜面找形时可将膜面与支承体系连接处取为固定边界。

膜面的形状必须满足边界条件和预应力自平衡条件。从结构角度,要求膜面预应力均匀、刚度均匀;从建筑角度,要求膜面曲面光滑和美观、易于排水。膜面有最小曲面和平衡曲面两种。最小曲面是指膜面预应力自相平衡且相等的曲面。在给定的几何边界下,这样的曲面具有最小的面积;平衡曲面是指膜面预应力仅满足平衡条件但不相等的曲面。最小曲面同时具有最佳的结构性能和建筑性能。

早期的膜面找形一般采用物理模拟法,物理模拟法包括肥皂膜和弹性尼龙膜模拟两类。图 33 为弹性尼龙膜模拟示意。

膜面找形的数值分析法有力密度法、动力松弛法和有限单元法等。找形分析时,可以将膜面划分为膜单元或索网,但基于索网找形完毕后必须将索网预张力转换为膜单元预应力以便进行膜结构的工作状态分析。

采用力密度法可以直接求解得到平衡曲面,也可迭代求解得到最小曲面。动力松弛法的优点是无须建立结构的平衡方程,因而无须分解矩阵求解方程,缺点是需迭代上千次才能得到收敛解。有限单元法的优点是迭代收敛快,但需要建立膜面的平衡方程,因而占用内存较多、求解方程较耗时。一般而言,动力松弛法和有限单元法的求解效率相当,基于索网的力密度计算效率最高。

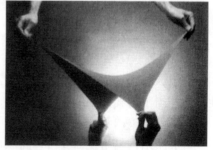

图 33　弹性尼龙膜找形示意

（2）膜面与支承体系的协同作用[24]

膜结构设计时，支承体系的几何位置一般是由建筑要求事先给定的。膜面找形完成后，与支承体系连接处的边界反力作用于支承体系上，支承体系的内力分布应根据给定几何和膜面反力进行确定，本质上属于给定几何的结构初始状态确定问题。从这一角度出发，可以根据支承体系类型按前述的索杆体系的矩阵方法或索梁体系的逆迭代法确定支承体系初始状态预张力分布，而支承体系零状态的几何可另行分析确定。这样的分析是建立在膜面与支承体系分离的模型基础上的。

近年来，有很多文献采用商用软件在给定的支承体系几何基础上，进行膜面与支承体系协同作用的分析，这样的分析固然是精确的。但必须谨记：支承体系几何满足给定的要求是基本的设计目的。所以，只有当支承体系具有足够刚度时，协同分析结果不会产生较大变形，这样的结果才能作为结构找形结果，而这种情况下分离模型的计算结果已经足够精确，所以协同分析是没有意义的。当结构较柔时，采用商用软件进行计算必须反复假定结构零状态几何进行分析直到得到与给定几何足够接近的支承体系与膜面内力分布，显得十分复杂和麻烦。事实上，在前述基于分离模型得到的支承体系初始状态基础上，对于索杆体系可以直接计算得到零状态杆长，而对于索梁体系可由逆迭代法得到零状态几何，然后再在零状态基础上进行协同分析以检验支承体系和膜面是否达到设计要求，这样的协同分析更有意义。

3.3.2 膜结构的裁剪分析

膜结构的膜面是通过找形得到的空间曲面，而膜材是具有一定幅宽的平面。裁剪分析要解决的问题是如何由膜材平面通过制造和安装生成膜结构曲面。显然，由平面生成曲面的唯一途径是：将曲面划分为一定尺寸的若干曲面片，寻找曲面片的近似平面片（即裁剪片），将裁剪片拼接后张拉成曲面。所以，裁剪设计包含了两个过程：（1）确定空间曲面的划分线（裁剪线）[26]；（2）将空间曲面上的曲面片展开成膜片或裁剪片[25]。图34为膜结构裁剪分析示意。

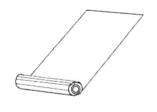
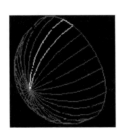

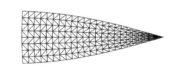

图34 膜结构裁剪分析示意

裁剪线的确定应满足两个要求：美观（规则排列）和经济（利用幅宽）。从美观角度出发，可以将空间平面与膜曲面切面的相交线取为裁剪线，按照美观要求控制裁剪线在曲面上的空间位置，但其缺点是难以把握每个裁剪片的幅宽、用材不经济。从经济角度出发，可以将膜曲面上的测地线取为裁剪线，因为测地线是曲面上两点之间距离最短的线段，对于可展曲面测地线展开后是直线，对于不可展曲面测地线展开后接近直线，这样可控制幅宽、节省材料，但其缺点是难以把握裁剪线在曲面上的空间位置。图 35 和图 36 分别为裁剪线为平面和膜面相交线的情况及裁剪线为测地线的情况示意图。

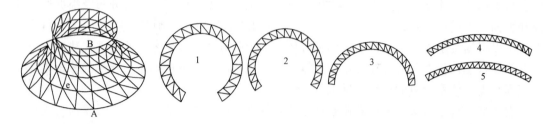

图 35 裁剪线为水平面与膜面的相交线及裁剪片展开图

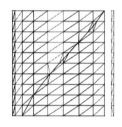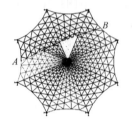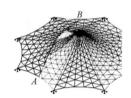

图 36 柱面和伞面上的测地线

在曲面片的展开方面，对于可展曲面（如柱面），可精确展开为平面；对于不可展曲面（如球面），只能近似展开为平面。近似展开方法可采用等效有限单元法、初始热应力法、单排三角形网格近似展开法等。这里，重要的是如何确定近似展开平面的精度判据。可以将空间曲面由三

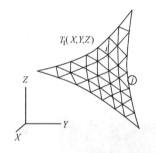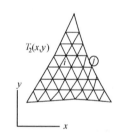

图 37 空间曲面展开示意

角形网格线描述（图 37），判据之一是：曲面网格线和相应的展开平面上网格线长度差的平方和最小；判据之二是：空间三角形面积和相应的展开平面上三角形面积差的平方和最小。

3.4 有待研究的相关问题

3.4.1 关于膜材的计算和设计参数

膜材是一种各向异性材料，在双向受力条件下，经纬向的弹性模量与两向荷载比例有关。目前一般取单向弹性模量测试的平均值作为膜材本构关系中经纬向的弹性模量，但精确的本构关系模型还有待于进行大量的膜材双向受拉实验和数值分析进行充分的研究。图 38 为国内研制的双向张拉设备及其测试结果示意。

第三章 索和膜结构在我国的应用、发展及存在问题

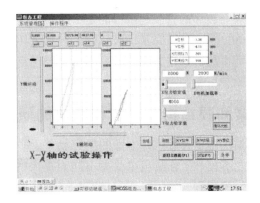

图 38 膜材双向张拉设备及其测试结果

此外，关于膜材强度的标准值和设计值的取用，还需针对不同厚度不同类型的材料进行大量的单向拉伸材性实验，在此基础上进行概率统计分析和研究。

3.4.2 膜面褶皱的数值模拟

国内相关规程规定：在永久荷载效应控制的组合作用下，不容许膜面松弛而退出工作；在短期荷载效应控制的组合作用下，容许一定面积的膜材松弛而退出工作。部分膜面退出工作后，一般膜单元有限元理论将失效，必须采用能考虑膜面褶皱的膜单元理论才能分析膜面其后的工作性能。在这一领域已有很多研究成果，但大部分还没有达到实用水平。

3.4.3 膜面破坏机理研究

膜面的破坏大都发生在极端条件如积雪积水、强台风作用等，图 39 为国内近年发生的膜结构工程破坏实例。

图 39 膜结构工程破坏实例

膜面破坏的直接起因可能是由于膜面应力超过设计强度，也可能是由于外部尖锐物的破坏等。不充分研究膜面破坏的机理和破坏过程就无法科学鉴定实际工程中发生的大量膜面破坏事故。在这方面的研究基本上还是空白。

4. 索和膜结构施工过程跟踪的研究进展

结构施工过程跟踪方面的研究始于上世纪80年代，当时主要针对钢筋混凝土框架在浇捣、拆模和工作阶段梁受力不同而提出的。到了上世纪90年代，大型钢结构在国内开始大规模建设，施工过程中与安装完成后钢结构构件的受力差异更大，因为忽略这样的差异而导致了重大工程事故，这样就提出了施工过程跟踪或所谓的施工力学的问题。

结构施工过程跟踪主要要解决两个方面的问题：一是能考虑结构体系中单元、节点、边界和荷载的变化，逐个阶段对这样的结构体系进行几何非线性计算分析；二是能考虑施工安装阶段结构构件和体系的机构运动和弹性变形的耦联作用问题。近几年，一些商用软件通过锁死和激活单元（节点）的方法提供了进行结构施工过程跟踪的可能性。但是，这些方法只解决了上述第一个方面的问题。运用商用软件进行框架和桁架结构的施工过程跟踪是比较方便的。但是，对于索结构和膜结构的施工过程，即使对于仅经历弹性变形的问题，一般商用软件就很难处理，更无法求解机构运动和弹性变形的耦联问题。

4.1 仅经历弹性变形的索结构施工过程跟踪理论[27]

4.1.1 考虑张拉过程的主动索计算模型

主动索在张拉过程中，张拉力 T 作用下的一端的索会拉出，从而发生长度改变，而另一端索会随所连其他结构部分发生变形，如图40所示。

第1至 $M-1$ 个索单元相当于被动索段，其预张力可按下式进行计算：

图40 考虑主动索张拉的力学模型

$$P_i = P_0 + \frac{L_i - L_{i0}}{L_{i0}} EA \qquad i = 1, 2, \cdots, M-1 \tag{7}$$

L_i 和 L_{i0} 分别为第 i 个索段张拉后的长度和零状态索段长度，L_i 根据单元两端节点位移计算得到；EA 为索的截面刚度；P_0 为索的零状态张力。

索端 M 单元为主动索段，形成单元刚度时零状态长度取：

$$L_{M0} = \frac{L_M}{(P_M/EA + 1)} \tag{8}$$

索端单元长度 L_M 根据迭代过程中支座节点和另一端节点坐标计算得到。

4.1.2 考虑安装过程中结构变形影响的几何确定方法

结构在安装过程中，因自重作用，已安装的结构部分会产生一定的变形，新安装构件的几何必须在变形后的已安装段的基础上予以确定。以 (o) 表示已安装结构段，(n) 表示新安装构件段，(n) 部分构件必须循 (o) 部分的切线、或直接连接两端的 (o) 部分、或取合龙连接点为各自 (o) 部分切线延伸线的平均值等，见图41。

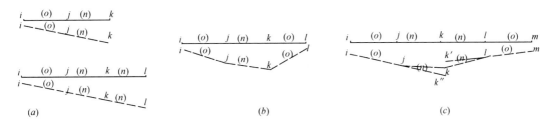

图41 考虑安装过程中结构变形的新增构件几何确定方法

考虑一般情况，新增节点 k 的坐标可表示为下式。

$$X_i^k = \frac{1}{N} \sum_{j=1}^{N} \left(X_i^j + \sum_{s=1}^{M} (l_i^j \times |\overrightarrow{JK_s}|) \right) \quad (9)$$

式中：X_i^k 为新节点 k 在 i 方向的坐标值；X_i^j 为相连的已有节点 j 在 i 方向的坐标值；l_i^j 表示相连的已有构件 IJ 在 i 向的方向余弦值；$|JK_s|$ 为第 s 根新构件的长度；N 为新增构件与已有构件相连的节点数；M 为与第 j 个已有节点相连的新增构件数。

4.1.3 考虑结构支撑拆卸过程中的特殊支承节点

结构支撑拆卸过程中，根据施工控制指令按给定位移卸载，这时支承节点是单向受力的，即只能受压，如计算位移超过给定值时支座脱开但不会受拉。数值模拟时需要推导给定位移下的单向接触单元。

结构给定位移的节点位移列阵为 $\{\Delta_2\}$，其他未知位移列阵为 $\{\Delta_1\}$，应将结构节点位移列阵分块写为：

$$\{\Delta\} = \begin{cases} \{\Delta_1\} & \text{未知位移列阵} \\ \{\Delta_2\} & \text{已知位移列阵} \end{cases} \quad (10)$$

与此对应，$[K]$ 与 $[P]$ 也作相应分块，结构平衡方程可按分块形式展开为：

$$\begin{aligned} [K_{11}]\{\Delta_1\} + [K_{12}]\{\Delta_2\} &= \{P_1\} \\ [K_{21}]\{\Delta_1\} + [K_{22}]\{\Delta_2\} &= \{P_2\} \end{aligned} \quad (11)$$

由式（11）第一项公式可得未知位移列阵为由：

$$[K_{11}]\{\Delta_1\} = \{P_1\} - [K_{12}]\{\Delta_2\} \quad (12)$$

与已知位移对应的约束反力矩阵可从式（11）中第二项公式求得：

$$\{X\} = [K_{21}]\{\Delta_1\} + [K_{22}]\{\Delta_2\} - \{P_{eq,2}\} \quad (13)$$

上式中 $\{P_{eq,2}\}$ 为与已知位移相应的固端反力向量。

考虑单向受力的连接单元，可以按下面的列式处理[28]。

$$\begin{cases} \alpha[[k_0]_i + [k_u]_i + [k_\sigma]_i]_t^{(t+1)} \{\Delta u\}_{t+1} = \alpha[\{f\}^{(n)} + \{\Delta f\} - \{f_R\}_t^{(n+1)}] \\ \alpha = \begin{cases} 1 & \sigma_{i,t} > 0 \\ 0 & \sigma_{i,t} \leq 0 \end{cases} (tension\ only) \quad \alpha = \begin{cases} 1 & \sigma_{i,t} < 0 \\ 0 & \sigma_{i,t} \geq 0 \end{cases} (compression\ only) \\ \Delta\varepsilon_{i,t+1} = [B_i]_t \{\Delta u\}_{t+1} \quad \varepsilon_{i,t+1} = \varepsilon_{i,t} + \Delta\varepsilon_{i,t+1} \\ \Delta\sigma_{i,t+1} = E\Delta\varepsilon_{i,t+1} \quad \sigma_{i,t+1} = \sigma_{i,t} + \Delta\sigma_{i,t+1} \\ \{u\}_{t+1} = \{u\}_t + \{\Delta u\}_{t+1} \\ \{f_R\}_{t+1}^{(n+1)} = \int_v [B_i]_{t+1} \sigma_{i,t+1} dv \\ t = 0,1,\cdots \end{cases} \quad (14)$$

上式中，$i=1, 2, \cdots, m$ 为单元数，t 为迭代步数；$\Delta\varepsilon$，ε，$\Delta\sigma$，σ 分别为增量应变、应变、应力增量和应力项；$[B_j]$ 单元几何矩阵；E 为弹性模量；$\{f\}$，$\{\Delta f\}$ 和 $\{f_R\}$ 分别为节点力、不平衡力以及内力向量。

4.2 膜结构计算设计误差的近似数值检验方法[29]

膜结构的计算设计遵循了由找形到裁剪的过程，但膜结构的实际制作安装是从裁剪片拼接完成到张拉成形的过程，正好与计算设计过程相反。因为裁剪片一般都是不可展曲面片，所以设计得到的裁剪片都是近似的。有这样近似的裁剪片拼接后张拉得到的实际膜曲面与找形得到的理论曲面之间必然存在误差。要检验这样的误差，必须完全模拟各裁剪片拼接后张拉成形的全过程，这在数值计算上是十分困难的。但是，可以忽略这样的过程，近似通过数值分析计算得到由裁剪片拼接张拉后的膜曲面，以检验其与理论曲面的误差。

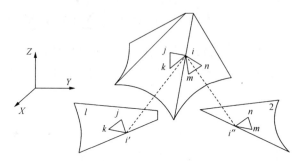

图42 裁剪片连接与张拉计算过程示意

图42 表示膜面上某一节点 i 位于裁剪片 1 和 2 拼接线处，对应于裁剪片 1 和 2 的节点分别为 i' 和 i''，而膜面上的单元 ijk 和 imn 分别与裁剪片 1 和 2 中的单元 $i'j'k'$ 和 $i''mn$ 相对应。显然，直接找形得到的膜面与裁剪片 1 和 2 拼接张拉得到的膜面间必然会存在差异。

可以以一个悬链面结构来说明这样的差异。图43 为悬链面的控制尺寸、找形结果（应力为 3N/mm² 的等应力曲面）及裁剪片形状。

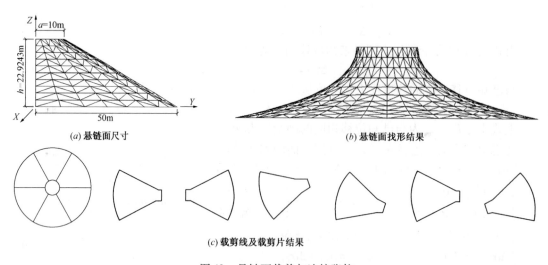

(a) 悬链面尺寸

(b) 悬链面找形结果

(c) 裁剪线及裁剪片结果

图43 悬链面裁剪与连接张拉

由裁剪片拼接张拉得到的实际膜面的最大和最小应力见图44 所示。

第三章　索和膜结构在我国的应用、发展及存在问题

(a) 实际结构最大应力显示

(b) 实际结构最小应力显示

图 44　由裁剪片拼接张拉得到的膜面最大和最小应力

4.3　考虑弹性变形和机构运动耦联作用的数值理论

日本东京大学的半谷彦教授最早借助电子理论中的广义逆矩阵方法导出了基于刚性构件几何长度不变的数值理论以模拟结构的机构运动，这一概念得到了其后很多学者的进一步应用和发展[30][31][32]。这里，介绍与这一思路不同的另外两种方法的研究成果。

4.3.1　基于广义逆矩阵的有限单元理论[33][34]

任意体系节点平衡方程的增量形式可写为：

$$[K]^{(n)}\{\Delta U\} = \{F\}^{(n+1)} - \{F_R\}^{(n)} \quad (15)$$

当体系为结构时，刚度矩阵不奇异，可遵循一般非线性有限单元的迭代格式求解上述方程得到给定荷载下的增量位移。

当体系存在机构自由度时，刚度矩阵奇异。根据广义逆矩阵理论，上述方程的解可写为：

$$\{\Delta U\} = \{\Delta U_G\} + \{\Delta U_S\}$$

$$\{\Delta U_S\} = ([K]^{(n)})^+ (\{F\}^{(n+1)} - \{F_R\}^{(n)})$$

$$\{\Delta U_G\} = ([I] - ([K]^{(n)})^+[K]^{(n)})\{\alpha\} \quad (16)$$

这里，$[K]^+$ 表示 Moore 广义逆矩阵；$\{\alpha\}$ 为运动约束方向矢量，$\{\alpha\} = \alpha_0 \{F\}^{(n+1)}$，$\alpha_0$ 为运动增量控制参数；$\{\Delta U_G\}$ 为方程的通解，其物理意义为在给定运动约束方向上的位移增量；

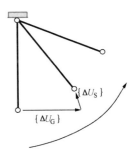

图 45　机构运动和弹性变形耦联问题中的位移增量

[ΔU_S]为方程的特解,其物理意义是体系为在新状态满足平衡条件所应作的位移增量修正。见图45所示。

考虑一个攀达穹顶的顶升过程,假定该穹顶顶升过程中侧向有支撑。顶升过程中体系经历了机构运动和弹性变形,计算跟踪结果见图46所示:

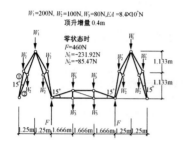 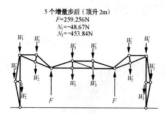 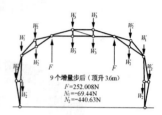

图46 攀达穹顶的顶升过程

公式(15)和(16)适用于任意结构体系,上述方法原则上也适用于索和膜结构施工过程的求解。但是,这一方法的成败取决于广义逆矩阵的计算精度和迭代求解的收敛效率。

4.3.2 基于结构运动分解方法的弹性体运动理论[35][36][37][38]

在每个弹性体单元上建立一个移动参考系,将该单元的运动分解为移动参考系相对于惯性参考系的刚体平动和刚体转动以及相对于移动参考系的弹性变形,如图47所示。

将单元上一点在惯性坐标系中的位置[24]表示为:$u = R + A(u_0 + \Phi f)$。这里,Φ 为弹性位移模式,移动坐标系内的矢量 $(u_0 + \Phi f)$ 通过旋转变换矩阵 $A = A(\varphi)$ 变换到惯性坐标系中,再与随动坐标系的平移 R 叠加得到其惯性坐标。设系统的广义坐标为 $q = \{R\ \varphi\ f\}^\mathrm{T}$,体系的支座约束和单元之间的连接由约束方程 $C(q, \dot{q}, t) = 0$ 表示。

计算系统的动能并带入含乘子的第二类拉格朗日方程得到系统的动力学控制方程[35][36]:

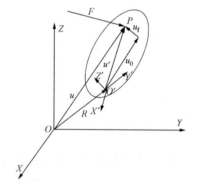

图47 用相对坐标描述的弹性体单元

$$M\ddot{q} + \dot{M}\dot{q} - \frac{\partial}{\partial q}\left(\frac{1}{2}\dot{q}^\mathrm{T} M\dot{q}\right) + Kq = Q_\mathrm{F} \tag{17}$$

解控制方程与约束方程的联立构成的微分-代数混合方程[39][40]可以得到系统的运动时程。

图48为某奥林匹克体育场屋盖平面钢桁架起吊翻身过程的跟踪仿真。起吊时在上弦节点设置两根吊索,计算得到的桁架翻身过程如图48所示。

图49所示为一找形所得的帐篷形索网结构[34],$EA = 3141.59\mathrm{kN}$,中心刚环直径1m,外径11m,中心环的高度为6m。以各索段相互连接放置于工作平台上作为零状态,此时所有索均处于松弛状态。通过竖直提升中心的刚性拉环使其达到预先设计的初始状态,提升过程如图49所示。

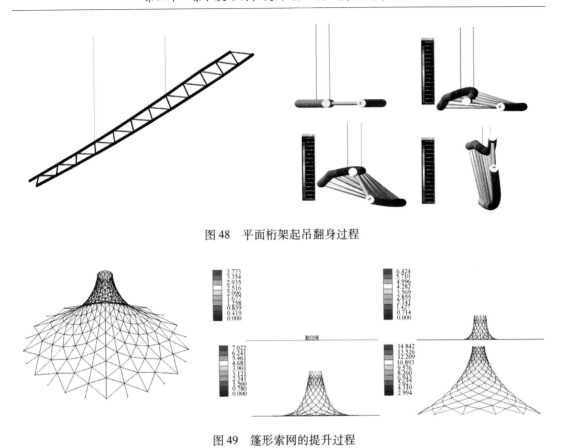

图 48 平面桁架起吊翻身过程

图 49 篷形索网的提升过程

跟踪仿真结果除了可以得到每一时刻的位移形状外，还可得到结构相应的内力分布及在这一过程中的稳定性能。

5. 索和膜结构工作应力的检测原理和方法

5.1 索工作应力的检测原理和方法

索力的检测不是一个新的研究课题，已有很多学者提出了很多方法，其中有些已成功地应用于实际工程中。其中，振动测试法在桥梁拉索索力检测中已有相当长的应用历史。对于幕墙结构中无护套且直径较小的索也可采用两端固定中间顶力的三点测力仪测量拉索内力。近年来，国内建筑索结构中索力检测多采用振弦式应变仪，但应用时必须特制一个钢的连接件以固定应变仪。因为应变仪只能测量表面应变，当连接件受力偏心时，索力的检测结果就会有误差。此外，使用时仍需十分小心，一旦碰倒应变仪，检测就告失败[41][42][43]。

这里，介绍一个简便易行的新的检测原理及其设备，即基于 EM 原理的索力检测方法。这一方法在发达国家业已开始实用，在国内也有望进入实用。这一检测方法不仅可以进行索力的检测，如需要也可进行索力的长期监测。

EM 基本原理为：铁磁材料的磁化性质可以通过材料的磁导率 μ 充分反映出来，而材料的磁弹性能 E_σ 与外加应力状态 σ 又息息相关，所以如果我们能得到材料的磁导率，就必然可以通过一定的方法得到材料的应力状态 σ[44][45]。对于一个线圈，内部有铁芯时的磁感应强度是无铁芯情况下的 μ 倍，这是磁导率 μ 的物理意义。据此可以通过在索上套线圈来测试磁导率 μ，进而测试索拉力。图 50 为线圈的基本构造图。

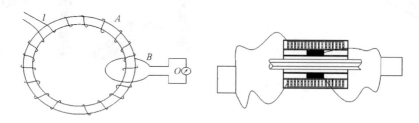

图 50　线圈基本构造图

根据 EM 原理，可以首先在实验室条件下标定与待测实际工程中拉索具有同等规格的索段，根据标定得到的输出电压与外加拉力的关系，得到实际工程中拉索的拉力。这一原理的基本前提条件是标定结果必须具有很好的准确性。图 51 表示实验时的输入电压、输出电压显示器和控制台。图 52 为针对三组同类索段标定所得的输出电压与索拉力关系，由图可见，标定结果具有较好的重复稳定性。

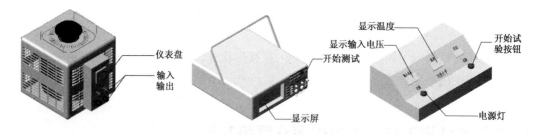

图 51　基于 EM 原理的输入电压、输出电压显示器和控制台

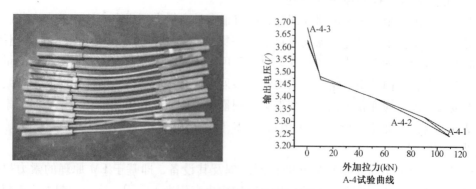

图 52　拉索试件及其标定结果

5.2 膜工作应力的检测原理和方法

在膜结构的计算设计时,膜面预应力是一个严格控制的参数。因为,膜面应力过低可能导致膜面松弛引起积水从而导致膜面破坏;应力过高可能导致风荷载作用下膜面应力超过其设计强度导致膜面破坏,也有可能导致边界刚性构件发生破坏。但是,实际施工张拉时,膜面应力完全是由工人根据经验进行控制的。这必然导致实际完成的膜结构完全不同于设计计算模型,也必然会给膜结构的安全性带来严重隐患[5]。

发达国家大型膜结构企业均研制并开发了具有专利权的膜面应力测试仪。其原理大致有振动原理和静力原理两大类,静力原理又有直接施加顶力的测试原理和利用真空泵施加吸力的测试原理[46]。同济大学研制开发了基于真空吸力原理的膜面应力测试仪,可以针对任意膜材料进行测试。如图 53 所示。图 54 为工程实测照片。

  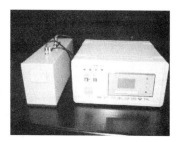

图 53 基于真空吸力原理的膜面应力测试仪

图 54 膜面应力测试仪

实验室测试表明,图 53 中所示膜面应力测试仪的误差小于 10%。目前,该仪器已应用于几个大中型膜结构的应力检测中,取得了很好的效果。

6. 结语

新型材料、先进计算理论和交叉学科知识的应用是推动新型结构体系出现和发展的三个重要因素。材料科学的发展,促成了玻璃结构、合金结构及索和膜结构的出现和发展;先进计算理论的完善为薄壁钢结构的应用提供了理论依据,也为索膜等非线性结构体系的发展提供了可能性;机电控制等学科在结构工程中的应用催生了大量满足各类建筑功能和结构功能的新型体系。

作为新型结构体系之一，索和膜结构在我国的应用历史只有短短十年不到。在这么短的时间内，我国索和膜结构的工程应用总量已经相当可观，已经建成和正在设计的结构体系十分丰富。我们经历了从最早的境外技术直接参与，到后来的国内企业模仿学习，再到最后的自主设计自主建造的快速发展的过程。在这一领域，国内的工程实践领先于理论研究。作为研究单位，必须虚心地学习工程单位的实践知识和经验，才有可能在此基础上发现客观规律和形成指导理论；而作为工程单位，也必须充分依靠研究单位的技术力量，才有可能真正提升自己的企业实力取得进一步发展。也就是说，只有研究单位和工程单位紧密配合和合作，才有可能尽快赶上发达国家的先进水平，并实现局部的突破和领先。这是索结构和膜结构在我国应用和发展历程给我们的启示。

尽管在短短几年的时间里，我们在索结构和膜结构领域取得了显著成就，但各方面的发展很不平衡。在设计计算理论方面，关于数值计算方法方面的研究很多，但对材料的本构关系、设计强度、温度膨胀系数等基础性数据的实验研究和统计分析鲜有人问津。在施工技术方面，工程单位更多地重视设备、依赖经验，而研究单位较多闭门造车、研究成果不能满足实际工程的需要。在检测验收方面，既缺少检测手段又没有检测规范，监理和管理单位都缺少必要的综合专业知识，导致工程检测和验收有形无实。这些问题使得我国在索和膜结构领域的整体技术水平仍然落后于国际先进水平，许多已建和待建工程还存在安全隐患，要解决这些问题尚待我国研究单位、设计单位、制作安装单位、监理检测单位及政府相关部门的密切合作、配合和工作。

参考文献

[1] 国家自然科学基金委员会工程与材料科学部. 学科发展战略研究报告（2006～2010年）. 建筑、环境与土木工程Ⅱ. 北京：科学出版社, 2006
[2] 丁大钧, 蒋永生. 土木工程概论. 南京：东南大学出版社, 1989
[3] 白丽华, 王俊安. 土木工程概论. 北京：中国建材工业出版社, 2002
[4] SEI Editorial Board, chairman: Geoff Taplin, Use and Application of High - Performance Steels for Steel Structures, Zurich of Switzerland: IABSE, 2005
[5] 张其林. 索和膜结构. 上海：同济大学出版社, 2000
[6] 陈务军, 张淑杰. 空间可展结构体系与分析导论. 北京：中国宇航出版社, 2006
[7] A. V. Srinivasan et al - Smart Structures: Analysis and Design, Cambridge of England: Cambridge University Press, 2000
[8] 韩林海. 钢管混凝土结构——理论与实践. 北京：科学出版社, 2004
[9] 陶忠, 于清. 新型组合结构柱——试验、理论与方法. 北京：科学出版社, 2006
[10] 石田保夫, 饭岛俊比古, 畔柳昭雄. Aluminium. 日本：大日本印刷株式会社, 2006
[11] 饭岛俊比古. Structural Design of Aluminium Architecture. 日本：鹿岛出版会, 2006
[12] J. Wardenier 著. 张其林, 刘大康译. 钢管截面的结构应用. 上海：同济大学出版社, 2004
[13] Tim Waite, NAHB Research Center, Steel - Frame House Construction, Carlsbad: Craftsman Book Company, 2005
[14] R. Hess, Material glass, 76-79, Structural Engineering International, J. IABSE, No. 2, 2004

[15] Bubner, Ewald, Membrane Construction, Essen of Germany: Druckerei Wehlmann GmbH, 1999

[16] Frits Scheubin, Arno Pronk, Adaptables 2006, International Conference on Adaptable Building Structures, Eindhoven The Netherlands: Eindhoven University of Technology, 2006

[17] 上海市工程建设规范. 建筑结构用索技术规程（DG-TJ08-019-2005）. 上海：上海市建设工程标准定额管理总站，2005

[18] 张其林，张莉等. 预应力索屋盖结构的形状确定. 同济大学学报，Vol. 28, No. 4, 2000, pp. 379-382

[19] 张其林，罗晓群等. 幕墙索杆桁架支撑体系设计计算中的若干问题研究. 工业建筑，Vol. 32 No. 1, 57-59

[20] 膜结构技术规程（DGJ08-97-2002）. 上海：上海市建设工程标准定额管理总站，2002

[21] 膜结构技术规程（CECS158：2004）. 北京：中国工程建设标准化协会，2004

[22] 张其林，张莉. 膜结构形状确定的三类问题及其求解. Vol. 25, No. 1, 2000, pp. 33-40

[23] Zhang Q. L., Zhang L., etc. Numerical method for form finding and cutting pattern of membrane structures, Computational Civil and Structural Engineering, P. 115-119, 2000

[24] Qi-Lin Zhang, Xiao-Qun Luo, Yan Nie. Numerical Models and Analysis Methods for Beam-Cable-Membrane Structures. Proceedings of the IASS International Symposium 2002 on sell & spatial structures. UK, September, 2002

[25] Zhang QL, Luo XQ. Finite element method for developing arbitrary surfaces to flattened forms. FINITE ELEM ANAL DES 39 (10): 977-984, JUL 2003

[26] Q. L. Zhang, X. Q. Luo. A FEM Model for Finding Geodesics on Discrete Surfaces Composed of Triangles. Journal of IASS, Vol. 47, No. 1, pp. 31-34, Apr. 2006

[27] 张其林，罗晓群等. 大跨钢结构施工过程的数值跟踪和图形模拟. 同济大学学报，Vol. 32, No. 10, 2004, pp. 1295-1299

[28] 张其林，罗晓群等. 只受拉单元的修正平衡迭代方法. 钢结构，Vol. 16, No. 52, 2002, pp. 62-64

[29] 张其林. 索和膜结构的施工过程模拟和检测验收标准研究技术报告. 上海市教育发展基金会曙光计划跟踪项目，2005

[30] Kawaguichi, K., Hangai, Y., Analysis of stabilizing paths and stability of kinematically indeterminate frameworks, Proceedings of the 3rd Summer Colloquium on Shell and Spatial Structures, 1990, pp. 195-204

[31] Tanaka, H., Hangai, Y., Rigid body displacement and stabilization conditions of unstable structures, Proc. IASS Symposium, Osaka, 1986, Vol. 2, pp. 55-62

[32] Kawaguichi, K., Hangai, Y., Nabana, K., Numerical analysis for folding of space structures, Space Structures, Vol. 4, 1993, pp. 813-823

[33] 张其林，罗晓群，杨晖柱. 索杆体系的机构运动及其与弹性变形的混合问题. 计算力学学报，Vol. 21, No. 4, 2004, pp. 470-474

[34] 杨晖柱，张其林，丁洁民等. 索杆体系机构展开问题的有限元求解方法. 同济大学学报，Vol. 31, No. 7, 2003, pp. 788-792

[35] 洪嘉振. 计算多体系统动力学. 北京：高等教育出版社，1999

[36] 陆佑方. 柔性多体系统动力学. 北京：高等教育出版社，1986

[37] 丁皓江，何福保等. 弹性和塑性力学中的有限单元法（第二版）. 北京：机械工业出版社，1996

[38] 王勖成，邵敏. 有限单元法基本原理和数值方法（第二版）. 北京：清华大学出版社，1999

[39] 林成森. 数值计算方法. 北京：科学出版社，2005. 212-299
[40] 潘振宽，洪嘉振，刘延柱. 多体系统动力学微分/代数方程组数值方法. 力学进展，1996，Vol. 26，No. 1：28-40
[41] 朱金国，陈宣民. 频率法测定拉索索力. 结构设计与研究应用，2003，2
[42] 魏建东. 索力测量技术. 中外公路，Vol. 21，No. 4，2001，8
[43] 林志宏，徐郁峰. 频率法测量斜拉桥拉索索力的关键技术. 中外公路，2003 年 10 月
[44] Sunaryo Sumitro, Monitoring based maintenance utilizing actual stress sensory technology. Smart Materials and Structures 14（2005）
[45] Sunaryo Sumitro. Sustainable structure health monitoring system. Structure control and health monitoring，2005，12
[46] Brian Forster, Marijke Mollaert. European Design Guide for Tensile Surface Structures. Tensinet，2004

第四章 | Chapter 4

新型结构材料的发展与应用

孙 伟

东南大学，江苏南京 210096　E-mail：Sunwei@seu.edu.cn

摘　要：本文主要介绍了新型结构材料的发展与应用。涵盖了四大系列结构材料，即绿色高性能与超高性能混凝土材料、环保型高性能与超高性能纤维增强水泥基复合材料的制备新技术、关键技术性能的形成机理及与矿物掺合料有关的收缩、徐变和疲劳方程和相应的机理模型。论述了充分和有效利用工业废渣取代水泥熟料提高结构混凝土耐久性和服役寿命是节能、节资、提高材料性能和保护生态环境的重要举措。本文还介绍了结构材料研究的新方向，即各系列智能混凝土材料和结构用 FRP 筋材、修补与加固材料，揭示了其技术与性能优势，提出了深化研究的核心问题。为适应和推进我国空前规模的基础工程建设和城市化推进，以此为巨大推动力，在我国发展更高层次的高性能、低收缩、低徐变、高抗裂、高耐久性、长寿命和智能化结构材料，并建立新的理论体系。

关键词：环保型，高与超性能混凝土材料，纤维增强水泥基复合材料，智能混凝土材料，纤维增强树脂基材料，关键技术性能，工业废渣

THE DEVELOPMENT AND APPLICATION OF NEW STRUCTURAL MATERIALS

W. Sun

Southeast University, Nanjing, 210096, P.R.C

Abstract: This paper mainly introduces the development and application of new structural materials, including new casting technology, key performance and mechanisms of green high performance and ultra-high performance concrete and high performance and ultra-high performance fiber reinforced cementitious composites. Related key properties involve shrinkage, creep and fatigue performance. It is revealed that replacing cement with waste solid residues is an efficient way to improve the durability and service life of structural concrete, besides that the utilization of waste residues is helpful to our ecological environment. New trends of structural materials, covering smart concrete, FRP bars, FRP repair and consolidation materials are introduced as well. The technical advantages of these new materials are described, relating key problems needing to be further investigated are proposed. To meet the requirements of the large scale constructions of infrastructures and to promote the progress of citification, we need to develop low shrinkage, low creep, high crack resistance, high durability and long service life high performance structural materials and smart structural materials and to develop new theories of these materials.

Keywords：Green, High performance and ultra – high performance concrete, Fiber reinforced cementitious composites, Smart concrete, Fiber reinforced plastics/polymer, Key property, Waste solid residues

1. 前言

土木工程结构材料的发展与应用起始于水泥问世以后，迄今已有160多年的悠久历史。发展至今，混凝土材料已是世界上研究最多、应用极广，其他材料所不能替代的主要土木工程结构材料之一。特别是随着材料科学与应用技术的不断发展与进步，高科技和重大工程的巨大推动力，混凝土材料的绿色化、复合化、智能化程度也逐年提升，特别是矿物掺和料、化学外加剂、各种纤维及其多元复合技术的有效和高校利用、混凝土科学与基本理论的不断发展，大大促进了混凝土材料组成与结构的优化，从本质上改变了混凝土材料结构形成与损伤劣化的规律与特点。通过多元复合技术，诸组分间扬其长、避其短、优势叠加与成分互补，从而使结构混凝土材料的关键技术性能不断得到相应提升。特别在我国重大基础工程、桥梁、铁道、港口、码头、机场、水电大坝、隧道、涵洞、国防防护、治山治水、治海治沙、南水北调、西气东输等重要和重大、大型和特大型工程的全面兴建，无一不与结构材料息息相关，不仅推动了结构材料用量巨幅提升，而且对材料性能的要求也越来越高，并各具特色，是结构材料发展的巨大推动力。总结发达国家基础工程过早失效造成巨大经济损失的教训，当今世界上混凝土科学与工程界对混凝土结构的耐久性和服役寿命给予了极大关注，从而对与混凝土耐久性和服役寿命密切相关的关键技术性能（如：各种收缩性能、开裂性能、长期徐变变形性能、动载作用下的疲劳性能及其控制和抑制开裂的各种技术）进行了全面研究，揭示了不同类型和不同掺量的工业废渣对关键技术性能的影响规律和影响机理。在大量科学研究和工程实践中人们充分认识到钢筋混凝土结构耐久性的优与劣、钢筋锈蚀速率的高与低、服役寿命的长与短，只要结构设计与施工技术有充分保证，那么关键因素就在于结构混凝土材料自身抵抗有害物质入侵和传输的能力。因此，结构混凝土材料对力学因素、环境因素及其耦合作用下的抵抗能力已是钢筋混凝土结构耐久性和服役寿命提高的基础、要害和核心。

众所周知，普通和高强结构混凝土材料是一种脆性易裂材料，且强度越高、脆性越大、越容易开裂，混凝土内部结构形成过程中裂缝尺度和数量，受力学或环境因素作用下裂缝引发与扩展程度都是影响混凝土结构耐久性下降和寿命缩短的隐患。因此，为提高结构混凝土材料的强度、韧性和阻裂能力，国际上纤维混凝土的发展已步入到深化阶段并形成研究热点[1]。在中国，纤维混凝土的强化与韧化技术也同步发展。至今不仅保持了钢纤维混凝土是研究最多、在重大工程结构中应用最广的水泥基复合材料，而且提高了纤维与混凝土基体界面粘结，提高了对基体强化、韧化与阻裂效应，纤维品种与外形不断增多和优化（平直型与异型纤维逐年增多）。特别是复合异型纤维（如哑铃与刻痕型复合）的出现，大大提高了钢纤维对混凝土强化与韧化的效率，而且通过复合异型纤维尺度的优化在保持高力学行为和物理特性的同时，有效解决了钢纤维混凝土超高泵送的技术难题[2]，并顺利泵送高度达 300m 以上，为在主要受力复杂的超高结构（如超高索塔锚固区）中应用开拓了新方向。由于钢纤维混凝土高动态效应的优势，它已成为国防防护工程首选的结构材料。当今，为抑制混凝土早期塑性收缩与开裂，缓解混凝土结构在高温和火灾中的爆裂，聚丙烯纤维（PPF）增强混凝土（PPFRC）得到广泛应用，发挥了极其重要的作用；为提高混凝土的延性、韧性，聚乙烯醇纤维（PVAF）增强水泥基复合材料（PVAFRCC）及工程纤维增强水泥基材料 ECC（Engineered fiber reinforced Cementitious Composites）这些年来发展迅速[3,4]；不同尺度不同性质的有机与无机、金属与非金属纤维混杂增强水泥基复合材料 HFRCC（Hybrid Fiber Reinforced Cementitious Composites）[5]，可使其在相应的尺度和性能层次上逐级发挥作用，大幅度提高了对混凝土基体阻裂与限缩的能力，在中国也得到了迅速发展。特别引人注目的是纤维增强水泥基复合材料已走上绿色化的道路，经纤维与活性掺合料有效复合，不仅大量节省了资源和能源，保护了生态环境，而且通过界面结构的优化与强化、界面效应和界面粘结的发挥与提高，又进一步强化了混凝土材料各项关键技术性能［如收缩率可降低 50% 以上，疲劳寿命可提高一个数量级（应力比相同时）］，大幅度提升了耐久性，延长了材料和结构的服役寿命。当今纤维增强水泥基复合材料已走上绿色化与高性能和超高性能化，其强度等级从 40MPa 到 200MPa 以上。

随着材料科学与技术的持续发展，混凝土材料走上了高科技与智能化[6]，通过自修复效应，把水泥基材料的裂缝消除在引发之初，使混凝土材料形成类似生物材料那样具有自感知、自诊断、自适应、自修复等特殊功能，对此，国内外都进行了大量研究，取得了显著效果，特别对提高和预测混凝土结构的耐久性和服役寿命，智能材料正发挥着重要作用。另一方面，由氯离子扩散引起钢筋锈蚀而导致混凝土结构过早损伤和失效问题，始终是被人们关注的重要难题，虽然混凝土科学与工程界应用了许多技术措施，以推迟钢筋锈蚀，延长钢筋混凝土结构的服役寿命，尽管也有相当效果，但在严酷环境条件下难以从根本上解决问题。因此，这些年来纤维增强树脂基复合材料（FRP）国内外又有了新发展，开拓了 FRP 研究与应用新方向，依据不同的环境特点和实际要求，可制成碳纤维增强树脂（FRP）、玻璃纤维增强树脂（GFRP）、有机纤维增强树脂（PFRP）和混杂纤维增强树脂（HFRP）筋材来取代钢筋，这一技术不仅提高了混凝土结构的耐久性和服役寿命，而且也大大减轻了结构重量。FRP 的新发展，不仅可用来取代钢筋，而且对混凝土结构损伤劣化过程中作为一种修复与加固材料可对混凝土结构已经损失的性能进一步提升，达到恢复和延长混凝土结构耐久性和服役寿命的目的。这些新技术，不管是智能化技术还是多重复合化技术都极大地推动了结构材料的发展，并从理论和技术双向推进。

2. 绿色化高性能混凝土结构材料的新突破

2.1 结构混凝土材料绿色化的重大意义

混凝土材料在研究、应用与发展过程中，特别在我国大规模基础工程建设推动下，人们充分认识到最大限度和高效利用工业废渣来取代更多的水泥熟料是节省资源、节省能源、保护生态环境、提高材料性能的重要举措，也是社会可持续发展的必由之路[7]。众所周知，我国是世界水泥大国，2003 年我国水泥总产量为 8.62 亿吨，占世界水泥总产量的

50%，到 2005 年就达到 10.6 亿吨（相应混凝土量约 80 亿吨），又占世界水泥总产量的 50%，预计到 2010 年我国水泥总产量要达到 12 亿吨，向世界水泥总产量的 2/3 逼近。但生产 1 吨水泥熟料要排放近 1 吨 CO_2 等有害气体，这就造成了严重的环境污染，严重危害着社会可持续发展的实现。并且我国工业废渣的储存量巨大，特别是粉煤灰、矿渣、钢渣、煤矸石、硅灰，还有各种尾矿等。虽然这些年来为满足重大工程的需求，作为矿物掺合料，其用量逐年增多，取代水泥熟料量也年有提升，但因工业的发展，其生产数量有增无减，且各自物理结构、化学成分、水化机理及掺量不同对混凝土性能也影响不一。其中有正效应，也有负效应。我国已采用多重复合技术，取其长、补其短、化"短"为"长"、化"负"为"正"，这种互补效应对混凝土性能的改善发挥了极大作用。当今磨细矿渣因其化学成分与水泥近似，自身有一定的水化能力，对氯离子结合能力也具有优势，所以它取代水泥熟料量可高达 80%。采用磨细矿渣已是较为成熟的技术，但由于必经磨细，同样要消耗能源，而且在大掺量时对混凝土变形性能仍有负面效应，因而在海工混凝土中为降低自由氯离子浓度和延缓钢筋锈蚀，磨细矿渣的应用量相对较多。

2.2 粉煤灰自身潜在的物理与化学优势

当今粉煤灰的利用从国内外大量报导和我们的科学研究与工程应用，充分显示了其自身具有潜在的物理与化学性能优势，又无须因粉磨而增大能耗，所以，各重大土木工程均把粉煤灰（特别是一级粉煤灰）作为首选的矿物掺合料，当今它已是在工程中应用最多、最广泛、潜力最大、优势最突出的取代水泥熟料的首选材料。早在若干年前，我国著名的粉煤灰混凝土专家沈旦申提出，粉煤灰有三大效应，即因颗粒外形是圆球而产生形态效应，有减水和提高工作性的显著效果；二是火山灰效应又称活性效应，经 $Ca(OH)_2$ 的激发可生成 C-S-H 凝胶，改善混凝土的微结构；三是微集料效应，因粉煤灰的颗粒分布属于连续级配，具有紧密填充效应，增进了水泥基材料致密性。此后，经大量科学实验，人们充分证实了粉煤灰对混凝土硬化前后性能提高的贡献。但这些年来，粉煤灰在重大工程如三峡大坝混凝土工程（三峡工程共用粉煤灰混凝土 2800 万 m^3，仅坝体用粉煤灰混凝土 1600 多万 m^3，其中粉煤灰用量在 100 万吨以上）、重大桥梁工程、隧道工程等中的应用和大量室内外试验都充分表明了，因粉煤灰的掺入有效降低了混凝土的干缩率，抑制了水泥基材料的开裂，减小了长期徐变值，提高了在疲劳荷载作用下抵抗疲劳损伤的能力，特别是提高了混凝土的耐久性和服役寿命（抗冻融和碳化除外），在用于预应力混凝土结构时，不仅不会增大徐变值，反而使预应力损失下降。这就解决和结束了长期以来预应力混凝土结构能否掺加适量优质粉煤灰的争议。

2.3 粉煤灰混凝土在重大工程应用实例

下面仅举有代表性的实例：粉煤灰混凝土在举世瞩目的三峡大坝混凝土工程（如图 1）、南京第二长江大桥工程（如图 2）、润扬长江公路大桥工程（如图 3）、具有世界四个第一的苏通大桥工程（如图 4）、杭州湾跨海大桥工程（如图 5）、南京地下铁道工程（如图 6）、西部电力工程（如图 7）以及厦门海底隧道工程（如图 8）等，都得到了广泛应用，充分发挥了粉煤灰及其与磨细矿渣复合的优势，充分发挥了粉煤灰的潜能与优势，取得了显著的技术、经济和社会效益。

(a)坝体　　　　　　　　　　　　(b)船闸

图1　长江三峡工程（全部采用粉煤灰混凝土）

图2　南京二桥工程（全部采用粉煤灰混凝土）

图3　润扬长江公路大桥工程（全部采用粉煤灰混凝土）

图4　苏通长江公路大桥工程（全部采用粉煤灰混凝土）

图 5　杭州湾跨海大桥工程（采用粉煤灰及粉煤灰与磨细矿渣双掺混凝土）

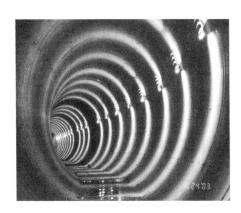 

图 6　南京地下铁道工程（采用粉煤灰及粉煤灰与磨细矿渣双掺混凝土）

(a) 高耐久抗腐蚀电杆根部　　　　　　　　(b) 高耐久抗腐蚀电杆

图 7　西部电力工程

图8 厦门翔安海底隧道工程

2.4 粉煤灰效应微观机理的剖析

大量试验表明，粉煤灰的火山灰效应及其反应程度仅有20%左右，最高也没有超过25%。为什么会对混凝土材料关键技术性能有那么大的贡献，究其原因，除了粉煤灰本身形态效应和火山灰效应之外，更重要的应归之于粉煤灰微珠中沉珠的物理与化学特性及特有的第二个微集料效应优势。大量试验表明，在粉煤灰体系中其颗粒组成主要有四种微珠，即漂珠、沉珠、复珠和磁珠，如图9所示。

漂珠：密度小于水，65%以上是中空颗粒，其中含有更小的玻璃微珠，又称子母珠或微珠，这种珠在粉煤灰中含量不高，约为原状灰的0.07%~0.50%，但活性很高，对混凝土性能影响很大。其平均粒径为2.2~22μm，壁厚为0.2~20μm。

沉珠：是粉煤灰中具有较大比例的微珠（约占90%以上），其比重大于水，故称沉珠。其力学行为主要是高弹高强，经纳米硬度测试，其弹性模量与水泥颗粒、磨细矿渣颗粒相比，有数量级的差别。而且它与水泥基和C-S-H凝胶的界面结合是强化学结合（如图10），其表面有不规则的突起点，又增进了与基体间界面的物理结合，这是微集料

图9 粉煤灰微珠的颗粒形貌 　　　　图10 粉煤灰颗粒与水泥基间的强化学结合

效应发挥的基础,也是强化界面结构的根本原因。壳壁上的气孔绝大多数为中空,内部也含有细小的玻璃微珠。

磁珠:是富铁微珠,在磁铁中 Fe_2O_3 的含量占55%,有一定的磁性,故称之为磁性微珠或磁珠。由此可见,沉珠是发挥双重微集料效应的关键。通过它与基体间的火山灰效应形成强化学结合与物理结合的强微界面效应的发挥,强化了三个层次界面结构(即粉煤灰颗粒~C-S-H凝胶、细集料~水泥基材料、粗集料~水泥砂浆),增进了界面粘结、强化了界面效应,从而有效发挥了更高层次的微集料效应,这是提高混凝土的阻裂能力、限缩能力、抵抗长期徐变能力、提高混凝土抗疲劳能力和耐久性能的关键和基础,进一步更科学和可靠地解释了粉煤灰潜在的优势,丰富与发展了粉煤灰效应的基本理论。

2.5 粉煤灰效应对提高混凝土关键技术性能的贡献

大量试验研究和巨量工程应用的结果充分显示,在低水胶比时,粉煤灰取代水泥熟料30%~40%干燥收缩可平均降低30%~35%,徐变值下降大于50%,疲劳寿命是不掺粉煤灰同强度等级混凝土的3倍以上(应力比相同时)。粉煤灰混凝土泵送高度可超过300m以上,在经过大量试验与应用的基础上,建立了与工业废渣特别是与粉煤灰掺量有关的(包括粉煤灰与磨细矿渣双掺)自收缩和干燥收缩方程、徐变方程和疲劳方程;经微观机理和理论剖析,建立了相应的收缩模型、徐变模型和疲劳损伤模型[8~10],在理论上有了新突破。这些方程经实际应用和完善,必将为工程设计提供理论依据。

到目前为止,大量工业废渣不同程度的取代了水泥熟料,粉煤灰已经广泛地应用于各类重大基础和建筑工程中,在大城市(如上海、北京、南京等)优质粉煤灰的用量已达到100%,全国平均粉煤灰用量为45%以上,由于地区不同,还有所差异。如果工业废渣取代水泥量平均能稳定地达到30%~50%,那么10亿吨水泥熟料就可生产14~20亿吨的水泥。

综上所述,粉煤灰的优势正得到充分揭示,粉煤灰对混凝土性能贡献的机理从结构形成和损伤劣化两个全过程中微结构的演变和微集料效应的有效发挥也足以论述明白。但事情总是一分为二,对大掺量粉煤灰后引起混凝土早期强度发展缓慢、推迟预应力筋张拉时间、抗碳化和抗冻融能力下降的问题,当今是采用限制粉煤灰最佳掺量或掺加功能组分通过物理、化学或物理化学原理来解决这些难题。对此尚应采取积极和科学的复合措施,在不同的时间历程中,全面发挥粉煤灰对混凝土各种性能的提升和强化作用。在特别情况下,还应对掺量做适当限制。

3. 纤维增强水泥基复合材料的绿色化及高与超高性能化

3.1 纤维增强水泥基材料的新进展

纤维增强水泥基复合材料在世界上的研究与应用已超过半个世纪的历史,我国基本上与国际上同步发展。这些年来我国在重大工程大规模兴建的推动下,为进一步提高纤维对混凝土增强、增韧与阻裂效应,充分发挥其潜在优势、提高性价比,并拓宽其应用领域,在优化基体组成与结构,扩大在基础工程和国防防护工程中应用的纤维品种、类型,从而在性能层次和技术水平上又不断有新进展。其中,钢纤维增强水泥基复合材料不仅是研究最多、应用最广，而且钢纤维与高效减水剂、矿物掺和料复合，又显示出新的复合优势，特别是又大幅度提高了纤维-水泥基间的界面粘结与界面效应、界面叠加与强化效应，从而纤维对混凝土增强、增韧与阻裂效应的优势得到了更有效的发挥，它已不仅用于土木、建筑、水利、交通、隧道、市政等各个领域，而且因其阻裂效应的特殊优势、动态效应的提高、破坏形态的转化，已成为国防防护工程中首选的高动态效应、抗爆炸、抗震塌与抗侵彻的结构材料[11,12]。

3.2 纤维增强水泥基复合材料主要特点

当今纤维增强水泥基复合材料发展的主要特点：一是绿色化。在水泥基体中有效发挥活性混合材的潜能与优势，强化与提高界面结构、界面粘结与界面效应，从而更大幅度提高了其增强、增韧与阻裂能力。其中阻裂能力是诸效应间的要害和基础，它直接影响到耐久性与服役寿命的提高，以及性价比的提升。二是纤维品种的增多与优化。对钢纤维混凝土，为提高纤维对混凝土基体间的粘结力与锚固作用，纤维外形正在不断优化（如图11所示），并由平直形、单一异形向复合异形方向发展。特别是纤维与绿色水泥基材料复合粘结力的大幅度提升，其粘结力提升幅度如表1所示。表1示出纤维与基体界面平均粘结强度。当90天龄期时，平直形与端钩形和哑铃形相比，其界面平均粘结强度的提高到2.43~2.55倍，平直形与复合异形相比，则可相应提高到2.95倍。如果在水泥砂浆基体中掺入30%粉煤灰来取代水泥，各界面平均粘结强度则有更大幅度的提高，平直形纤维与复合异形纤维相比，与基体界面粘结强度的提高可达3.27倍以上。而且界面粘结刚度和纤维脱粘与拔出时所做的功也以同样比例提升，这无疑在促进其增强、增韧，特别是阻裂效应提高方面具有新突破。

(a)钢锭铣削异形

(b)钢丝切端端钩形

(c)钢丝切断哑铃形

(d)钢板切削盾铃形

图11 钢纤维外形

纤维品种与外形对界面粘结力的影响　　表1

界面粘结性能	FA取代水泥量（%）	测试龄期（d）	纤维外形			
			平直形	端钩形	哑铃形	哑铃与刻痕复合形
界面平均粘结强度（MPa）	0	90	3.37	8.19	8.61	9.93
	30	90	3.70	9.82	10.5	12.11
纤维脱粘与拔出做功（N·m）	0	90	0.72	1.74	1.83	2.12
	30	90	1.12	2.97	3.17	3.43

3.3 自流平与高泵送化

长期以来钢纤维混凝土因工作性特别是流动性差而难以走上商品化,更不能做到高泵送化,这个问题始终是国内外混凝土工程界没有解决的技术难题,超高泵送更是难以实现。随着高效外加剂的高速发展,又与粉煤灰形态效应复合,结合近年来重大桥梁工程中受力复杂的超高索塔锚固区高抗裂、高抗疲劳与高耐久性的特殊要求,对纤维混凝土基体组成结构和集料品种和尺度、钢纤维外形与尺度进行了全面优选与优化,充分发挥高效减水剂和粉煤灰的复合效应,从而大大提高了钢纤维混凝土的工作性,结束了钢纤维混凝土流动性差、泵送难的历史判断。在确保钢纤维混凝土各项物理、力学性能和耐久性能有增无减的前提下,现在钢纤维混凝土可顺利泵送高度达300m以上,这为拓宽钢纤维混凝土的应用领域,有效发挥钢纤维混凝土用于工程的高空部位,充分发挥其高抗裂、高抗疲劳和高抵抗力学因素与环境因素耦合作用下损伤劣化的能力,为保证高空特殊结构的高耐久要求开拓了新方向。

3.4 最大限度发挥纤维对混凝土基体的阻裂效应

由于中国在重大工程建设中开始越来越多地强调提高工程结构的耐久性和服役寿命,人们已充分认识到提高结构材料的耐久性,延长服役寿命是最大的节约。如果混凝土结构和结构混凝土的服役寿命由50年提高到100年,甚至提高到200年,那么材料用量则相应减少到50%和25%。由此,极大推动了纤维增强技术的发展,而且把有效发挥纤维对混凝土的阻裂效应已排到首位。强调混凝土材料在结构形成过程中,应最大限度减少裂缝源的尺度和数量,最大限度降低影响结构耐久性和寿命的潜在隐患,特别在复杂力学与环境因素耦合作用下的损伤劣化过程中,还要最大限度抑制各种裂纹的引发与扩展,充分发挥不同性质、不同尺度的纤维,在相应尺度层次、性能层次和时间层次上充分发挥作用,在水泥基材料结构形成与损伤劣化两个过程中抑制各种收缩变形和开裂,最大限度抑制裂缝引发与扩展。不同尺度钢纤维的作用是:不仅提高了混凝土增强与增韧的效果,而且由于裂缝引发和扩展受到纤维的控制,从而使裂缝的尺度变小、数量减少,有效提高了抵抗有害物质入侵的能力。表2所示为当今常用于增强混凝土的几种纤维及其尺度和力学特性。

工程用纤维品种与基本性能 表2

纤维品种	密度	抗拉强度(MPa)	弹性模量(GPa)	极限延伸率(%)	长度(mm)	直径(μm)	价格(万元/t)
钢纤维	7.8	380~2000	200~210	3.5~4.0	14~35	150~500	0.5~1.8
PVAF	1.3	1200~1600	30~40	5~17	6~12	12~39	7~8
PPF	0.9~0.91	300~660	3.5~4.8	15~20	19~50	26~62	1~2
碳纤维	1.3~1.9	1400~3500	230~300	0.40	任意	7~10	20~30

当今扩大纤维品种,提高应用水平和性能层次,用来增强混凝土基体的增强体,除

钢纤维外，合成纤维应用也日益拓宽（见表2）。如表2所示，钢纤维（SF）强度等级的变化范围是380~2000MPa以上。钢纤维是研究最多、应用极广，对混凝土增强、增韧与阻裂效应提高的幅度也最大；聚丙烯纤维（PPF）的作用主要是抑制混凝土早期塑性阶段的收缩和由此引起的混凝土开裂，它还能抑制高强高性能混凝土在高温和火灾中的爆裂，所以在中国不少商品混凝土公司将PPF增强混凝土（PPFRC）列为主要产品供工程应用，已走上了商品化。这些年来，由于聚丙烯醇纤维（PVAF）的强度与弹性模量比PPF高一个数量级，所以成为制备高延性纤维增强水泥基材料的首选而被广泛应用。

3.5 高延性纤维增强水泥基材料的新进展

国际上快速发展的工程纤维增强水泥基复合材料 ECC（Engineered fiber reinforced Cementitious Composites）就是典型的代表[3]。ECC是采用了大掺量PVAF来增强环保水泥基材料，其组成配比如表3（a）所示。其高延性特征及荷载~挠度曲线如图12和图13所示。

国外 ECC 的配合比与材料组成　　　　　　　　　　　表3（a）

材料组成	水泥 P.O42.5	Ⅰ级粉煤灰	集料（最大粒径100μm）	高效外加剂	水	PVA 纤维	水胶比
掺量（kg/m³）	583	700	467	19	298	26	0.23

本课题组制备 ECC 的配合比与材料组成　　　　　　　表3（b）

材料组成	水泥 P·O42.5	Ⅰ级粉煤灰	集料（天然砂，最大粒径3000μm）	高效减水剂	水	PVA 纤维	W/B
掺量（kg/m³）	580	698	464	20	294	26	0.23

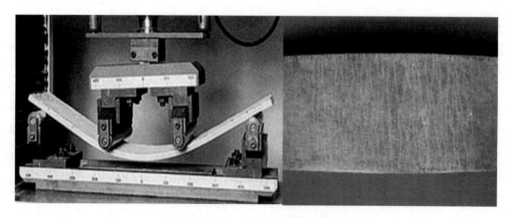

图12　ECC 材料的多缝开裂形貌[3]

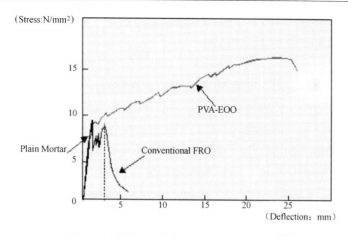

图 13　国外 ECC 材料的荷载—挠度曲线[3]

如表 3（a）所示，ECC 组成材料的特点是高掺量 PVAF，粉煤灰取代水泥量远大于 50%，具有明显的绿色特征。特别是荷载—挠度曲线具有明显的多缝开裂阶段和特征。ECC 在美国、日本等国家已在基础与建筑工程中应用，研究与应用均显示了这种复合材料物理与力学性能的独特优势，并在可用作桥面铺装材料、抗震结构材料、快速修补材料、高速公路罩面材料、抗冲击材料与抗裂材料等方面显示了很广的应用前景和潜力。实验结果表明，当 PVAF 掺量为 26kg/m³ 时，混凝土最大裂宽为 0.06mm，抗弯强度 15MPa，抗压强度 70MPa，极限拉伸应变 2.5%～3.5%，最大挠度达 10mm～25mm，冻融循环 300 次动弹模不下降。特别是在荷载－挠度曲线中出现了多缝开裂阶段，充分揭示了 ECC 具有高延性与高韧性的本质，也是优异性能产生的根源和基础。

近年来，为提高其性价比和降低能耗，我们适当调整了其组成材料与配比（如表 3（b）所示）。胶凝材料与 W/C 基本与国外相当，其主要不同点是用粒径相同为 100μm 的工业废弃物尾砂来取代磨细石英砂，或用最大粒径为 2000μm 的天然砂取代粒径为 100μm 的磨细石英砂来制备 ECC 材料。实验结果表明，其抗压强度达 65MPa 左右，最大裂宽也控制在 0.06mm，抗弯强度达 10～15MPa，弯曲挠度为 1.7cm，其荷载—挠度曲线如图 14 所示。同样也出现了多缝开裂阶段和明显的高延性特征,但挠度最大值尚达不到35mm。与

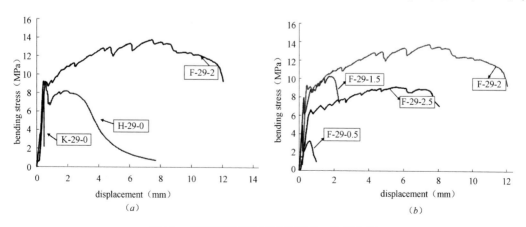

图 14　本课题组制备 ECC 的荷载—挠度曲线

国外的 ECC 相比,一是集料粒径增大了 20 倍或粒径相同（100μm）,但采用工业废弃物尾砂取代了磨细石英砂,大大节约了能耗。

3.6 高性能钢纤维增强混凝土已是防护工程的首选材料

大量室内外动态效应试验表明,高性能与超高性能纤维增强水泥基复合材料具有独特的动态效应优势,以及极高的抗爆炸与抗侵彻能力,彻底改变了防护工程在强动载下的破坏特征。依据动态荷载大小与作用方式不同,材料抗力不一,出现了裂而不散、微裂而不散、甚至不裂也不散的鲜明特征,其具体情景如图 15 所示。

(a)钢筋混凝土与HPFRCC侵彻后的形态对比

(b)钢筋混凝土与HPFRCC爆炸后的形态对比

图 15 高性能与超高性能纤维增强水泥基复合材料抗爆炸与抗侵彻能力

3.7 纤维增强水泥基材料的绿色化与超高性能化

众所周知,对纤维复合材料而言,不论增强什么基体,其与基体界面粘结性状始终是影响复合材料性能最为重要的关键。当今不论哪种纤维增强水泥基复合材料均掺入不同品种和掺量的矿物掺和料,大大优化了界面结构、提高了界面粘结,强化了界面效应,充分发挥了纤维与工业废渣间的物理与化学复合效应。当今中国的纤维增强水泥基复合材料其强度等级有中、高和超高三个层次（CF40～CF200 以上）。自 1992 年至今我国已颁布了两本国家规程,即《钢纤维混凝土结构设计与施工规程》和《纤维混凝土应用技术规程》,但均未包含超高性能水泥基复合材料。发展至今,为满足大跨与超大跨薄壁结构和高抗动力荷载的特殊要求,在中国发展超高性能纤维增强水泥基复合材料已是科学研究的新动向。众所周知,多少年前,国内外曾出现了MDF、DSP等超高性能水泥基材料,这些材料具有很高的强度性能优势,但终因制备工艺的复杂性、长期性能的不稳定性、价格的昂贵性,而难以向工程应用转化。因此,这些材料因发展前途渺茫而逐渐被人们淡忘。随着高效减水剂的高速发展和品位的大幅度提高,在20世纪90年代中期,国际上出现了活性粉末混凝土RPC(Reactive Powder Concrete),由于具有自流平特征且制备工艺相对比较简单而引起国内外极大关注,并开始在典型工程和典型制品中应用,加拿大首先建立了一座步行桥(walk bridge)。但值得关注的是,其粉体材料和集料均需经过磨细,导致能耗大幅度增大,影响到性价比的提高和大面积推广应用。当今在中国制备超高性能水泥基材料时,改变了一种研究思路与方法[13]。首先充分发挥超细工业废渣及其复合的潜能,取代超磨细石英粉和大掺量硅灰,采用最大粒径为3000μm的天然砂取代100μm的磨细石英砂,采用标准养护和热水养护取代能耗大的蒸压处理,并通过优选外加剂使超高性能纤维增强水泥基复合材料具有自流平密实成型的特征。其配比组成见表4。采用的钢纤维抗拉强度为1800MPa,$d_f=0.175$mm,$l_f=14\sim15$mm,$l_f/d_f=75$。其界面粘结强度见表5,与国际上RPC200的对比见表6。在材料组成方面,超细工业废渣取代50%~60%的水泥,从而因诸废渣间的成分互补与粒径叠加效应,充分发挥了各组成材料自身的优势与潜能,大大强化了界面结构和界面粘结。

超高性能纤维增强水泥基复合材料组成配比 表4

基体	水泥(%)	硅灰(%)	粉煤灰(%)	磨细矿渣(%)	减水剂(%)	水胶比	灰集比
M1	50	0	25	25	1.7	0.15	1:1.2
M2	50	10	0	40	1.7	0.15	1:1.2
M3	40	10	25	25	1.7	0.15	1:1.2

纤维—基体界面平均粘结强度 表5

养护龄期	M_1	M_2	M_3
28	7.35	8.44	8.91
90	13.56	14.88	15.55
180	15.06	16.62	18.01

如表5所示,纤维与M_1、M_2、M_3基体的界面粘结,当标准养护时,M_3与纤维的界面粘结强度,随龄期增长而大幅度提高,龄期由28d到180d,纤维与M_1、M_2、M_3的粘结强度分别提高到196.9、197和202.1倍。平均界面粘结强度最高可达18.01MPa,与纤维和普通水泥基粘结强度相比,有数量级的差异。

经过材料组成与结构的变化,现将其各项力学性能及长期耐久性能进行对比,对比结果见表6。在表6中示出了RPC200与超高性能水泥基复合材料的抗压强度基本相同,其各项力学性能指标也都接近,其中抗弯强度、断裂能和耐久性均高于RPC200的数值。但超高性能水泥基材料因用大掺量超细工业废渣取代了磨细石英粉,用粒径为3000μm的天然砂取代磨细石英砂,在力学行为相当时,大大优化了材料的组成与结构,增进了界面粘结,高度强化了界面效应,在保持自流平特征的前提下,大幅度节省了能源和资源,是一种很有应用前景的大跨、薄壁结构材料,也是能满足特高要求的防护工程材料,应用前景十分广阔。

与国外 RPC200 的性能对比　　　　　　　　　　　　　　　　　　　表6

性 能	RPC200	超高性能水泥基材料	性 能	RPC200	超高性能水泥基材料
抗压强度（MPa）	170~230	171~228	抗冻性	好	更好
抗弯强度（MPa）	30~60	38~65	抗腐蚀	好	更好
断裂能（J/m^2）	20000~30000	27000~34000	抗渗透	好	更好
弹性模量（GPa）	50~60	52~56	动态效应	—	优异

另一个不能忽视的方面,是超高性能水泥基材料具有高耐久的重要特征。就是这种材料在冻融 700 次时 RPC 基体动弹模量的下降 <2%;掺入钢纤维后相对动弹模量的损失几乎为零。如图 16 和图 17 所示。碳化 90 天的碳化深度几乎接近为零。在严酷的西部盐湖卤水中浸 700 天未发现有害离子向内部入侵。显示了这种材料具有超高的耐久性,必将有长的服役寿命,为此这种材料已被土木工程界、防护工程界给予了极大的关注。

综上所述:采用超细复合工业废渣取代 60% 水泥并取代磨细石英粉和大掺量硅灰、用天然砂取代磨细石英砂、增大集料粒径由 100μm 到 3000μm 来制备超高性能纤维增强水泥基复合材料,当纤维尺度与掺量、减水剂的减水剂效应相同,在标准养护 90d 或热水养护条件下,其各项力学性能指标与国际上报导的 RPC200 均处于相近、相等或更高,在抗弯强度和断裂能两项力学行为均超过了 RPC200,且其耐久性十分优异、性价比明显提高,可节省资源、节省能源,具有明显的绿色与环保特征,因其自流平特征显著,故与 RPC200 具有等同工作性优势,是一种发展前景广泛的超高性能水泥基材料。产生优异性能的超高性能主要来自于两个方面:一是来自纤维间距影响,二是来自基体组成结构优化,这都是产生超高性能的机理。

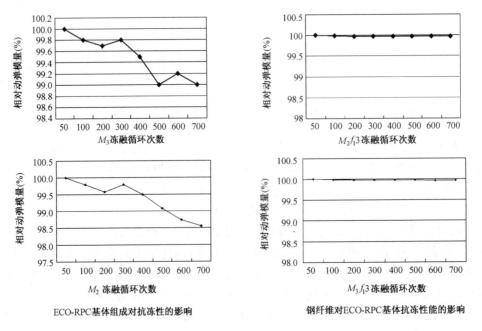

ECO-RPC基体组成对抗冻性的影响　　　　钢纤维对ECO-RPC基体抗冻性能的影响

图 16　抗冻性能

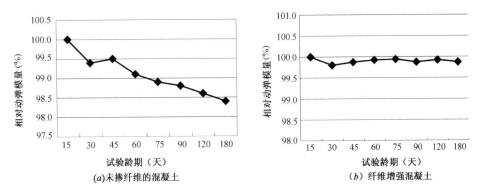

图 17　抗多种有害离子腐蚀的能力

纤维增强不同层次的水泥基材料在中国已有长足的发展。在今后随着国家基础工程、国防工程建设和城市化的推进，这一材料必将有更大的突破和发展。

3.8　纤维混凝土高强水泥基材料的新进展

综上所述，纤维增强水泥基材料在我国已有充足的发展，但在大量的研究工作中我们可以发现，单一性能和尺度的纤维增强水泥基材料由于自身尺度和性质尚难以适应多尺度非均匀水泥基材料在结构形成与损伤劣化过程中在不同结构层次和时间层次上抑制收缩与开裂的要求，往往会产生因此失彼的情景。因此，不同尺度的纤维混杂会在混凝土相应结构层次上发挥作用。同样，不同性质的纤维也会在相应的性能层次上有影响。因此，纤维混凝土增强效应是单一尺度和单一性质的纤维所无法比拟的增强、增韧与阻裂效果。不仅尺度和不同性质纤维混杂混凝土增强，其阻裂效应会贯穿于结构形成和损伤劣化的全过程，它还是提高结构混凝土抗裂、抗渗，提高耐久性和寿命的重要举措。

图 18 和表 7 示出纤维混杂增强对混凝土收缩与抗渗的影响。如图 18 和表 7 所示，不同尺度和不同性质的纤维混杂增强，有效抵制了不同阶段混凝土裂缝的引发与扩展，减少与缩小了裂缝源的尺度和数量，从而有效改善了混凝土的孔结构。不同尺度与不同性质的纤维混杂增强，均比单一尺度和单一纤维增强，其收缩值可下降 50%（与不掺纤维相比）和 70%（混杂增强与不掺纤维相比）；其最大渗水高度和相对渗透系数可下降到 1/4（由 4.3cm 下降到 1.1cm）和 1/22（由 2.9×10^{-7} cm/h 下降到 0.127×10^{-7} cm/h），为 SFC 的发展提供了更科学、更有效的多元复合技术。

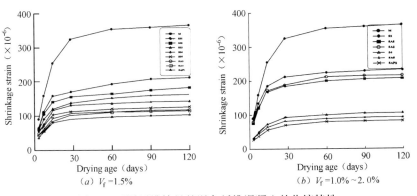

图 18　不同纤维掺量的混杂纤维混凝土的收缩特性

纤维混杂增强对渗水高度的影响　　　　　　　　　　表7

系列	最大渗水高度（cm）	平均渗水高度（cm）	相对渗透系数（$\times 10^{-7}$cm/h）	系列	最大渗水高度（cm）	平均渗水高度（cm）	相对渗透系数（$\times 10^{-7}$cm/h）
M	4.3	2.9	2.9	SA4	1.9	1.6	0.88
MU	2.5	2.2	1.7	SAU4	1.2	1.0	0.34
S31	2.8	2.4	2.0	SAP1	1.5	1.3	0.59
S3U1	1.8	1.7	1.0	SAPU1	1.1	0.7	0.127

随着我国重大工程规模空前的建造，纤维增强水泥基复合材料的发展，必将在理论和技术上有更大突破；必将成为延长混凝土结构服役寿命的重要举措；必将成功地走向高科技和智能化的道路；必将是节省资源、节约能源的新一代土木工程材料。

4. 智能混凝土

4.1 智能材料简介

近年来，智能材料（Smart/Intelligent Materials）与结构的发展引起了人们的极大关注。所谓智能材料是指模仿生命系统、能感知环境变化，并能实时改变自身的性能参数，做出人们所期望的、并能与变化后的环境相适应的复合材料或材料的复合[14]。在工程结构中复合智能材料，可以使其能感知和处理信息，并执行处理结果，对环境的刺激可做出自适应响应，这种结构称为智能结构（Smart Structures）。智能材料的使用能改善结构的性能，提高结构对环境的敏感程度和自适应能力，以实现动态或在线状态下的自检测、自诊断、自监控、自修复及自适应等功能。从某种意义上来说，结构中使用智能材料能提高结构自身的"智商"[15]。

智能材料是20世纪90年代迅速发展起来的一种新型材料，属于特殊功能材料中的一种。由于这种材料不是过去常见的、单一的、简单的结构，因此常被称为智能材料系统（Smart/Intelligent Material System）[15]。20世纪50年代，人们提出了智能结构，当时把它称作自适应系统（Adaptive System）。在智能结构的发展过程中，人们越来越认识到智能结构的实现离不开智能材料的研究与开发。智能材料的概念最早是由美国弗吉尼亚理工学院及州立大学的Rogers教授1988年9月首次在美国"Smart Materials, Structure and Mathematical Issue"的专题研讨会上提出的。我国学者认为智能材料是一种能从自身的表层或者内部获取关于环境条件及其变化的信息点，然后进行判断、处理和做出反应，以改变自身的结构与功能，并使之很好地与外界相协调的具有自适应的材料系统[16]。智能材料的主要特点有[17]：

（1）具有感知功能，能检测并可识别外界（或内部）的刺激强度，如应力、应变、热、光、电、磁、化学或核辐射等；
（2）具有驱动特性及响应环境变化功能；
（3）能以设定的方式选择和控制响应；
（4）反应灵敏、恰当；

(5) 外部刺激条件消除后，能迅速回复到原始状态。

可见，应用于智能结构中的智能材料应具备感知、处理和驱动三个基本要素[5]，但由于现有的智能材料多为单一均质材料，难以具备多功能的智能特性，因此，一般需要两种或几种材料的复合，构成一个智能材料体系。

智能材料可以分为本征型和集成型两类[18]。本征型智能材料主要由具有类似生物细胞性能的材料组成，例如，将具有信息处理功能的蛋白质和执行控制功能的蛋白质相混合，合成具有信息处理和控制功能的新型蛋白质，这就意味着要进行原子或分子级的材料设计与合成。这无疑是相当复杂和困难的工作，目前尚未制成这种智能材料。集成型智能材料是由传感器、控制器和执行器等嵌入结构中而集成的功能系统，有时又称之为智能结构。另外，根据智能材料在结构中对环境的响应情况，它又分为主动式智能材料和被动式智能材料[19]。被动式智能材料能响应环境的变化，通常有自修复或承受环境突然变化的能力；主动式智能材料能感知环境的变化，利用反馈系统产生响应。

集成型智能材料在目前研究较多，且得到了一定的应用。它一般由基体材料、敏感材料、驱动材料和信息处理器四部分组成。基体材料担负着承载的作用，一般宜选用轻质材料；敏感材料担负着传感的任务，其主要作用是感知周围环境的变化（包括压力、应力、温度、电磁场、pH值等）；驱动材料担负着响应和控制的作用，在一定条件下，驱动材料可以产生较大的应变和变形；信息处理器对传感器输出信号进行判断处理。基体材料、敏感材料、驱动材料和信息处理器构成了一套完整的智能体系，从而使材料具有感知和适应周围环境的能力。

智能材料的工作原理[20]：敏感材料感知到外界环境的变化（比如压力、应力、温度场、电磁场、pH值等发生改变），将此变化信息传送到信息处理器，信息处理器对信息进行判断处理，再将信号传送给驱动材料，驱动材料根据传送的信号控制基体材料，改变其状态，从而使材料适应周围的环境。

虽然智能材料的提出还不到二十年，且目前所研制的各种智能材料及结构距离人们的设想还有一定的距离，其设计的理论和方法还不够完善，但由于其功能优越，实用范围广，而受到机械、航空航天、医药卫生、土木建筑、船舶、服装纺织、仿生学等领域研究人员的重视[14,21~23]，并已初步应用在这些行业中，产生了一定的经济效益和社会效益。

4.2 智能混凝土

智能混凝土（smart concrete）作为智能材料的一个研究分支，是近年来国内外正在研究的热点问题。它是以水泥、砂浆或混凝土为基体，其中复合智能型组分（比如形状记忆合金、电流变体、压电陶瓷、光纤材料等），或其他材料（如碳纤维、高分子材料等），使混凝土材料成为具有一定自感知、自适应和损伤自修复等智能特性的多功能材料[24,25]。利用智能混凝土的这些特性可以有效地预报混凝土材料内部的损伤，满足混凝土结构自我安全监测的需要，防止混凝土结构潜在的脆性破坏，同时根据监测的结果，实现材料及结构的自动修复，以提高混凝土结构的安全性和耐久性。智能混凝土最早可追溯到20世纪60年代，当时苏联学者尝试以碳黑石墨作为导电组分制备水泥基导电复合材料；20世纪80年代末，日本土木工程界的研究人员设想并着手开发构筑高智能结构的所谓"对环境变化具有感知和控制功能"的智能建筑材料；20世纪90年代，日本学者H. Hiarshi采用在水泥基材内复合内含粘结剂的微胶囊（称为液芯胶囊）制成具有自修复智能混凝土[26]，美国伊利诺伊斯大学的 Caolyn Dry 采用在空心玻璃纤维中注入缩醛高分子溶液作为粘结剂，研制成一种可自行愈合的混凝土，以实现建筑结构的自动加固功能[27]。1993 年美国创办了与土木建筑有关的智能材料与智能结构的工厂，从而初步实现了智能材料与结构新技术的产业化。总之，自 20 世纪 90 年代以来，国内外对混凝土在智能化方面作了一些有益的探讨，并取得了一些阶段性的成果。相继出现了损伤自诊断智能混凝土、自适应自调节智能混凝土混凝土、自修复智能混凝土、具有反射和吸收电磁波功能的智能混凝土、温度自监控智能混凝土等。

4.2.1 损伤自诊断智能混凝土[28~39]

普通混凝土本身并不具备自感知能力，但在混凝土基材中复合导电、传感器等其他材料组分使混凝土本身具备自诊断和自感知功能，从而成为自诊断混凝土。所谓自诊断，是指混凝土能够通过自身物性的变化，来反应外界环境对自身的作用情况，并能做出材料安全与否的判断，这种判断通常是通过光、电、声等信号的变化来反映出来。在智能混凝土中，常用的材料组分有：聚合物类、碳类、金属类和光纤等，其中最常用的是碳类、金属类和光纤。目前，正在研制的损伤自诊断智能混凝土有碳纤维智能混凝土、光纤智能混凝土和压电智能混凝土等。

碳纤维是一种高强度、高弹性且导电性能良好的材料。在水泥基材料中掺入适量碳纤维不仅可以显著提高强度和韧性，而且其物理性能，尤其是电学性能也有明显的改善，可以作为传感器并以电信号输出的形式反映自身受力状况和内部的损伤程度。碳纤维智能混凝土是以水泥砂浆或是普通混凝土为基材，在其中添加碳纤维等导电组分及其他外加剂，从而使混凝土成为具有一定机敏性和温敏性的智能材料（称为 Smart Carbon Fiber Concrete 或是 Carbon Fiber Reinforced Cement or Concrete 简称 SCFC 或 CFRC）。为了探讨碳纤维对混凝土性能的影响，美国的 D. D. L. Chung 教授将一定形状、尺寸和掺量的短切碳纤维掺入到混凝土材料中，通过对材料的宏观行为和微观结构进行观测，发现材料的电阻变化与其内部结构变化是相对应的，即：碳纤维混凝土构件在弹性受力阶段，其电阻变化率随内部应力线性增加，当构件处于非弹性变形时，电阻的变化率出现了非线性增加，这表明材料结构内部出现了损伤，而当接近构件的极限荷载时，电阻增大幅度较大，预示构件即将断裂破坏。而普通材料的导电性在加载过程中几乎无变化，直到临近破坏时，电阻变化率剧烈增大。这一现象表明，碳纤维混凝土电阻率的变化规律可反映材料结构内部处于安全、损伤或破坏的哪一阶段。碳纤维可使混凝土具有自感知内部应力、应变和损伤程度的功能。根据纤维混凝土的这一特性，通过测试碳纤维混凝土所处的工作状态，可以实现对结构工作状态的在线监测。当结构内部应力接近损伤区或破坏区时，即可自动报警。同时利用复合材料的敏感性可有效地监测拉、弯、压等情况以及在静态或动态荷载作用下材料的内部情况。比如当在水泥净浆中掺加 0.5% 体积分数的碳纤维时，它作为应变传感器的灵敏度可达 700，远远高于一般的电阻应变片。除了损伤自诊断外，碳纤维自诊断混凝土可用于确定公路路上的车辆方位、载重和速度等参数，为交通管理的智能化提供材料基础；还可用于机场跑道等处的化雪除冰、钢筋混凝土结构中的钢筋阴极保护、住宅建筑的电热结构等。

光纤传感智能混凝土是另一种自诊断智能混凝土，它是在混凝土结构的关键部位埋入光纤维传感器或其阵列，探测混凝土在碳化以及受载过程中内部应力、应变变化，并对由于外力、疲劳等产生的变形、裂纹及扩展等损伤进行实时监测。图19所示为将光纤传感器埋入混凝土中以测量混凝土裂纹尖端扩展位移（CTOD）。为了提高光纤传感器的灵敏度，将直线光纤［图19（a）］改变为环状［图19（b）］以增加其长度，准确测量微裂纹开展的规律。通过埋有环状光纤传感器的混凝土简支梁受载实验，结果表明，光纤传感器能测量位移为5μm的裂纹尖端扩展位移，这是传统测试方法难以达到的。它是利用光在光纤的传输过程中受温度、压力、电场、磁场等外界环境因素对光强度、相位、频率、偏振态等光波量的影响规律而开发的一种传感技术。近年来，国内外进行了将光纤传感器用于钢筋混凝土结构和建筑监测这一领域的研究，开展了混凝土结构应力、应变及裂缝发生与发展等内部状态的光纤传感器技术的研究，这包括在混凝土的硬化过程中进行监测和结构的长期监测。光纤在传感器中的应用，提供了对土建结构智能及内部状态进行实时、在线无损检测手段，有利于结构的安全监测和整体评价和维护。目前，光纤传感器及光纤传感智能混凝土已用于多个工程，主要有加拿大Calgary建设的一座名为Beddington Trail的一双跨公路桥内部应变状态监测、美国Winooski的一座水电大坝的振动监测、江苏润扬长江公路大桥长期监测与安全评估系统、重庆渝长高速公路上的红槽房大桥监测系统等。

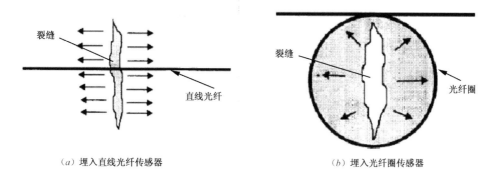

图19　光纤传感器测量混凝土裂纹尖端扩展位移（CTOD）

压电智能混凝土是将具有压电特性的敏感元件（如压电石英、压电陶瓷）埋入混凝土中，并按一定的排列方式构成阵列网络，由于压电效应，敏感元件可以无源、不间断地工作，直接将结构的应力、应变变化转换成电信号输出。由于无源工作的压电敏感元件只能传感结构的动态变化，而

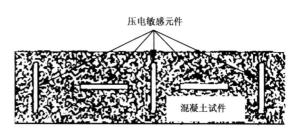

图20　压电智能混凝土试件

土木工程结构受环境影响的变化往往是十分缓慢的，结构自身的缺陷在载荷的作用下，逐渐发展并最终导致结构的失效或损坏。因此通常对土木工程结构的监测，更主要的是对其静态（或准静态）情况的评估、监测。为了利用埋入混凝土中的无源压电敏感阵列网络，实现对结构的静、动态监测，在结构中安置了多个与压电敏感元件的相同压电元件（如图20），使它们与混凝土结构复合形成主动式的压电智能混凝土结构，利用压电敏感元件所具有的逆压电效应，在交变电压的作用下，使压电元件产生声振动，声波在结构中传播，

结构的静、动态特性将对声波产生调制，阵列网络中的压电元作为敏感元件，将声波信号转换成电信号，从而实现静、动态参数的监测。

4.2.2 自修复智能混凝土[40~47]

混凝土结构在使用阶段，由于受到各种荷载及外部环境的作用下，不可避免地发生各种形式的损伤甚至开裂，特别是所产生的裂缝会加速有害物质的传输并加速氯离子扩散和钢筋锈蚀，从而影响结构的安全性。因此，对混凝土结构的修复就成为人们非常重视的一个工

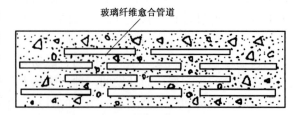

图21 裂缝自愈合智能混凝土

程问题。自修复智能混凝土是研究人员模仿生物组织对受创伤部位能自动分泌某种物质，从而使受创伤部位愈合的机理而研制的。它是在混凝土材料中加入某些特殊的成分（主要在受拉方向上），如内含粘结剂的空心胶囊、空心玻璃纤维或液芯光纤（如图21），当混凝土材料或构件在受到损伤或开裂时，部分空心胶囊、空心玻璃纤维或液芯光纤也会破裂，粘结剂流到损伤处，从而弥补混凝土内部的缺陷，使混凝土的裂缝重新愈合。

常用的修复剂有两种：一种修复剂是本身就具有粘结基体材料的功能。Day C M 将装入化学药品的多孔玻璃纤维放置在混凝土中，如果混凝土因地震或其他应力而发生破裂，空心玻璃纤维就会破裂，释放出一种粘结剂阻止进一步的破裂。日本学者三桥博三教授将内含粘结剂的空心胶囊掺入混凝土材料中，一旦混凝土材料在外力作用下发生开裂，部分空心胶囊就会破裂，粘结剂流向开裂处，可使混凝土裂缝重新愈合（如图22）。在1994年，美国伊利诺伊斯大学的 Carolyn Dry 将内注有缩醛高分子溶液作为粘结剂的空心玻璃纤维埋入混凝土中，使混凝土产生了自愈合效果（如图23）。另一种是修复剂本身不具备粘结基材的功能，但是当其与另外的物质相遇时，能够发生化学反应，生成具有粘结功能的物质，从而具有粘结基材的功能，实现混凝土裂缝的修复。如采用磷酸钙水泥为基体材料，在其中加入多孔编织纤维网，在水泥水化和硬化过程中，多孔纤维释放出引发剂，引发剂与单聚物发生聚合反应生成高聚物。这样，在多孔纤维网的表面形成了大量有机及无机物质，它们互相穿插粘结，最终形成了与动物骨骼结构相类似的复合材料，具有优异的强度和延展性、柔韧性等性能。在混凝土材料使用过程中，如果发生损伤，多孔纤维就会释放高聚物，自动愈合损伤。

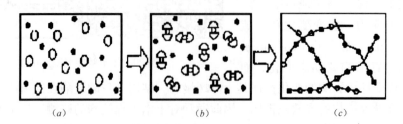

图22 内置胶囊仿生自愈合混凝土自愈合机理示意

(a) 内含修补剂的胶囊事先被埋入混凝土内；(b) 裂缝的发生使胶囊破裂，修补剂流出；
(c) 流出的修补剂修复了裂缝

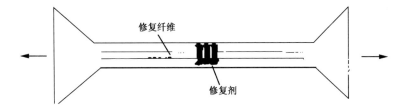

图 23　内置封闭式空心玻璃纤维仿生自愈合混凝土自愈合机理示意

在混凝土中埋入粘接剂后，不仅可实现混凝土裂缝的自愈合或损伤自修复，还可提高开裂或损伤部位的强度，增强材料或构件的延性弯曲的能力。同时，埋入混凝中的低模量粘接剂可以改善建筑结构的阻尼性能，提高结构的抗风抗震能力。

4.2.3　自调节智能混凝土[48~51]

混凝土结构除了正常负荷外，人们还希望它在受到地震、强风等自然灾害时，能够调整结构的承载能力和减轻结构的振动。混凝土本身是惰性材料，要达到自调节的能力，就需要在混凝土中复合具有驱动功能的组合材料。常见的驱动功能的材料有形状记忆合金（SMA）和电流变/磁流变体（ER/MR）等。

形状记忆合金是具有形状记忆效应和超弹性效应等多种性能的一种智能材料。形状记忆效应是指具有热弹性马氏体相变的材料能记忆它在高温奥氏体下的形状。当环境温度 $T<A_f$（A_f 为材料马氏体逆相变结束温度）时，在外应力作用下，产生了一定的残余变形；当加热到 A_f 以上时，残余变形消失，材料能恢复到加载前的形状和体积。而超弹性效应是指环境温度 $T \geqslant A_f$ 以上一定温度区间内，外加应力超过弹性极限，产生塑性变形后，卸载时即使不加热，应变也会随外应力的下降而下降，且外应力为零时，应变也恢复到零，应力应变曲线呈现出滞回效应。也就是说，形状记忆合金是一种对温度和应力非常敏感的智能材料。20 世纪 90 年代初，日本建筑省建筑研究所曾与美国国家科学基金会合作研制了具有调整建筑结构承载能力的自调节混凝土材料。其方法是在混凝土中埋入形状记忆合金，利用形状记忆合金对温度的敏感性和不同温度下恢复相应形状的功能，在混凝土结构受到异常荷载干扰时，通过记忆合金形状的变化，使混凝土结构内部应力重分布并产生一定的预应力，从而提高混凝土结构的承载力。

电流变液/磁流变液是一种可控流体，它是用不导电的母液（常为硅油或矿物油）和均匀散布其中的固体电解质颗粒或磁性颗粒制成的悬浮液。在电场或磁场的作用下，电流变液和磁流变液中的固体颗粒会形成一束束纤维状的链，横架于电场的正负两极或磁场的两极之间。这样，对于平行于两极的剪切力而言，电流变液和磁流变液在电场或磁场的作用下就能从流动性良好的具有一定黏滞度的牛顿流体转变为有一定屈服剪应力的黏塑性体，即所谓的"固化"，当外界电场或磁场拆除时，仍可恢复其流变状态。这种特性就称为电流变或磁流变效应。利用电流变体的电流变效应，在混凝土中复合电流变或磁流变体，当混凝土结构受到台风、地震袭击时，可调整其内部的流变特性，改变结构的自振频率、阻尼特性以达到减轻结构振动的目的。

有些混凝土结构对湿度的要求比较高，这就要求混凝土结构具有能够根据周围环境的状况自动调节自身湿度的能力。研究发现，在混凝土中加入调湿性组分天然沸石所组成的调湿混凝土，能够根据周围环境湿度的变化改变自身湿度。调湿混凝土的吸、放湿与温度有关，当温度升高时放湿；温度降低时吸湿。

4.2.4 温度自控智能混凝土[52~56]

混凝土结构在温度的作用下，容易发生膨胀收缩变形，产生内应力，这对建筑结构造成很大危害。因此对于那些对温度要求比较严格的建筑结构，就需要对温度进行实时监测。

碳纤维混凝土具有很好的温敏性。一方面，含有碳纤维的混凝土会产生热电效应(Seebeck 效应)，即温度变化引起电阻变化（温阻性）及碳纤维混凝土内部的温度差会产生电位差的热电性。试验表明，在最高温度为70℃，最大温差为15℃的范围内，温差电动势（E）与温差 Δt 之间具有良好稳定的线性关系。当碳纤维掺量达到某一临界值时，其温差电动势率有极大值，且敏感性较高。如在普通硅酸盐水泥中加入碳纤维，其温差电动势率可达 18μV，将微细钢纤维混凝土和碳纤维混凝土联结形成的水泥基热电偶，其敏感度可以达到 70μV/℃。因此利用这一效应可以把碳纤维智能混凝土制成热电偶，埋入混凝土结构中，实现对混凝土结构内部和建筑物周围环境的温度分布及变化的监控；也可以实现对大体积混凝土的温度自监控以及用于热敏元件和火警报警器等，可望用于有温控和火灾预警要求的智能混凝土结构中。另一方面，当对碳纤维混凝土施加电场时，在混凝土中会产生热效应，引起所谓的电热效应，利用电热效应，可以把碳纤维智能混凝土制成机场跑道、桥梁、道路路面等以实现自动融雪和除冰的功能。

4.2.5 具有反射与吸收电磁波功能的智能混凝土[57~60]

随着电子信息时代的到来，各种电器电子设备的数量爆炸式地增长，导致电磁泄露问题越来越严重，而且电磁泄露场的频率分布极宽，从超低频到毫米波，它可能干扰正常的通信、导航，甚至危害人体健康。因此电磁污染是影响我国城市化可持续发展的灾害之一。

在普通混凝土中加入导电增强介质诸如短切碳纤维、钢纤维制成的智能混凝土能够反射和吸收电磁辐射，利用智能混凝土的这种特性可以研制 EMI 防护罩。鉴于电磁辐射已经成为社会污染的一个严重问题，EMI 防护罩的重要性已经是显而易见的。在混凝土中掺入体积含量为 1.5%、直径为 0.1μm 的短切碳纤维，其对 1GHz 电磁波的反射强度为 40dB，要比普通混凝土对 1GHz 的电磁波的反射强度高 10dB，且其反射强度比透射强度高 29dB，而普通混凝土反射强度比透射强度低 3~11dB。研究表明，对碳纤维微丝经臭氧处理后再掺入混凝土中，不但能提高混凝土反射电磁波的能力，而且能提高混凝土的抗拉强度。采用这种混凝土作为车道两侧导航标记，可实现自动化高速公路的导航。汽车上的电磁波发射器向车道两侧的导航标记发射电磁波，经过反射，由汽车上的电磁波接收器接收，再通过汽车上的电脑系统进行处理，即可判断并控制汽车的行驶线路。采用这种混凝土作导航标记，其成本低，可靠性好，准确度高。

4.3 小结与展望

随着材料科学的发展和工程应用的需要，各种形式、不同功能的智能混凝土还将不断地被研制出来。而前面所述的自诊断、自调节和自修复等各种混凝土是智能混凝土研究的初级阶段，它们所具有的智能程度还较低，一般只具备智能混凝土的某一些基本特征，是一种简单形式的智能混凝土。为了进一步提高智能混凝土的性能，实现土木工程结构向"高智商"的智能结构迈进，必须要从材料学、物理学、化学、力学、电子学、人工智能、信息技术、计算机技术、生物技术、控制论、仿生学和生命科学等众多学科中开展交叉研究。目前，智能混凝土的研究与应用虽处于初级阶段，但它拥有巨大的工程应用前景和社会经济效益。而作为一种新型的土木工程材料，如果要被广泛地应用于工程实际，还有很多工作需要进一步研究：

（1）进一步研究各种智能混凝土的工作机理；

（2）开发适合于智能混凝土的新型智能控制材料，为研制新的智能混凝土提供基础；

（3）积极研发具有集传感、处理、控制于一体的多种功能智能混凝土，提高智能混凝土的集成化及小型化程度；

（4）开展智能混凝土材料—工程结构—智能体系的一体化研究，推动混凝土在工程中应用的进度；

（5）开展现有各种智能混凝土材料及结构的优化设计研究，如碳纤维混凝土的电极布置方式、耐久性等；光纤混凝土的光纤传感阵列的最优排布方式；自愈合混凝土的修复粘结剂的选择、封入的方法以及愈合后混凝土耐久性能的改善等；

（6）开展智能混凝土复合材料中各种材料组分之间的相互作用的微观分析研究，建立相应的分析理论和分析方法，为智能混凝土材料与结构的宏观设计提供基础；

（7）广泛开展智能混凝土材料及其结构的实验研究，特别是将智能混凝土复合于结构中所进行的结构及体系的实验研究；

（8）建立智能混凝土材料及结构的设计理论和设计方法。

5. 纤维增强树脂（FRP）

5.1 纤维增强树脂的新进展

纤维增强树脂（FRP）从 20 世纪 40 年代问世，并在航空、航天、船舶、汽车、化工、机械等领域得到广泛应用，当时在中国被称之为"玻璃钢"。当今 FRP 加固、增强和取代钢筋用于承担钢筋混凝土中的拉应力，这种材料是有机与无机、金属与非金属、纤维型与颗粒型材料多重复合的新技术。

氯盐环境下混凝土结构中钢筋锈蚀问题十分突出。因钢筋锈蚀导致许多桥梁、建筑物体前破坏，造成巨大经济损失。有资料显示因钢筋锈蚀导致混凝土结构提前破坏而造成的经济损失将可能成为制约未来经济发展的重要因素。在美国，每年因钢材腐蚀造成的工程结构损失高达 700 亿美元，近 1/6 的桥梁因钢筋锈蚀而严重损坏。加拿大用于修复因老化损坏的工程结构的费用高达 490 亿加元，我国目前因钢材锈蚀而造成的损失也在逐年增加[61~64]，每年造成的经济损失也高达 1000 亿人民币。为防止钢筋锈蚀，各国都采取了相应的措施，主要是严格保证混凝土的施工质量，限制最大水灰比和最小水泥用量，规定最低混凝土强度等级、最小保护层厚度和最大允许裂缝宽度等要求；同时提高钢筋的抗锈蚀能力，如在钢筋表面镀锌或在钢筋表面喷涂环氧树脂形成环氧涂层钢筋等，但这些措施仍不能从根本上解决钢筋混凝土结构的耐久性问题。自上世纪 80 年代中期以来，欧美及日本等国陆续采用连续性纤维增强树脂材料（Continuous Fiber Reinforced Plastics/ Polymer，简称 FRP）来代替钢筋，这种材料不仅轻质高强，而且在恶劣环境中具有高抗腐蚀能力，有效延长了恶劣环境下混凝土结构的耐久性和服役寿命，降低结构寿命周期维修费用和成本，是一种前景广阔的新型结构材料。二十多年来，世界各国对 FRP 材料的原材料、配合比、加工工艺、力学性能与结构性能进行了系统研究。在混凝土结构中应用的 FRP 复合材料包括片材、网材和筋材等品种，片材主要是通过树脂粘结剂粘贴于混凝土结构受拉表面起补强作用，网材可以代替钢丝网或钢筋网对混凝土结构起到增强作用，也可以通过树脂粘结剂将其贴在被补强的结构上起加固和提升性能的作用，筋材主要是替代传统钢筋用于混凝土结构或预应力混凝土结构增强。许多国家先后颁布了 FRP 筋混凝土结构设计与施工规程，与此同时世界各国采用 FRP 材料进行修补、加固和新建的工程也日益增多。为克服 FRP 筋刚度小、结构变形大的缺点，并充分利用混凝土徐变和收缩对低弹模筋不会引起大的预应力损失的特点，人们对 FRP 筋施加预应力，许多跨海大桥、大跨度重载建筑结构也开始采用预应力 FRP 筋增强混凝土结构，从而 FRP 材料与结构的研究已成为材料科学与土木工程界研究的一个热点。我国对于 FRP 增强混凝土的研究时间不长，但一开始我国的混凝土科学与混凝土工程专家就对 FRP 材料的制备技术及其增强混凝土结构的静、动力学特性和各种环境条件下的耐久性给予了高度重视，在不长的时间历程中进行了大量的研究工作，取得了一系列的新进展。尽管 FRP 材料强度较高，但当今为解决 FRP 材料极限延伸率较低（1~5%）、脆性破坏、FRP 结构和 FRP 组合结构及其耐久性的研究工作还在不断深化。

5.2 FRP 材料的组成结构与性能

FRP 材料与传统钢材有很大不同。它是以纤维为增强，以合成树脂为基体，并掺入适量辅助剂，经不同特点的成型工艺制成的一种新型复合材料。FRP 筋中纤维体积含量可达到 60%，具有抗拉强度高、质量轻的特点，FRP 重量约为普通钢筋的 1/5，强度为普通钢筋的 6 倍以上，高抗腐蚀、低松弛、非磁性、抗疲劳等许多钢筋不具备的优点都集中在 FRP 复合材料中；FRP 材料的物理力学性能与纤维种类、纤维含量、粘结基体种类、用量、表面处理以及成型工艺等因素密切相关，不同的材料组成的 FRP 结构性能有很大差别。此外，在 FRP 复合材料中通常还加入一些填充物和添加剂，可以起到减少收缩、降低成本、改善加工性能以及提高物理力学性能的作用。树脂基体主要有热固性树脂和热塑性树脂两大类，通常采用不饱和聚酯树脂、乙烯基树酯、环氧树脂和酚醛树脂等，而常用的增强纤维材料主要有玻璃纤维、碳纤维、芳纶纤维及其他新型高性能有机合成纤维等，FRP 复合材料的性能取决于纤维和聚合物基体类型、

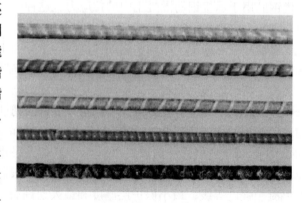

图 24　FRP 筋的外形[69]

纤维分布状态及其与聚合物间的界面结构。由于聚合物基体弹性模量较低，制成的FRP复合材料的弹模小于纤维的弹性模量，FRP材料的强度也小于用来制造它们的纤维的强度。为改善FRP的力学性能，克服一些无机纤维如碳纤维和玻璃纤维没有屈服点呈脆性破坏的缺点，常常采用不同性质纤维混杂与性能互补技术以达到增强和增韧的双重目的。根据采用纤维的不同，制成的FRP分别称为碳纤维增强聚合物（CFRP）、玻璃纤维增强聚合物（GFRP）、芳纶纤维增强聚合物（AFRP）和混杂纤维增强聚合物（HFRP）。常用的FRP主要包括筋材、片材和网材（或格状材）等品种。筋材主要用来取代钢筋增强混凝土，片材和网材主要用来制备FRP供工程修补和加固用。FRP筋的外形又包括光圆、螺纹、矩形、工字形等，外径一般在3~40mm之间。FRP的组成与性能与树脂基体及增强纤维密切相关。常用的筋材外形如图24所示。

5.3 FRP用增强体纤维及其发展方向

作为FRP复合材料的重要组成部分，增强纤维决定了复合材料最终的机械性能。一般要求增强纤维具有较高的强度和弹性模量，以及优异的热稳定性和耐高温性能。目前玻璃纤维、碳纤维以及聚酰胺纤维和聚乙烯醇纤维等高性能有机纤维是制造高性能FRP的主要纤维品种，表8所示为常见纤维的主要技术性能，可以看出所有纤维的抗拉强度都比钢筋高，而极限延伸率则均较低。与钢材与铝材相比，纤维材料强度为钢材的20~50倍，具有显著高强轻质的性能。碳纤维模量为钢材的5~10倍，芳纶纤维为2~3倍，玻璃纤维与钢材相近。经大量研究与实践表明，碳纤维有优势，但其极限延伸率低。

用于FRP的纤维性能[65]　　　　　　　　表8

纤维品种 技术参数	碳纤维				聚酰胺纤维 （芳纶纤维）		玻璃纤维		聚乙烯醇 高强纤维
	丙烯腈		沥青树脂		Kevlar49	Tech-nora	E-玻纤	抗碱玻纤	
	高强度	高弹模	高强度	高弹模					
抗拉强度（MPa）	3430	2450~3920	764~980	2940~3430	2744	3430	3430~3528	1764~3430	2254
弹性模量（GPa）	196~235	343~637	37~39	392~784	127	72.5	72~73	68~70	59.8
延伸率（%）	1.3~1.8	0.4~0.8	2.1~2.5	0.4~1.5	2.3	4.6	4.8	4.52~3	5.0
密度（g/cm³）	1.7~1.8	1.8~2.0	1.6~1.7	1.9~2.1	1.45	1.39	2.6	2.27	1.30
直径 μm	5~8		9~18		12		8~12		14

碳纤维有两大系列，即沥青基和丙烯酸腈（PAN）基制成，其性能因石墨晶体的结合情况而异，根据其制备原料碳纤维可分为沥青基碳纤维和丙烯酸腈碳纤维两种，一般来说PAN基碳纤维能提供高强度，而沥青基碳纤维的模量较高。碳纤维具有高强度、高弹模量、耐高温和导电等性能，同时与玻璃纤维相比碳纤维具有很高的抗腐蚀能力和较高的比强度（抗拉强度/表观密度），这使得它在目前的实际工程中有较多应用。由于碳纤维的极限延伸率在所有纤维中最低，使得碳纤维复合材料的韧性下降，从而用CFRP材料难以承受冲击荷载。通过控制微观结构缺陷、结晶取向、杂质和改善工艺条件，利用PAN或沥青基碳纤维均可获得高强高模量的FRP复合材料[65]，目前CFRP的研究主要围绕改善韧

性、提高模量和强度以及降低成本。

玻璃纤维是在商业上最为成功、也是应用最多的一种增强材料,根据组成的不同可分为 E—玻璃纤维、C—玻璃纤维、S—玻璃纤维和 D—玻璃纤维。用于复合材料生产的玻璃纤维大部分是 E—玻璃纤维,也有部分抗碱玻璃纤维,前者含有大量的硼酸和铝酸盐,而后者则掺有一定量的氧化锆,可以有效提高玻璃纤维的耐碱腐蚀性[66]。聚酰胺纤维是一种高度定向的有机纤维,市售商品有 Kevlar、Twaron 和 Tech - nora 三种。Kevlar 和 Twaron 均为芳香族聚酰胺纤维,其晶核由直线苯核结合;Tech - nora 为芳香族聚醚酰胺纤维,其晶核由乙醚结合[5]。丙烯酸醇纤维是在生产时采用了高聚合化丙烯酸醇,纤维在生产过程中经缠绕而增加了强度和弹性。聚乙烯醇纤维质量较轻,并具有较好的延性,但与其他纤维相比,抗拉强度与弹性模量则相对较低。目前国内外正大力开展对高性能有机纤维的研究,高性能有机纤维包括具有柔性链结构的超高强度聚乙烯纤维(UHTPE)、芳纶纤维(PTAA)和刚性链结构的聚对苯撑苯并二恶唑纤维(PBO)。UHTPE 纤维密度低、拉伸强度和模量极高,采用等离子表面处理方法解决了它和基体粘结差的问题,使之应用越来越广。然而 UHTPE 纤维在 150℃熔融和在室温下会出现蠕变这一缺点,严重阻碍了它作为结构材料的应用。目前 PTAA 纤维的拉伸模量已达 $100 \sim 200\text{GPa}$、断裂强度达 $2 \sim 4\text{GPa}$、密度为 $0.97 \sim 1.47\text{g/cm}^3$。PTAA 纤维的最大缺点是压缩和横向拉伸性能差,复合材料生产中的热收缩应力可能导致纤维劈裂,水分会沿着劈裂的纤维进入复合材料而加速复合材料的失效。PBO 是聚对苯撑苯并双恶唑纤维的简称,被誉为 21 世纪超级纤维,其商品名为柴隆(Zylon)。PBO 具有十分优异的物理机械性能和化学性能,其强度、模量为 Kevlar(凯夫拉)纤维的 2 倍,并兼有间位芳纶耐热阻燃的性能,而且物理化学性能完全超过迄今在高性能纤维领域处于领先地位的 Kevlar 纤维,PBO 纤维的耐冲击性、耐摩擦性和尺寸稳定性也很优异[66]。

5.4 FRP 用树脂基体及其发展方向

FRP 复合材料除上述不同增强材料纤维外,优选不同类型的树脂作为基体也十分重要,只有二者通过有效的界面结合,才能达到传递纤维承受的荷载,保护纤维免受周围环境影响的目的。由于树脂的灌注、浸渍工艺对复合材料性能有重要影响,如方法不当会在复合体(Matrix)内引起应力集中而使抗拉强度大大降低。常用的树脂有热固性和热塑性两大类,常用的热固性树脂有酚醛树脂、环氧树脂、不饱和聚酯树脂和有机硅树脂等,而这类树脂首次加热时软化,可塑造成型,但固化后再加热时将不再软化,也不溶于溶剂。热塑性树脂加热时则软化和熔融,可塑造成型,冷却后即成型并保持即得形状,这一过程具有重复性。国内外常用的热塑性树脂主要有:聚乙烯(PE)、聚丙烯(PP)、尼龙(PA)、聚酰胺—酰亚胺(PAI)、丙烯腈—丁二烯—苯乙烯(ABS)、聚酰亚胺(PI)、聚醚醚酮(PEEK)、聚醚砜(PES)和聚醚酮(PEK)等。常见树脂基体的性能见表9。热固性树脂基 FRP 材料有一个共同点,即最终固化成型必须通过熟知的化学交联反应才能实现,化学交联度不同,材料的热稳定性和脆性也不同,交联度提高了热稳定性但脆性也随之增大,从而导致材料的极限延伸率和韧性均下降,此外热固性树脂还具有不溶解、不熔化、不可焊接以及在高温下炭化的特点。针对热固性树脂基加固材料存在硬脆、耐冲击性差、固化成型周期长、成本高、有机溶剂挥发不环保、难以回收再利用等诸多缺陷。因此，以改性低熔融黏度热塑性树脂作为基体，制备 FRP（简称 FRTP），与用热固性树脂制备 FRP（简称 FRSP）相比，前者具有韧性较高、成型加工周期较短、可重复使用、有类似于金属的加工特性、成本低、维修方便等优点，所以 FRTP 越来越受到各国重视，研究应用也十分活跃[67]。

典型常见树脂基体性能[66]   表9

名　称	热变形温度（℃）	拉伸强度（MPa）	延伸率（%）	压缩强度（MPa）	弯曲强度（MPa）	弯曲模量（GPa）
环氧树脂	50～121	98～210	4	210～260	140～210	2.1
不饱和聚酯	80～180	42～91	5	91～250	59～162	2.1～4.2
酚醛树脂	120～151	45～70	0.4～0.8	154～252	59～84	5.6～12
聚酰胺	—	70	60	90	100	—
乙烯基树脂	137～155	59～85	2.1～4	—	112～139	3.8～4.1

发展高性能 FRTP 材料对树脂的耐热性和机械强度都有较高的要求，考虑土木工程应用时的性价比、对环境有无污染以及热变形温度、加工温度等因素，聚丙烯、尼龙等可以作为合适的树脂基体进行 FRTP 筋、板、管的试验研究，而一些对耐火及特定强度有高要求的结构可以考虑使用成本较高的高性能热塑性树脂。同时考虑添加适当的添加剂和其他树脂组分对树脂基体进行改性，以提高树脂自身的流动性和易成型性。目前世界各国争相开发各种高强度、高耐热的热塑性树脂基体，如英国 ICI 公司和美国 DuPont 公司开发的聚醚醚酮（PEEK）树脂（其熔点高达 334～380℃，长期使用温度为 240～260℃）；德国开发的聚醚酮（PEK）树脂，美国 Phillips 开发的聚苯硫醚（PPS）等均属高性能树脂基体，我国也开发了聚芳醚砜酮（PPESK）、聚芳醚酮酮（PPE2KK）等新型高性能工程树脂基材料[66]。表10 列出了常用的高性能树脂的热性能及力学性能。总之，对 FRP 材料的基体除要求复合材料中的热塑性树脂有良好的机械性能、高热稳定性、耐化学腐蚀性外，另一关键技术在于其加工性能，一般高性能热塑性树脂大都难溶或难融，这就给复合材料的树脂浸渍和成型加工造成了困难。加工温度越高，生产过程中树脂越容易热氧化、降解，因此要选择合适的树脂，避免生产时对设备过高的要求。如何降低成本已成为发展高性能 FRTP 材料需要解决的关键技术难题。

新型高性能热塑性树脂基体性能[66]   表10

高性能树脂基体	T_g（℃）	抗张模量（GPa）	抗张强度（MPa）	延伸率（%）
PEKK	156	4.50	102	4
PEEK	144	3.79	103	11
HTX	205	2.48	86	13
PAS-2	215	3.28	101	8
PPS	85	3.91	80	3
PEI	217	2.96	104	60
PES	260	2.41	76	7

续表

高性能树脂基体	T_g（℃）	抗张模量（GPa）	抗张强度（MPa）	延伸率（%）
PAI	283	3.30	135	25
PES-C	260	3.20	89	3
PEK-C	228	3.50	103	3.3
PPESK	284	1.41	90.7	11.2

5.5 FRP材料的组成结构与性能

纤维增强树脂（FRP）主要由树脂基体和纤维增强材料按照一定比例组合而成，根据需要纤维含量在55%~80%，树脂基材在20%~45%。树脂基体赋予FRP变形性能、耐化学腐蚀性、热性能和加工性能，而增强纤维则在很大程度上决定了复合材料的机械性能，纤维含量越高制成FRP材料的强度也越高，但成型也越困难，当树脂少到不足以包裹纤维表面并填充纤维空隙时，FRP复合材料的各项性能必然要下降，相反树脂用量越高FRP复合材料的韧性提高，但强度下降、成本增加。因此控制好纤维与树脂间的比例对于形成致密的组织结构、优异的力学性能、高性价比的FRP复合材料十分重要。

FRP复合材料虽自身有明显优势，但目前大量应用的FRP复合材料还存在诸多不足，主要有应力—应变呈线性关系、在达到极限抗拉强度之前无塑性变形特征，其极限应变比钢筋小得多，断裂时呈脆性破坏，有关FRP筋与普通钢材的性能比较见表8[64,68]。此外FRP材料整体抗剪强度及层间剪切强度低，造成连接件设计困难[64,69]；FRP筋的轴向热胀系数较小，而横向热胀系数较大，在温差变化较大的环境下使用FRP筋有可能造成与混凝土的粘结破坏并有可能引起混凝土开裂，而影响到混凝土结构的耐久性能；FRP材料热稳定性较差，当超过温度范围时FRP材料的抗拉强度会明显下降，这些也是应用到实际工程中不可忽视的方面。

综观上述，目前正从优选与优化纤维和树脂基体性能出发，大力研制新型纤维和高性能树脂，同时采用不同纤维混杂和性能互补技术，以提高其物理与力学性能，通过改进工艺和使用功能型外加剂以改善树脂和纤维的界面粘结和界面结构。HFRP筋的混杂方式见图25。如图25所示，中央有玻璃纤维，而周围则采用了聚酰胺纤维；目前，用于建筑结构加固的HFRP主要是将碳纤维和玻璃纤维作为混杂纤维增强相而制成的复合材料，充分发挥这两种材料的优点，从而有效提高复合材料综合性能。有研究表明[62]，GFRP与FRP体积比为2的混杂纤维复合材料的断裂应变比CFRP高30%~50%，在CFRP中加入15%的GFRP，抗冲击强度可增加2~3倍，且性价比高。通过调整这2种纤维的比例，还可使

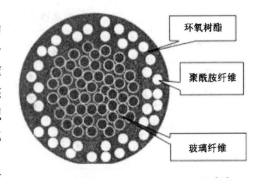

图25 混杂纤维FRP复合材料结构[65]

混杂纤维复合材料的强度和弹模在 GFRP 与 CFRP 的强度及弹模之间变化，扩大了材料与构件设计的自由度[67]。HFRP 的强度和弹性模量等力学性能与其组成材料的性能和比例有关，根据纤维与树脂基体的性能和体积分数按照混合率法则（Rule of mixture）进行评价[65]。根据纤维混杂的原理也可以对树脂基体进行改性或混杂作为基体增强相，近年来研制成功的 PBO-FRP 除具有与高强 CFRP 相近的力学性能外，还表现出更好的柔韧性，实质上就是通过改善基体相而达到高性能的[70]。

5.6 FRP 复合材料制备工艺进展

纤维增强树脂的成型方法有 FRP 筋成型方法和 FRP 修补与加固材料成型方法两大系列，成型工艺是保证纤维与基体共同工作的基础。主要有手糊工艺、喷射成型工艺、团状模塑料（Bulk Molding Compound，BMC）工艺、片状模塑料（Sheet Molding Compound，SMC）工艺、层压工艺、树脂传递模塑成型（Resin Transfer Molding，RTM）工艺、缠绕成型工艺、反应注射成型工艺和拉挤成型工艺等。连续纤维增强塑料的成型一般与制品的成型同步完成，再辅以少量的切削加工和连接即成成品；随机分布短纤维增强塑料可先制成各种形式的预混料，然后进行挤压、模塑成型[71,72]。

根据增强纤维长度的不同（短纤维、长纤维、连续纤维），FRTP 的基体浸渍工艺主要包括：溶液法、熔融法、薄膜镶嵌法、流化床法、混合纱法、悬浮熔融法、Atochem 法等[73,74]。成型工艺主要包括：隔膜成型工艺、热压成型、树脂注入成型、拉挤成型、缠绕成型等[67,75,76]。在诸多的成型工艺当中，拉挤成型是适应土木工程用制品生产（如：筋、板、管）最合适的成型工艺。FRTP 的拉挤成型工艺与 FRSP 基本相似。只要把进入模具前的纤维浸渍工艺加以改造，生产热固性玻璃钢的设备便可使用。由于热塑性树脂熔体的黏度大，并且表面极性低，造成纤维浸渍困难，因此国内外科研人员研究工作的重点集中在浸渍技术方面[77,78]，各种不同拉挤工艺的根本区别也在于浸渍方法和浸渍工艺的不同。Twintex 纤维是在玻纤拉丝的同时用挤出机将树脂（如聚丙烯）通过模头形成有机纤维，两种纤维掺混排列在一起形成一种混合无捻粗纱，在这种混合粗纱中，玻纤纱含量可达 60%~75%。这种无捻粗纱可以织成方格布或制成针刺织物，通过加热使有机纤维熔化成树脂基材，从而使玻璃纤维与树脂固结在一起，通过模压工艺将其制成模压制品，混合无捻粗纱也可以用拉挤工艺制成型材[79]。目前，法国 Vetrotex 与德国 JM 两家公司能够生产 Twintex，这种被称为"革命性的热塑性 FRP"，已经在欧美成功地应用于建筑、汽车和船艇等设备上[67]。

5.7 FRP 复合材料的劣化机理

FRP 最大的优点是其在恶劣环境特别是在氯盐环境中的高抗腐蚀性能，但 FRP 存在其他劣化机理，如环境介质的高碱性可能是导致 FRP 材料性能劣化的重要原因之一，其他如水和热的作用、冻融循环、紫外线、干湿交替、荷载作用等都会加速其劣化进程。FRP 性能劣化不仅与构成 FRP 的纤维及树脂的品种、纤维与树脂间的界面结构等密切相关，还在很大程度上受腐蚀介质的种类及浓度、环境温度与热作用、紫外线以及所施加的应力水平与作用方式等外在因素的影响；因此 FRP 材料的耐久性是材料性能及其与环境和应力相互作用的函数。

各种FRP材料在不同环境中劣化机理不同，玻璃纤维的价格最低、用量也最大，一般玻璃纤维的强度和弹性模量均较高，但其耐碱性很差，特别是在有拉应力和碱溶液的环境中玻璃纤维性能劣化十分明显，它比碳纤维和芳纶纤维更易于遭受应力腐蚀，有研究表明GFRP筋在强碱性环境中持续工作6个月其抗拉强度下降可达25%左右[80]。人们对GFRP在水和碱溶液中的耐久性有较多研究，不同的研究者对FRP劣化提出了不同的机理，Bank and Gentry 1995[81]认为水分能使聚合物链间的范德华力破坏，这使得FRP的弹性模量、强度、韧性等发生变化，Hayes等（1998）[82]认为水分引起膨胀应力使基体开裂、水解以及纤维与基体脱粘等。一般认为GFRP的劣化是由于玻纤中的S-O键分别与水分子和碱中的氢氧根的离子交换或化学反应使S-O键断裂所致，其反应分为离子交换和碱液水解两个过程。最终其劣化机理可归集到水分使树脂膨胀或由于其中的碱对玻璃纤维的腐蚀作用所致。

CFRP耐环境腐蚀能力极强，这主要由于碳纤维不吸水，其抗酸、碱及多种化学腐蚀能力都很强，此外CFRP受环境的影响程度也很小，但是由于碳纤维导电CFRP可能会产生电流腐蚀。芳纶纤维（芳族聚酰胺）易吸水，吸水后弯曲强度有较大降低，尤其当温度升高强度下降则越显著；芳纶纤维在高应力下易徐变断裂，此外芳纶纤维耐高温性能较差。

在FRP中树脂的作用是将纤维胶结在一起形成整体的复合材料，同时保护纤维免遭有害介质腐蚀，但树脂种类不同其耐腐蚀性也有很大差异，一般环氧树脂和乙烯酯抵抗水分和碱侵蚀能力较强，而聚酯树脂较差，如果水在聚合物基体中的扩散用Fick定律来描述，那么水分在乙烯酯树脂的扩散系数小于在聚酯树脂中的扩散系数；温度越高树脂的抗化学腐蚀能力下降幅度越大，树脂的耐热作用能力也较差。一般环氧树脂和聚酯树脂持续工作温度最高不超过120℃和105℃；Katz等[83]研究了提高温度对FRP与混凝土粘结性能的影响，发现直到100℃粘结强度的损失与传统变形钢筋相似，但当温度再提高时粘结强度就会大幅地降低，当温度提高到200~220℃时剩余粘结强度约为室温时的10%，粘结强度的大幅度降低是由于此时温度已超过树脂的玻璃化温度，一般用聚酯树脂制成的GFRP的玻璃化温度很低，约在80℃左右，用乙烯酯和环氧树脂制成的GFRP的玻璃化温度分别为145℃和165℃。此外树脂与纤维结合方式和结合行为不同，制成的FRP耐久性也不同，这与树脂注入纤维时有无孔隙存在关系很大，孔隙在拉应力作用下会产生应力集中并可能导致树脂开裂，因此尽管树脂极限应变高于纤维，但往往在纤维破坏前基体树脂已经开裂，FRP筋外层很薄的树脂基体一旦开裂将会加速侵蚀介质的渗入，好的挤压成型工艺与设备将会有助于消除复合材料中的空隙与缺陷、提高结构的密实度；对FRP表面进行涂敷处理也将会增加纤维与基体的粘结，提高复合材料的耐久性。

归纳起来，FRP复合材料腐蚀机理主要是：（1）环境介质渗入材料内部与大分子发生化学反应使大分子主价键破坏和裂解；（2）环境介质渗入材料内部破坏大分子次价键并使大分子溶解或融胀；（3）环境介质渗透扩散引起复合材料基体与界面脱粘；（4）环境介质与应力共同作用或应力单独作用使树脂产生银纹并生长成裂缝直至脆性断裂；（5）复合材料经受温度、湿度、应力共同作用引起损伤与开裂使材料性能下降[84]。

5.8 FRP 在工程结构中的应用进展

FRP 增强和加固混凝土是有机与无机、纤维型与颗粒型、多重混杂的复合过程。不管是做成筋材取代钢筋,还是对受损结构加固和修复,都有性能提升的过程。目前国内外 FRP 在工程结构中的应用研究主要集中在如下几方面:(1) FRP 材料制备与特性;(2) 采用 FRP 片材进行受损混凝土结构维修加固;(3) FRP 配筋混凝土结构;(4) FRP 型材结构以及与钢木组合结构;(5) FRP 预应力结构;(6) FRP 拉索。其中尤以 FRP 对现有结构维修加固研究逐年增多,工程应用实例也最为广泛。FRP 筋具有良好的耐腐蚀性,FRP 片材最早用于破损混凝土结构的加固,之后应用于环境恶劣地区的混凝土结构以替代钢筋或预应力筋,并可应用于岩土工程中的加筋土中。由于 FRP 具有轻质高强优势,如 CFRP 拉索可极大地增大缆索承重桥梁的跨越能力,使桥梁结构更加优化、设计更加灵活。CFRP 吊杆因其良好的抗疲劳性能在大跨吊桥、系杆拱桥中有广泛的应用前景;FRP 作为筋材在美国、加拿大、日本应用已有 500 多工程实例(如图 26 ~ 图 28 所示)[64],FRP 作为预应力筋主要用于桥梁结构,采用 FRP 预应力筋修建桥梁已有 50 多座,其中尤以日本最多。

图 26　GFRP 筋在 Crowchild 桥面板中的应用
(Calgary, Alberta, 1997)[69]

图 27　FRP 增强桥面板
(Lima, Ohio, 1999)[69]

图 28　GFRP 筋在 Ohio 的 Salem Avenue
桥面修复中的应用(Ohio, 1999)[69]

我国较早开展了 FRP 片材的加固研究和工程应用,我国已颁布了《碳纤维片材加固混凝土结构技术规程》(CECS146:2003)[64,85];薛伟辰在 1997 年开展了 FRP 筋预应力混凝土梁的试验研究[86,87],高丹盈从 1998 年开展 FRP 及其混凝土结构受力性能的系统研究[88,89],方志于 2000 年开始对配置 CFRP 的预应力混凝土梁、CFRP 吊杆和系杆混凝土拱桥及 CFRP 锚具进行研究和开发[90~92],吕志涛等也较早开展了对 FRP 配筋结构、FRP 预应力结构的研究,并应用于实际工程,国内首座 CFRP 拉索斜拉桥就是由东南大学主持设计和研究，锚具采用新研制的套筒粘结型，该桥的建成为今后采用 CFRP 建造特大跨度的海峡大桥等工程作出了成功的探索和实践[93,94]。但是与传统结构材料不同，FRP 制品通常为各向异性，特别是 CFRP 往往沿纤维方向的强度和弹性模量较高，而垂直于纤维方向的强度和弹性模量很低，此外它在受力性能上有许多不同于传统结构材料的现象，这些都增加了 FRP 结构的分析与设计难度。FRP 结构的设计通常由变形控制，可通过设计 FRP 构件的截面、合理地与混凝土等材料组合以及采用预应力等方法控制结构的变形，补偿刚度的不足。FRP 材料的剪切强度、层间拉伸强度和层间剪切强度仅为其抗拉强度的 5%～20%，由此使得 FRP 构件的连接成为突出的问题。FRP 结构可采用铆接、栓接和粘接，但不管哪种连接方式，连接部位往往都容易成为整个构件的薄弱环节[95]。此外 FRP 材料防火性能较差，主要是由于多数树脂在高温下会软化，在树脂达到软化温度时力学性能会大大降低。因此，往往在 FRP 树脂基体中掺入适量阻燃剂可提高抗火性能。

5.8.1 FRP 片材在加固工程中的应用进展

将 FRP 片材粘贴在构件表面可以增强构件的抗拉性能。早在 20 世纪 80 年代，FRP 片材加固混凝土结构技术在欧洲、日本、美国和加拿大等国得到迅速发展，并在实际工程得到较多的应用，特别是美国北岭地震和日本阪神地震后，FRP 加固技术的优越性在已损坏结构的快速修复加固中得到了很好的验证，目前，这些国家和地区先后颁布或出版了 FRP 加固混凝土结构设计规程或指南。FRP 片材加固技术在我国的多项工程实践中也曾尝试过，比如南京长江大桥引桥就采用了环氧树脂粘贴玻璃布进行了加固[96]。我国在 1998 年后才对 FRP 加固技术开展了系统研究，使得这一技术得到推广，并在一些重大工程如人民大会堂、民族文化宫的加固改造中得到了应用。提升和延长了这些受损结构的服役寿命。我国在 2000 年完成了首部 FRP 片材加固设计与施工技术规程[85]。近年来我国还开展了 FRP 片材加固砌体结构和钢结构的研究和应用，对于隧道、涵洞、烟囱、壳体等特种结构的加固，采用粘贴 FRP 片材的技术在我国也有开展，目前虽然 FRP 片材加固混凝土结构已得到广泛应用，但仍有许多问题值得深入探讨，主要包括界面粘结性能、抗疲劳能力、防火问题、耐久性能等[95]。

5.8.2 FRP 筋增强混凝土结构应用进展

在混凝土结构中采用 FRP 筋代替钢筋可避免由于钢筋锈蚀对结构所带来的损害，大大降低结构维护费用，通过对 FRP 筋表面砂化、压痕、滚花以增强与混凝土的粘接力；在桥梁工程中采用 FRP 索可用作悬索桥的吊索及斜拉桥的斜拉索以及预应力混凝土桥中的预应力筋，用作预应力 FRP 筋的索一般较柔软，具有一定的韧性。20 世纪 70 年代末 FRP 筋开发成功以来，FRP 配筋和 FRP 预应力筋混凝土结构的研究和应用发展较快，特别是在北美、北欧等西方国家，为解决除冰盐对桥梁结构中钢筋腐蚀所带来的严重问题，较多地采用了 FRP 筋混凝土，并在桥梁结构和建筑结构中得到了较多的应用。随后德国、日本相继建成了 FRP 预应力混凝土桥[83]。目前我国已初步制备出 FRP 筋产品和预应力锚夹具。同时针对 FRP 筋与混凝土之间的粘结问题、FRP 筋埋长、FRP 筋外部约束和表面变形以及混凝土保护层厚度对 FRP 筋与混凝土粘接性能的影响等问题进行了诸多探索性研究[63]，对 FRP 筋和预应力 FRP 筋混凝土构件受力性能进行了试验研究[96]。此外 FRP 筋及预应力 FRP 筋在挡土墙、地基锚杆及喷射混凝土筋等工程中也得到了广泛应用[95]。但由于 FRP 筋拉断前一直表现为线弹性，没有碳素钢的屈服平台，因此构件多为脆性破坏，必须针对其性能采用合理的设计方法；此外像箍筋等形状较复杂的配筋，需事先设计好后由工厂加工或采用高性能热塑性 FRTP 筋才更恰当[97]。

5.8.3 FRP 结构及其组合结构的应用进展

FRP 结构是指用 FRP 制成各种基本受力构件所形成的结构，如可以利用 FRP 拉挤型材单向受力性能好的特点，制做成工形、槽形、箱形等型材，组成 FRP 框架或桁架结构。FRP 组合结构则是指，将 FRP 与传统结构材料如混凝土和钢材通过受力形式的组合共同工作承受荷载的结构形式，如 FRP—混凝土组合梁、板就是一种合理的 FRP 与混凝土组合的结构形式，其设计概念与钢—混凝土组合梁、板相同，即上部为混凝土主要受压，下部为 FRP 构件主要受拉，它们之间通过剪力连接件使两者协同工作，同时 FRP 构件可兼做模板，便于施工。世界各国学者对于将混凝土浇入预制的 FRP 管中形成的 FRP 管混凝土的研究进行得比较深入，FRP 管对内部混凝土起约束作用，兼作模板又可以极大地提高混凝土的强度和变形能力，同时混凝土也可防止 FRP 管的屈曲破坏。FRP 管混凝土受力性能好，施工方便并具有很好的耐腐蚀性，优势非常明显，因此在结构工程中得到了较广泛的应用。此外，FRP 与钢材组合可发挥出钢材的高弹性模量和 FRP 耐腐蚀、耐疲劳性能好的优势，达到互补的效果。可在拉挤 FRP 型材时直接将钢筋和钢丝嵌入型材中成型，也可在钢结构外部采用 FRP 型材封闭，一方面防止钢结构锈蚀，另一方面可与钢结构共同受力[95]。

在 FRP 结构设计与施工规程方面，目前日本、加拿大和英国已制定了 FRP 设计指南，美国和挪威也制定了临时性设计指南，各国指南规定了与耐久性有关的环境和长期应力对抗拉强度降低系数的取值[98]，但是目前的设计指南还基本只是限制使用 FRP 类型、规定不同环境和应力下的强度降低系数等经验方法，远没有上升到基于概率的结构设计。FRP 具有抗腐蚀、抗疲劳、强度高、重量轻、非电磁性等优点，在建筑、桥梁、地下工程、海洋工程、隧道以及有特殊要求的土木工程中，应用前景十分广阔，已成为一个重要的研究领域和新兴产业[95,99]。

5.9 小结与展望

目前许多国家对 FRP 材料的开发和应用研究十分重视，因而使得 FRP 成为一个非常活跃的研究领域。对于 FRP 材料制备技术的开发主要围绕增强和增韧开展工作，比如混杂纤维复合增强增韧技术、高性能有机纤维制备技术、高性能热塑性树脂改性技术以及先进成型工艺技术等方面，且已经取得了许多重要进展。对于 FRP 材料的应用研究，也从早期的主要采用片材补强加固混凝土结构为主转向以研究 FRP 筋增强混凝土结构、预应力 FRP 结构、FRP 复合结构的静、动力学特性研究，同时逐步开展了对 FRP 结构的耐久性及使用寿命预测的研究。在应用领域尽管采用 FRP 结构或其组合结构具有许多优势，但从价格上看，FRP 结构和 FRP 组合结构与钢筋混凝土结构相比目前的竞争力不明显，考虑到自重轻以及 FRP 材料耐腐蚀能力强、维护费用低等因素，其综合效益值得深入研究和分析。随着技术进步和应用规模的不断扩大，FRP 的性价比可望进一步提高，FRP 结构服役寿命将进一步延长，使这种新型结构材料给土木工程结构的改革与创新带来新的活力和动力。

6. 结语

综上所述,这些年来由于我国重大基础工程建设的推动,新型结构材料的研究与应用在技术、理论和应用三个方面均有很大发展。随着重大工程建设的多样化、环境条件的复杂化、耐久性和服役寿命的高要求,要求环保节能型结构材料必须在性能上、技术上有更大突破,理论上有更大创新,智能化程度有更大幅度提高,工业废渣的潜能更有效发挥及应用水平再上一个高档次。为了适应各类工程不同的特殊要求,结构工程材料必须具有低收缩、低徐变、高抗裂、高耐久和长寿命的特点。寿命预测理论与方法应由单一环境因素向力学与环境耦合作用转变,并必须与工程实际密切结合,以增进寿命预测的安全性、准确性和可靠性。特别是要经过大量科学研究和现场数据的积累,建立可靠的数据库和寿命预测专家系统,以更好地指导工程应用。对具有自感知、自诊断、自修复的智能结构材料和新型高性能 FRP 复合材料而言,在理论和研究两个方面应有更大的跨越。深信通过我国混凝土科学与工程界的不懈努力,各类新型结构材料在理论和技术两个方面均将取得更大的突破和更高水平的科研成果,确保我国重大土木工程沿着高耐久、长寿命、智能化和高科技化的方向发展。

参考文献

[1] Luo X., Sun W., Chan Y. N. Steel fiber reinforced high performance concrete: A study on the mechanical properties and resistance against impact [J]. Materials and Structures, 2001, 34 (237): 144-149

[2] 蒋金洋,孙伟,张云升,陶剑飞. 超高索塔锚固区用高性能纤维混凝土泵送性能研究 [J]. 桥梁建设, 2006 (4): 1-4

[3] Lim, Yun Mook; Li, Victor C.. Durable repair of aged infrastructures using trapping mechanism of engineered cementitious composites [J]. Cement and Concrete Composites Volume: 19, Issue: 4, 1997, pp. 373-385

[4] Kamada, Toshiro; Li, Victor C.. The effects of surface preparation on the fracture behavior of ECC/concrete repair system [J]. Cement and Concrete Composites Volume: 22, Issue: 6, December, 2000, pp. 423-431

[5] 孙伟,钱红萍,陈惠苏. 纤维混杂及其与膨胀剂复合对水泥基材料的物理性能的影响 [J]. 硅酸盐学报, 2000, 28 (2): 95-99

[6] Pierre-Claude AÔÈtcin. Cements of yesterday and today, Concrete of tomorrow [J]. Cement and Concrete Research, 2000 (30): 1349-1359

[7] 唐明述. 提高基建工程寿命是最大的节约 [J]. 中国建材资讯, 2006 (2): 24-25

[8] 田倩. 低水胶比大掺量矿物掺合料水泥基材料的收缩及机理研究 [D]: [博士学位论文], 东南大学, 2005

[9] 赵庆新,孙伟,郑克仁等. 粉煤灰掺量对高性能混凝土徐变性能的影响及其机理 [J]. 硅酸盐学报, 2006 (4): 446-451

[10] 郑克仁. 矿物掺合料对混凝土疲劳性能的影响及机理 [D]: [博士学位论文], 东南大学, 2005

[11] Luo X., Sun W., Chan Y. N.. Characteristics of high performance steel fiber reinforced concrete subject to

high velocity impact [J]. Cement and Concrete Research, 2000, 30 (6): 907-914

[12] 焦楚杰，孙伟等. 钢纤维混凝土抗冲击试验研究 [J]. 中山大学学报（自然科学版），2005，44 (6): 41-44

[13] 赖建中，孙伟等. 生态型 RPC 材料的动态力学性能 [J]. 工业建筑，2004，34 (12): 63-66

[14] 杨大智等. 智能材料与智能系统 [M]. 天津：天津大学出版社，2000

[15] Shahinpoor m. Intelligent materials and structures revisited [C]. Proceeding of SPIE, 1996, 2716, 238-279

[16] 杨亲民. 智能材料的研究与开发 [J]. 功能材料，30 (6), 1999: 575-581

[17] 杨大智，魏中国. 智能材料——材料科学发展新趋势. 物理，1997，26 (1): 6-11

[18] 陶宝祺. 智能材料结构 [M]. 北京：国防工业出版社，1997

[19] 姚武，吴科如. 智能材料和智能建筑 [J]. 上海建材，2000 (3): 16-18

[20] 姚康德，成国祥. 智能材料 [M]. 北京：化学工业出版社，2002

[21] 赵连城，郑玉峰，蔡伟. 合金的形状记忆效应与超弹性 [M]. 北京：国防工业出版社，2002

[22] 黄尚廉. 智能结构系统——减灾防灾的研究前沿 [J]. 土木工程学报，33 (4), 2000: 1-6

[23] Inderjit Chopra. Review of state of art of smart structures and integrated systems [J]. AIAA Journal, 40 (11), 2002: 2145-2187

[24] P. W. Chen And D. D. L. . Chung. Concrete as a new strain/stress sensor. Composite: Part B, 27B, 1996: 11-23

[25] 姜德生，claus Richard. 智能材料器件结构与应用 [M]. 武汉：武汉工业大学出版社，2000

[26] Hiraishi H. . Smart structure system [J]. Concrete journal, 1998, 36 (1): 11-12

[27] Dry Carolyn. Smart earthquake resistant material (using time released adhesives for damping stiffening, and deflection control) [M]. Proceeding of SPIE – The international society for optical engineering, 1996, 2779: 958-967

[28] N. Muto, Y. Arai, S. G. Shin, H. Matsubara, H. Yanagida, M. Sugita, T. Nakatsuji. Hybrid composites with self – diagnosing function for preventing fatal fracture. Composites science and technology, 61, 2001: 875-883

[29] Xuli Fu and D. D. L. Chung. Self – monitoring of fatigue damagein carbon fiber reinforced cement [J], Cement and Concrete Research, 26 (1), 1996: 15-20

[30] P. W. Chen, D. D. L. Chung. Concrete reinforced concrete as a smart material capable of non – destructive flaw detection, Smart Material and structure, 2, 1993: 22-30

[31] Chen p w, Chung D D L. Carbon fiber reinforced concrete as intrinsically smart concrete for damage assessment during dynamic loading. Journal of Caramics Society, 78 (3), 1995: 412-418

[32] 姚武，吴科如. 智能混凝土的研究现状及发展趋势. 建筑石膏及胶凝材料 [M]. 10, 2000: 22-24

[33] C. K. Y. Leung. Fiber optic sensors in concrete: the future? [J]. NDT&E international 34, 2001: 85-94

[34] Insang Lee, Yuan Libo. Farhad Ansari and Hong Ding. Fiber – optic crack – tip opening displacement sensor for concrete [J]. Cement and concrete composites, 19, 1997: 59-68

[35] 巴恒静，冯奇. 光纤传感智能混凝土的研究与现状 [J]. 工业建筑，32 (4), 2002: 45-49

[36] Ansari F. . State – of – the – art in the applications of fiber – optic sensors to cementitious composites [J]. Cement and concrete composites, 19, 1997: 3-19

[37] Wen Yu – Mei, Li Ping, Huang Shang – Lian. Study on the readout of piezoelectric distributed sensing newt – work embedded in concrete [J]. Proceeding of SPIE, 3330, 1998: 67-74

[38] 孙明清,李卓球. 压电陶瓷-混凝土机敏结构研究 [J]. 混凝土,172 (2),2004:5-9
[39] 欧进萍. 土木工程结构用智能感知材料传感器与健康监测系统的研发现状 [J]. 功能材料信息,2 (5),2005:1-6
[40] Dry Carolyn And Mindy Corsaw. A comparison of bending strength between adhesive and steel reinforced concrete with steel only reinforced concrete [J]. Cement and concrete research, 33, 2003: 1723-1727
[41] Dry Carolyn, Nancy Sottoe. Passive Self-Repair in Polymer Matrix Composites Materials, Conference of Adaptive Materials, Albquerque, New Mexico. 1993 January
[42] Day C M. First European Conference on smart structures and Materials. Glasgow UK. May, 1992
[43] Jacobse N S, Sellevold E J. Self healing of high strength concrete after deterioration by freeze/thaw [J]. Cement and Concrete Research, 26 (1), 1996: 55-62
[44] 张雄,习志臻,王胜先,姚武. 仿生自愈合混凝土的研究进展 [J]. 混凝土,3,2001:10-13
[45] 三桥博三. 強度の自己修復機能を有するイリテリソコソトコソクリートの開発に関する基礎的研究,コソクリート工学年次論文報告集,Vol. 11, No. 2, 2000
[46] Brown E N, Sottos N R. Performance of embedded microspheres for self-healing polymer [J]. Society of experimental mechanic IX international congress on experimental mechanics, 8 (5), 2000: 563-566
[47] Brown EN, Sottos N R. White S R. Fracture testing of a seir-healing polymer composite [J]. Experimental mechanics, 42 (4), 2002: 372-379
[48] 邓宗才,李庆斌. 形状记忆合金对混凝土梁驱动效应分析 [J]. 土木工程学报,35 (2),2002: 41-47
[49] Maji A K, Negret I. Smart prestressing with shape memory alloy [J]. Journal of Engineering Mechanics, 124 (10), 1998: 1121-1128
[50] 左晓宝,李爱群,倪立峰,陈庆福. 超弹性形状记忆合金丝（SMA）力学性能的试验研究 [J]. 土木工程学报,37 (12),2004:10-16
[51] 瞿伟廉,李卓球,姜德生,官建国,袁润章. 智能材料结构系统在土木工程中的应用 [J]. 地震工程与工程振动,19 (3),1999:87-95
[52] S. Wen, D. D. L. Chung. Effect of fiber content on the thermoelectric behavior of cement. Journal Of Materials Science. 2004 (39)
[53] S. Wen, D. D. L. Chung. Enhancing the Seebeck effect in carbon fiber-reinforced cement by using intercalated carbon fiber. Cement and Concrete Research. 30 (8), 2000: 1295-1298
[54] 侯作富,李卓球,胡胜良. 融雪化冰用碳纤维导电混凝土的电阻变化特性研究. 混凝土. No. 3, 2003
[55] 孙明清,李卓球,沈大荣. 碳纤维水泥基复合材料的 SEEBECK 效应 [J]. 材料研究学报,12 (1),1998:111-112
[56] Y. S. Xu, D. D. L. Chung. Cement of high specific heat and high thermal conductivity, obtained by using silane and silica fume as admixtures [J]. Cement and concrete research. 30 (7), 2000: 1175-1178
[57] Luo Xiangcheng and Chung D. D. L. Electromagnetic interference shielding using continuous carbon fiber carbon matrix and polymer matrix composites [J]. Composites Part B: Engineering, 30 (3), 1999: 227-231
[58] L. Gnecco. Building a Shielded Room is not Construction 101. Evaluation engineering. 1999, 38 (3)
[59] Wen S. and Chung D. D. L. Electromagnetic interference shielding reaching 70 db in steel fiber cement [J]. Cement and concrete research, 34 (2), 2004: 329-332

[60] 张跃, 职任涛, 朱逢吾, 肖纪美. 碳纤维 (LCF) —无宏观缺陷 (MDF) 水泥基复合材料电学性能的研究 [J]. 材料研究学报, 4, 1992: 12-17

[61] Nanni A. FRP reinforcement for concrete structures [M], Elsevier Science Publishers, 1993

[62] 洪乃丰. 钢筋混凝土基础设施腐蚀与耐久性 [M] // 陈肇元, 土建结构工程的安全性与耐久性, 北京: 中国建筑工业出版社, 2003: 76-83

[63] 薛伟辰, 刘华杰, 王小辉. 新型 FRP 筋粘结性能研究, 建筑结构学报, 56 卷 5 期

[64] 方志, 杨剑. FRP 和 RPC 在土木工程中的研究与应用, 铁道科学与工程学报, 2005.8

[65] 朱航征. 纤维增强聚合物 (FRP) 用作混凝土配筋材料的开发与应用 (上), 建筑技术开发, V (31), No (11), Nov. 2004

[66] 陈平, 于祺, 孙明等. 高性能热塑性树脂基复合材料的研究进展, 纤维复合材料, No12 (52), 2005 年 6 月

[67] 王川, 张志春, 周智, 欧进萍. FRTP 材料在土木工程制品中的应用研究, 纤维复合材料, No12 (35), 2006 年 6 月

[68] American Concrete Institute. State-of-the-Art Report on Fiber Reinforced Plastic (FRP) Reinforcement for Concrete Structures (ACI440R-96) [R]. ACI Committee 440, 1996

[69] American Concrete Institute. Guide for the Design and Construction of concrete Reinforced with FRP Bars (ACI440.1R-01) [R]. ACI Committee 440, 2001

[70] 曹运红, 谢雄军, 刑娅. 21 世纪新型材料-PBO 纤维 [J], 飞航导弹, 2002 (6): 59-61

[71] 高丹盈, 李趁趁, 朱海堂. 树脂基复合材料在混凝土结构中的应用, 纤维复合材料, No2 (37), 2002.6

[72] 黄晖, 马翠英, 王福生. 车用纤维增强塑料 (FRP) 的成型工艺及其应用, 拖拉机与农用运输车, Vol. 33. No. 3, 2006.6

[73] 杨岭, 何华珍, 顾海麟, 王东川. 玻璃纤维增强热塑性复合材料及其应用. 汽车工艺与材料, 2002 (6): 28-31

[74] 孙宏杰, 张晓明, 宋中健. 纤维增强热塑性复合材料的预浸渍技术发展概况, 玻璃钢/复合材料, 1999 (4): 40-44

[75] Dae Hwan Kim, Woo HLee. A model for a thermoplastic pultrusion process using commingled yarns. Composites Science and Technology, 2001 (61): 1065-1077

[76] P Mitschang, MBlinzler, A Processing technologies for continuous fibre reinforced thermoplastics with novel polymer blends. Composites Science and Technology, 2003 (63): 2099-2110

[77] 章亚东, 段跃新, 左璐等. 经编织物法制备连续纤维增强热塑性复合材料的微观形貌和浸润过程分析, 复合材料学报, 2004, 21 (6): 63-69

[78] RMarissen, L Th van der Drift, J Sterk. Technology for rapid impregnation of (r) bre bundles with a molten thermoplastic polymer, Composites Science and Technology, 2000 (60): 2029-2034

[79] Christophe DUCRET. TWINTEX (r) PULTRUSION: 6th World Pultrusion Conference "A stronger profile for the future" development, 2002.4

[80] 阮积敏, 王柏生, 张奕薇. FRP 筋特点及其在混凝土结构中的应用 [J], 公路, 2003, 3: 96-99

[81] Bank L C, Gentry T R, Accelerated test methods to determine the long-term behaviour of FRP composite structure: environmental effects, Journal of reinforced plastic and composites [J], June, 1998 (14): 558-587

[82] Hayes M D, Garcia K, Verghese N, Lesko J J, The effects of moisture on the fatigue behavior of a glass/vinyl ester composite, Fiber composite on fiber composites in infrastructure ICCI98 [C], Tucson, 1998

(1): 1-13

[83] Katz A, Berman N, Bank L C, Effect of cyclic loading and elevated temperature on the bond properties of FRP rebars, Proceedings from the first international conference on durability of fiber reinforced polymer composites for construction, Sherbrooke, August, 1998: 403-413

[84] 吴培熙,沈健主编. 特种性能树脂基复合材料 [M],北京:化学工业出版社, 2003

[85] 碳纤维片材加固混凝土结构技术规程 [M],北京:中国计划出版社, 2003

[86] 薛伟辰,康清梁. 纤维塑料筋在混凝土结构中的应用 [J],工业建筑, 1999, 29 (2): 19-21

[87] 薛伟辰. 新型FRP筋预应力混凝土梁试验研究与有限元分析 [J],铁道学报, 2003, 25 (5)

[88] 高丹盈,赵军,Brahim B.. 玻璃纤维增强聚合物筋混凝土梁裂缝和挠度的特点及计算方法 [J],水利学报, 2001 (8): 53-58

[89] 高丹盈,BrahimB. 玻璃纤维聚合物筋混凝土梁正截面承载力的计算方法 [J],水利学报, 2001 (9)

[90] Zhi, Campbell T I. General and Simplified Models for the Analysis of Partially Prestressed Concrete Beams Containing FRP Tendons of Arbitrary Bonded Condition [A], Proceedings of the International Conference on FRP Composites in Civil Engineering [C], Hong Kong 2001

[91] 杨剑,方志. 配置CFRP预应力筋混凝土T梁受力性能研究 [A],工业建筑(增刊)第三届全国FRP学术会议论文集 [C], 2004, 255-258

[92] 方志,梁栋. 单根碳纤维(CFRP)预应力筋粘结式锚具的试验研究 [J],南华大学学报, 2004, 18 (1): 35-37

[93] 梅葵花,吕志涛. CFRP斜拉索的静力特性分析 [J],中国公路学报, 2004, 17 (2): 43-45

[94] 吕志涛. 高性能材料FRP应用与结构工程创新,建筑科学与工程学报,No. 1 (22) 2005. 3

[95] 叶列平,冯鹏. FRP在工程结构中的应用与发展,土木工程学报, 2006. No. 3

[96] 曾宪桃,车惠民. 复合材料FRP在桥梁工程中的应用及其前景 [J],桥梁建设, 2000 (2)

[97] 赵洪凯,钱春香,周效谅等. 拉挤工艺成型连续纤维增强热塑性FRP的性能研究,化学建材, 2006年第22卷第1期

[98] Dejke Valter, Durability of FRP reinforced in concrete, Degree of Licentiate of Engineering, Chalmers University of technology, Sweden 2001

[99] 卢亦焱,黄银,张号军等. FRP加固技术研究新进展,中国铁道科学, Vol. 127. No. 13, 2006, 5

第五章 | Chapter 5

THE DEVELOPMENT TREND OF ADVANCED BUILDING MATERIALS

Z. J. Li[1] and H. S. Chen[2]

1. Department of Civil Engineering, The Hong Kong University of Science and Technology, China. E-mail: zongjin@ust.hk
2. College of Materials Science and Engineering, Southeast University, China

Abstract: The development trend of building materials in this century has three important characteristics: functional development, sustainable development, and multiple disciplinary integrated. This paper first introduce functional building materials, including cement-based piezoelectric composite, thin-walled low frequency sound shielding materials, controllable heat insulation materials based on phase change mechanism and electromagnetic wave shielding and absorbing materials. Secondly, sustainability of building materials is discussed, including recycling of various industrial waste (coal combustion products, plastics, glass cullet, tire, construction and demolition waste, cement kiln dust, steel slag, phosphorus slag), application of geopolymer and magnesium phosphate cement (MPC). Finally, the multi-disciplinary development of building materials is presented.

Keywords: Building material, developing tendency; functionality; sustainability, multiple disciplinary integration.

INTRODUCTIOM

With the development of technology, improvement of people's living condition, various new building materials come to the market. People also propose higher requirement on the properties and functions of available materials. The development trend of building materials in this century has three important characteristics: functional development, sustainable development, and multiple disciplinary integrated.

The functional building materials introduced in this paper include controllable heat insulation materials based on phase change mechanism, thin-walled low-frequency sound shielding materials, cement-based piezoelectric composite and wave absorption materials. To produce the controllable heat insulation materials, the by product with low melting point from chemical engineering is first intruded into porous light aggregate and then sealed. The sealed aggregates are then mixed with binder and produced using extrusion technique. By utilizing the latent heat stored due to phase changing, the inside temperature can keep in a comfortable range and the energy for air conditioning can be reduced. The thin-walled low-frequency sound shielding material is a composite. Its elastic constant can be adjusted to block the low frequency noise under 500 Hz. The newly developed cement-based piezoelectric material are made by piezoelectric ceramic powder, cement, iron powder and other additives. It has good compatibility with Portland cement concrete and suitable for application of health monitoring in civil engineering and dynamic response measurement and vibration control. The wave absorption material will be developed based on the mechanism of bio-structural coloring using magnetic coating.

The sustainable development of building materials has caught more and more attention world wide. Take China as an example, the cement production in 2005 has reached 1.05 billion tons. In the near future, the need for cement will continue increase. It not only creates a big pressure to the raw materials and energy, but also pollutes our environment seriously. On the other hand, the by products from other industries such as coal ash, (coal) gangue, CKD, phosphate and steel slag, and construction waste bring more burden to our environment. For this reason, the sustainable building materials should have energy saving, fully using resources, reducing environment pollution, and be able to be recycled and reutilized. In this paper, geopolymer, MgO based cement, and recycling of industry waste will be introduced.

Multi disciplinary development of building materials is a power for new structure development. Nanotechnology and solar energy is gradually entering our life. So in this section, the potential application nanotechnology in building materials is introduced firstly. Then dynamic shading window system is presented. Nowadays, fibrous reinforced composites have been introduced into civil engineering. Their high strength, light weight and high corrosion resistance are beneficial to the durability of the structure. The third part will present FRC's characterization, including fiber reinforced concrete, fiber reinforced polymer, ECC, RPC et al. Due to the lower noise-level, less labor involved, self-compacting concrete (SCC) is well developed and get wide application in western country. So, the forth part of this section will present SCC's characterization. The integration of organic and inorganic materials will overcome the brittleness of concrete and aging problem of organic and will be explained.

1. FUNCTIONAL MATERIALS

1.1 Cement-based Piezoelectric Materials

With the advanced development of smart or intelligent structures in civil engineering, the health monitoring and active vibration control of structures are being introduced (Chang 1999; Tzou and Guran 1998; Aizawa et al 1998). In a smart structure, sensors and actuators are essential components for sensing and controlling. Compared with the achievements in controller design, the study and design of sensors and actuators are lagging (Brennan et al. 1999). In addition, the sensors and actuators suitable for applications in the field of mechanical engineering, such as space structures, may not be applicable in civil engineering structures due to differences in the properties between a sensor or actuator and the host structures. For example, the acoustic impedance, temperature coefficient, and shrinkage and creep characteristics of concrete, which is the most popular material in civil engineering structures, are quite different from those of the metal and alloy that are frequently used in mechanical engineering structures. New sensors and actuators should, therefore, be developed to meet the requirements of civil engineering applications. Among the techniques used in sensors and actuators, piezoelectricity has proved to be one of the most efficient mechanisms for most applications in smart structures (Banks et al. 1996; George et al. 1997; Tao 1997). Moreover, piezoelectric materials have attracted great attention from researchers (Xu 1991; Furukawa 1989; Okazaki 1985; Banno 1988; Safari et al. 1992; Latour et al 1994).

Piezoelectric materials belong to a class of dielectrics which exhibit significant material deformations in response to an applied electric field as well as product dielectric polarization in response to mechanical strains. Piezoelectric materials can be classified in three categories: *piezoelectric ceramics*, *piezoelectric polymers*, and *piezoelectric composites*. The advantages of piezoelectric ceramics include high electromechanical coupling coefficient (K), high piezoelectric strain coefficient (d33), and very fast response. However, the relatively heavy density, high acoustic impedance, brittleness, and inconvenient machinability of piezoelectric ceramics restrict their applications. In contrast, piezoelectric polymers have low acoustic impedance, but also a much lower piezoelectric factor and electromechanical coupling coefficient. To overcome these deficiencies, researchers (Banno 1988; Safari et al. 1992; Mazur 1995) have developed piezoelectric composites by incorporating particles or other forms of piezoelectric ceramics into a polymer matrix, such as epoxy, rubber, or some types of piezoelectric polymer, such as polyvinylidene floride (PVDF).

Piezoelectric composites can be classified according to the connectivity of piezoelectric ceramics and matrix phases. There are ten connectivity patterns, designated as 0-0, 0-1, 0-2, 0-3, 1-1, 1-2, 1-3, 2-2, 2-3, and 3-3 (as shown in Figure 1) (Safari et al. 1992; Dias et al. 1996) where the first number denotes the connectivity of the ceramics phase, and the second refers to that of the matrix. The "0" means that material is in the form of particles. It is understandable that 0-3 composites can be more easily fabricated in complicated shapes than other forms of composites. Piezoelectric ceramic/polymer 0-3 composites have been studied in detail in previous investigations (Banno 1998; Safari et al. 1992; Mazur 1995; Levassort et al. 1996; Reed et al 1990; Zheng et al 1998; Furukawa et al. 1979; Olszowy 1996; Wang et al. 1999). However, a problem remains in ceramic/polymer piezoelectric composites regarding the difficulty of poling piezoelectric ceramic powders due to the large differences in resistivity and permittivity between polymer matrix and ceramic powders (Dias et al. 1996).

Lead zirconate titanate (PZT) and lead lithium niobate-lead zirconate-lead titanate (PLN)

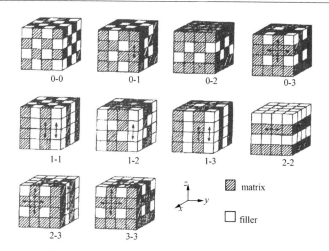

Figure 1 Connectivity patterns of two-phase piezoelectric composites (adapted from Dias et al. 1996)

have a good pyroelectric and piezoelectric properties and are widely used in various applications (Zheng et al. 1998; Furukawa et al. 1979). However, their incompatibility with the construction materials limits their use in the civil engineering. To develop a new kind of smart materials suitable for building and infrastructures, various types of cement-based piezoelectric composites incorporating PZT have been fabricated and studied (Li et al. 2001, 2002, 2005; Shen et al. 2006; Dong, 2005; Huang 2004, 2005; Lam et al. 2005). Due to the complexity of fabrication process, three types of connectivity patterns, i.e. 0-3 (Li et al. 2002, 2005; Shen et al 2006; Dong, 2005; Huang 2005), 1-3 (Huang 2005; Lam et al 2005) and 2-2 (Li et al. 2001), have been considered for cement-based piezoelectric composites in literatures. And the types of cement used include white cement, sulphoaluminate cement and Portland cement. Compared with other two types of connectivity pattern, the fabricating process of 0-3 piezoelectric composite is simpler. Therefore, this section will use 0-3 cement-based piezoelectric composites incorporated with PZT as an example to present the current study on cement-based piezoelectric composites. Since it is the first stage of the study for such kind of piezoelectric composites, the poling behavior and piezoelectric properties of the composites are focused. And this section will investigate the influence of various factors, including poling voltage, duration and PZT content on poling behavior and piezoelectric properties of composites. As to compatibility between cement-based piezoelectric composites and concrete, with regard to acoustic impedance, temperature coefficient, shrinkage and creep etc., it will be considered in the future.

1.1.1 Preparation of piezoelectric composites and sensor

In this study, lead zirconate titanate (PZT) ceramic Powder (Shanghai Keda Electronic Ceramics Co. Ltd.) and white Portland cement (H. S. L. Enterprises Co. Ltd.) were used to prepare the 0-3 type cement-based piezoelectric composites. The properties of The PZT ceramic are listed in Table 1. The particle size distributions of PZT ceramic and white Portland cement were measured by a laser particle size analyzer (LS 230, Coulter Corporation,

USA). Figure 2 shows that the mean diameter of PZT ceramic is 153.6 μm with median of 83.52 μm, the mean diameter of white Portland cement is 15.32 μm and median of 12.55 μm.

Properties of PZT powder, hardened Portland cement paste, and normal concrete   Table1

	PZT ceramics	Cement paste	Concrete
Piezoelectric strain factor d_{33} (10^{-12}C/N)	513	—	—
Piezoelectric voltage factor g_{33} (10^{-3}Vm/N)	15.9	—	—
Dielectric constant ε_r (at 1KHz)	3643	~56	—
Electromechanical coupling coefficient K_P (%)	67	—	—
Mechanical quality Q_m	43	—	—
Elastic compliance s_{33} (10^{-12}m^2/N)	16.7	~72	~30
Density ρ (10^3kg/m^3)	7.5	~2.0	~2.4
Acoustic velocity V (10^3m/s)	2.83	~2.64	~3.73
Acoustic impedance ρV (10^6kg/m^2·s)	21.2	~5.3	~9.0

PZT ceramic powder and white Portland cement are mixed together to make the 0-3 type piezoelectric composites. In order to improving the fluidity of the fresh mixture, a superplasticizer (W19, W. R. Grace) was used. Using a normal mixing (mixing duration is about 2 min) and vibrating (using small shaking table) method (vibrating duration is about 1 min), up to 50% volume content of PZT ceramic powder could be incorporated into the composites. The mixing proportions are listed in Table 2.

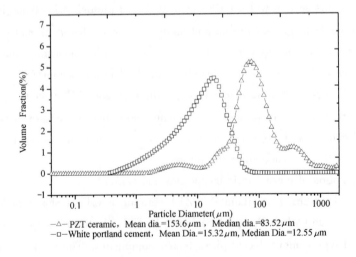

Figure 2 Particle size distribution of PZT powder and white Portland cement

The mixing proportions of cement-based PZT composites			Table 2
	C18	C30	C50*
Ceramic (g)	100	100	100
Cement (g)	76.76	41.33	17.71
Water (g)	38.38	20.67	8.85
ADVA** (g)	0.64	0.34	0.30

Note: C18, C30 and C50 are used to identify the specimens of cement-based piezoelectric composites. The numbers (18, 30 and 50) are the indication of volume fraction of PZT powder in whole system;

* For C50 specimens, the ratio (w%) of ADVA/Water = 1%; other two kinds of specimens, the ratio (w%) of ADVA/Water = 0.5%;

** A type of superplasticizer (water solution, the solid content is 30w%).

To achieve a uniform mixture, cement and PZT particles were mixed thoroughly first, then water and superplasticizer were added into the mixture. The mixing process was continued until the mixture became uniform. Then the mixture was compacted into the glass (or plastic) model. After casting, the specimens were put in the curing room with a temperature of 65 ℃ and relative humidity of 98% for 24 h. After curing, the specimens could be polarized immediately. Before polarization, the specimens were cut into square slices of $8 \times 8 \times 1.5$ mm^3 (or cube of $8 \times 8 \times 8$ mm^3, shown in Figure 3). Then the surfaces of the slices were polished and coated with silver paint (AGAR Scientific Ltd.). In order to ensure the reliability of experiment results, three specimens were tested for every kind of composites and took the average value of test results as its property index.

Figure 3　The specimens of 0-3 type cement-based PZT composites

A key technique for fabrication of piezoelectric materials is poling procedure. Poling was carried out in a silicon oil bath with a temperature of 160℃. After polarization the specimens were immersed in cold silicon oil to for fast cooling in order to maintain the status of polarization. Then the polarized specimens were wrapped with aluminum foil to eliminate remnant charge generated during poling process.

Piezoelectric strain factor d_{33} was measured with a piezo d_{33} meter (model ZJ-3B, fabricated by Institute of Acoustics of Academia Sinca). The impedance spectra were measured with a Hewlett Packard impedance/gain-phase analyzer (model 4194A) at 1 kHz. The entire tests were carried out at 22 ℃ and R. H. = 50%.

1.1.2 *Poling behavior of cement-based PZT composites*

Many factor may affect the poling behavior of cement-based PZT composites, such as the properties of piezoelectric ceramic powder, particle size distribution of piezoelectric ceramic powder and cement, aging, poling voltage, poling duration, and PZT content. This section will focus on the influence of aging, poling voltage, poling duration, and PZT content.

The dependence of piezoelectric strain factor d_{33} on aging for ceramic/cement composites and ceramic/polymer composites are shown in Figure 4. It can be seen from Figure 4 that d_{33} value of cement-based PZT composites improved with the curing time before reaching saturation. On the other hand, ceramic/polymer composite presented opposite trend. Its d_{33} value decreased with the aging procedure. And the final d_{33} value of ceramic/cement composites is much higher than that of ceramic/polymer composites (Mazur 1995). This interesting phenomenon shows that the ceramic/cement composites may have a coupling effect between ceramic powder and cement particle.

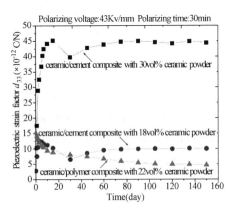

Figure 4 Aging influence on piezoelectric factor d_{33} for PZT/cement composites and PZT/polymer composites

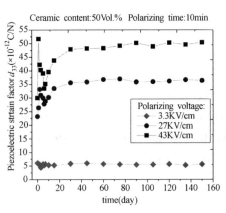

Figure 5 The effect of poling voltage on d_{33} for 0-3 type cement-based PZT composites

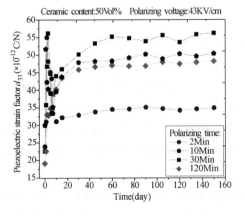

Figure 6 The effect of poling duration on d_{33} for 0-3 type cement-based PZT composites

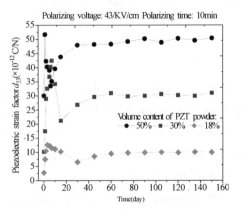

Figure 7 The effect of PZT content on d_{33} for 0-3 type cement-based PZT composites

Figure 5 shows the poling voltage dependence of piezoelectric strain factor d_{33} for 0-3 type cement-based composites (C50) with 50% volume content of PZT ceramic powder. It can be seen in Figure 5 that with the voltage increasing, d_{33} rises greatly. Low voltage cannot polarize the composites effectively. For example, when the voltage is about 3.3 kV/cm, d_{33} value is quite low. The peaks appeared in the figure are the results caused by remnant charge in the composites during the poling procedure. It is clear that the larger the poling voltage, the higher the peaks in the figure. After removing the remnant charge, d_{33} values are reduced. However, with aging, the d_{33} values go up again. For the whole process, the specimen polarized with higher voltage shows superior d_{33} values to that with lower voltage.

Apparently, variation of the poling duration also affects the piezoelectric properties of the composites. Figure 6 displays the results of time dependence of d_{33} for 0-3 type cement-based PZT composites. It reveals a similar tendency as given in the Figure 4. It can be found in Figure 6 that for samples polarized 2min, 10min and 30 min, longer poling duration leads to higher d_{33} value. However for samples polarized for 120 min, their d_{33} values are lower than those polarized for 10 min. This phenomenon may be attributed to a partial breakdown of the saturated piezoelectric materials

Figure 7 shows the poling behavior of 0-3 type cement-based piezoelectric composites with different content PZT ceramic powder. It can be found that the higher the content of PZT ceramic, the larger the d_{33} value. It can be easily understood that as a functional phase in the composites, higher PZT ceramic content is beneficial to the piezoelectric performance of 0-3 type cement-based PZT composites.

1.1.3 *Piezoelectric properties of cement-based PZT composites*

By utilizing the impedance analyzer, the capacitance of the cement-based piezoelectric composites can be determined. After obtaining the capacitance, relative dielectric constant can be calculated according to the plate condenser Eq. 1:

$$C = \frac{\varepsilon_r \cdot \varepsilon \cdot S}{d} \tag{1}$$

where S is the area of the specimen and d is thickness of the specimen. ε is the vacuum dielectric constant ($\varepsilon = 8.855 \times 10^{-12}$ F/m). The relative constant values calculated by Eq. 1 are listed in the Table 3.

Capacitance and relative dielectric constant of the composites　　Table3

Specimen codes	C18	C30	C50
Size of Specimen	$8 \times 8 \times 1.5$ mm^3	$8 \times 8 \times 1.5$ mm^3	$8 \times 8 \times 1.5$ mm^3
Capacitance (pF)	38.4	63.8	112.5

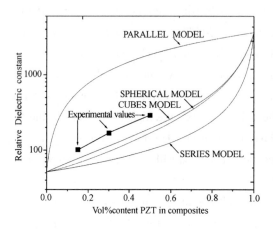

Figure. 8 Experimental and theoretical values for dielectric constants of composites with various PZT contents (Theoretical values are calculated based on parallel, series, cubes and spherical models).

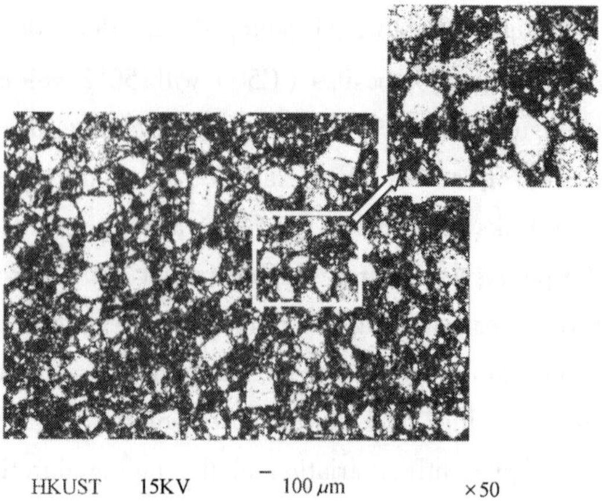

Figure. 9 Backscatter SEM image of 0-3 type Cement-based PZT composites (with 50% volume fraction of PZT particle)

Basically, ε_r increases with PZT contents. In the Figure 8, the dielectric constants for samples polarized for 30min are plotted against PZT content in the composites. In the figure, the theoretical predictions are also provided based on different models. In the calculation, it is assumed that the polarization of PZT particle is saturated. It is clear that the experimental results are close to the theoretical value of the cubes models, which means that the PZT particles in the composites are well dispersed. This conclusion is confirmed by the result of SEM image for the composites. From the backscatter scanning electronic microscopy (SEM) image (Figure 9), uniform distribution of PZT particles can be observed.

Figure 10 displays the relationship between experimental values of piezoelectric strain factor d_{33} of the composites polarized for 30min versus PZT content. For comparison purpose, the theoretical curves based on different models (parallel mode, series model, cubes model) are aloe given. The experimental results confirm that the poling behaviors of the 0-3 type

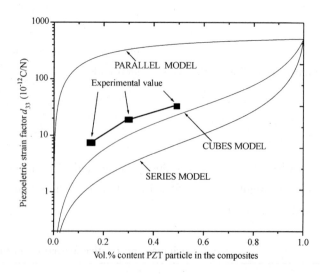

Figure. 10 Experimental and theoretical values for piezoelectric strain factor d_{33} of composites with different PZT content (Theoretical values are calculated based on parallel, series and cubes models)

cement-based PZT composites can be well described by the cubes model. The theoretical formulae for serial model and parallel model are known. What follows gives the theoretical solutions of cubes model for the relative dielectric constant (ε_r) and piezoelectric strain factor (d_{33}) in two-phase systems (Banon 1988; Mazur 1995):

Theoretical formula of cubes model for the dielectric constant:

$$\varepsilon_{33} = \frac{{}^1\varepsilon_{33} \cdot {}^2\varepsilon_{33}}{({}^2\varepsilon_{33} - {}^1\varepsilon_{33}) \cdot {}^1v^{-\frac{1}{3}} + {}^1\varepsilon_{33} \cdot {}^1v^{-\frac{2}{3}}} + {}^2\varepsilon_{33} \cdot (1 - {}^1v^{\frac{2}{3}}) \tag{2}$$

Theoretical formula of cubes model for the piezoelectric strain factor:

$$d_{33} = {}^1d_{33} \cdot \frac{{}^1v}{{}^1v^{\frac{1}{3}} + (1 - {}^1v^{\frac{1}{3}}) \cdot \frac{{}^1\varepsilon_{33}}{{}^2\varepsilon_{33}}} \cdot \frac{1}{1 - {}^1v^{\frac{1}{3}} + {}^1v} \tag{3}$$

Where ε_{33} is the relative dielectric constant of composites; ${}^1\varepsilon_{33}$ is the relative dielectric constant of PZT ceramic; ${}^2\varepsilon_{33}$ is the relative dielectric constant of cement paste; ${}^1v$ is the volume ratio of PZT particle in the composites; d_{33} is the piezoelectric strain factor of the composites; ${}^1d_{33}$ is the piezoelectric strain factor of PZT ceramic.

Relative dielectric constant and piezoelectric strain factor for various PZT composites   Table 4

Material	Ceramic PZT	PZT/PVDF	PZT/Rubber	PZT/POM	PZT/Cement
ε_r	~3000	~120	~55	~95	~300
d_{33} (10^{-12} C/N)	~500	~20	~35	~17	~55

Compared to the PZT/polymer composite, it is easy to find from Eq. 3 that the larger d_{33} value of the PZT/cement composite is attributed to the larger dielectric permittivity of cement. For the sake of comparison, Table 4 displays the relative dielectric constant and piezoelectric strain factor both for PZT/polymer composites and for PZT/cement composites, it can be seen that the piezoelectric properties of PZT/cement composite seems to be significantly different from those of PZT/polymer composites.

1.1.4 Sensor application of cement-based piezoelectric composite in civil engineering

Fabrication of aggregate-like sensors So far, all discussion only focuses on the properties of piezoelectric composites and its influencing factors. To realize the function of the above cement-based piezoelectric composites as sensor, cement paste is employed as a "host" for piezoelectric composites instead of the hermetic metal housing used in traditional and commercial available sensors (as shown in Figure 11 (a)). The new sensor was made by a sandwich method. It was formed by jointing two pieces of cubes made of cement paste with a piece of PZT sensing element. A new adhesive was developed for bonding the cuboids of hardened cement paste and piezoelectric ceramic plate. The adhesive had high stiffness, high resistivity and was water-proof. It was made by mixing dry white cement powder and epoxy. When hardened, the stiffness of the adhesive was much higher than hardened pure epoxy and close to that of the hardened cement. This mechanical match enabled the effectiveness and accuracy of the transfer of the stress from cement cubes. The resistivity of the adhesive was as high as 10^{16} Ωcm just like pure epoxy, which assured the electrical stability of them. In addition, the adhesive could prevent the water from penetrating into the sensing element that was sensitive to moisture. The fabricated sensor elements are shown in Figure 11 (b).

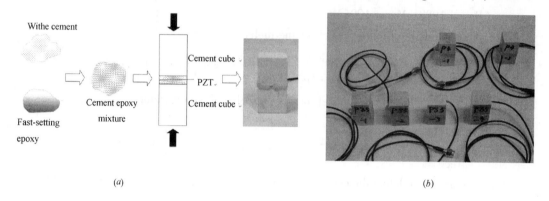

Figure 11 Fabrication of cement-based piezoelectric composite sensor

1.1.5 Properties of sensor

Figure 12 shows the setup of experiments for measuring the basic properties of the aggregated-like sensor. Compression plate of MTS is driven by MTS actuator to generate various load patterns such as sine wave, square load, random load and complex load. Under the compression of MTS, the sensor can generate charge that is amplified by charge amplifier signal. The charge amplifier also converts the signal to a voltage that can be acquired by any readout or data acquisition devices such as data log and Oscilloscope. Figure 13 shows the measuring system used in experiments, which is called charge amplified system. It is suitable for the sensors with artificially polarized polycrystalline ceramics. Usually the amplifying circuit or signal conditioning electronics are built in the sensors. However, for a laboratory study, an integrated electronics is not convenient due to its fixed range, e.g. lack of range adjustability. An external charge amplifier is usually used in studying a new design of piezoelectric sensors because of its flexibility in choosing different measuring scales and ranges.

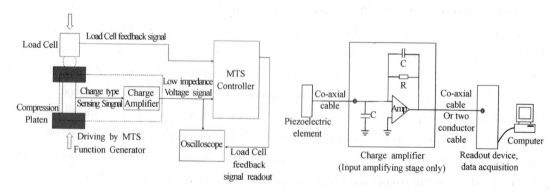

Figure 12 Setup of experiments   Figure 13 Measuring system of experiments

(1) Frequency response

Before the results of frequency response experiment are presented, the loading scheme used in the experiment is introduced, which is shown in Figure 14. This load scheme is named frequency sweep, in which the amplitude of input sine wave force is kept as a constant while the frequency of input force is changed from 0.01Hz to 40 Hz, the upper frequency limit of MTS we used. The mean of input force is chosen as 3000N for simulating the preloading condition of sensor embedded in structure. Figure 15 shows the frequency response of P85 series sensors. From Figure 15 (a) it can be seen that the amplitude of output is almost a constant in the whole frequency range. In other words, the relationship between amplitude of output and the frequency of input force is almost a horizontal straight line in the intended frequency range. From Figure 15 (b) it can be seen that the phase shift between output and input of P85 series sensors is very close to zero. These two characters can prevent distortion of the output signal when the input force is a complicated load with components of different frequencies and amplitudes. The frequency characteristic of P43 series sensors is similar to P85 series sensors and thus is not shown in this paper for the sake of the conciseness.

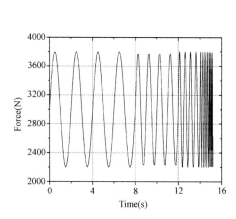

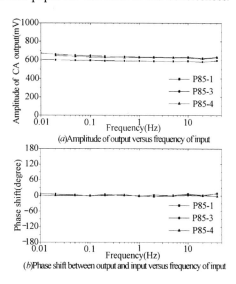

Figure 14 Load scheme: frequency sweep

Figure 15 Frequency response of P85 series sensors

(2) Linearity

To measure the relation between amplitude of input and sensor output, another load scheme named amplitude sweep as shown in Figure16 was applied. In this amplitude sweep, the frequency of input sine wave force was kept as a constant, 1Hz in the experiments, and amplitude (from peak to mean) of input force was changed from 100N to 1900N. The mean value of the input force was also chosen at 3000N compressive force for simulating the preloading condition. Figure 17 shows the P85 sensors' responses to the amplitude sweep load. From Figure 17 (a), a linear relation between amplitude of sensor output and input can be easily seen. The responses of different sensors in P85 series were similar. However, the experimental fitting lines showed a slightly different slope. This might be due to the minor differences in the dimension of the ceramic plate in each sensor that led to a different charge generated. From Figure 17 (b) it can be seen that the phase shift between output of sensors and input amplitude sweep load is near zero. The linear relationship between output and input is preferred in application of sensor. It is possible for these sensors to be conveniently applied in health monitoring of structures.

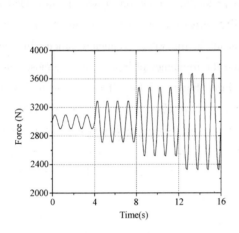

Figure 16   Load scheme: amplitude sweep

Figure 17   Response to amplitude sweep load

(3) Response to complex, random and square loads

After the basic properties of these sensors had been determined, more complicated loads were applied to these sensors for further evaluation. The first one was a complex load that was generated by five components with different frequencies and different amplitudes as shown in Table 5.

Components of complex load Table 5

No.	Frequency (Hz)	Period (s)	Mean compressive load (kN)	Peak to mean amplitude (kN)	Phase (degree)
1	0.05	20	3.0	1	0
2	0.2	5	3.0	0.4	60
3	1	1	3.0	0.3	120
4	5	0.2	3.0	0.2	180
5	20	0.05	3.0	0.2	240

The frequency range of the complex load was within the frequency range in pervious experiments. The amplitude of the complex load reached to the maximum amplitude used in the experiment of linearity check. Figure 18 shows the complex load and the output of the sensor. The output of sensor corresponded to the input complex load very well. Both global and local responses of the sensor were almost identical to the shape of the input as shown in Figure 18. Figure 19 shows the sensor's response to a random load that can better simulate the realistic condition that sensor may encounter during its application. From Figure 19, it can be found that the wave shape of sensor output is almost same as that of input load. Figure 20 shows the sensor's response to a square load with 3000N of mean, 0.1Hz of frequency and 400N of amplitude. There is not any overshot happened during the up-step loading and down-step loading, which is important to dynamic measurement.

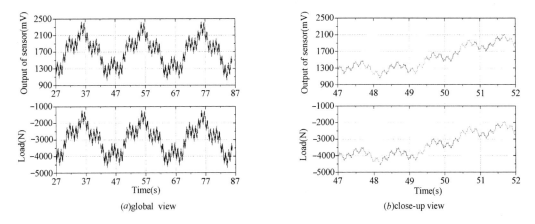

Figure 18 Response to complex load

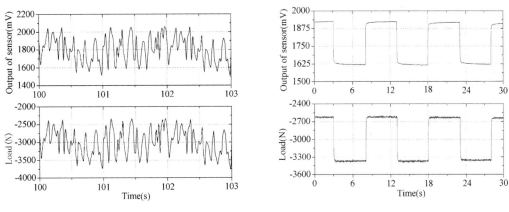

Figure 19 Response to random load   Figure 20 Response to square load

The above information demonstrates that cement not only provides excellent chemical and physical stability for the composite, but also simplifies poling conditions. Consequently, cement-based PZT composites have great potential as sensor, especially in civil engineering.

1.2 Thin-Walled Low Frequency Sound Shielding Material

The study of acoustic insulation of a wall panel is generally based on the consideration of sound transmission loss. Many researchers have investigated the transmission loss (TL) for building products, such as gypsum and concrete walls. Green and Sherry (Green et al 1982) attempted to determine the TL of gypsum partitions with steel studs. Litvin et al (Litvin et al 1978) examined the TL for concrete. Actually, most of the works have been carried out to increase the TL by different structural designs. Only a few paid attention to enhance the TL by incorporating smart material. In this research, a new concrete is developed using the idea of the local resonant unit.

Liu et al (Liu et al. 2000) has developed such a local resonant unit, a lead ball coated with silicone rubber, for the purpose of sound shielding. They arranged the lead balls into an epoxy matrix and proved that the lead balls are able to reduce the transmission coefficient of the epoxy composite at a certain frequency range by local resonance. In this study, we incorporated the coated lead balls into a short fiber reinforced cementitious composite panel to improve TL as an extension of this invention. Two samples are made for the investigation. One has been incorporated with the lead balls in the cement-based matrix whereas the other one is a pure cement-based material as a reference. To perform a reliable sound TL test while avoiding building an expensive reverberation room (ASTM 1991a; ASTM 1991b), a novel and simple test method was developed. The method adopted an automatically controlled sound generation system and sand bath set up. By using such a method, the sound transmission loss has been measured for the two types of specimens at 1/3 octave frequency band. The experimental results show that the new concrete specimen has much better sound proofing capability in the low frequency range of 100-200Hz. The exciting results reveal that the new concrete has good potential in application of low frequency sound shielding.

1.2.1 Test Method

Figure 21 indicates the experimental set up and the flow chart of the new test method for sound TL measurement.

By using the programmable port of the HP function generator (RS232 communication), a program in QuickBasic is written to control the frequency output of the function generator. Then the function generator can produce sounds at different frequencies. The power amplifier is connected to a speaker and used to adjust the amplitude of the sounds so that the signals detected by the microphone 1 and 2 are intact without distortion. Therefore, the sounds can be displayed as sine waves by monitoring the amplitude. It is necessary to keep the same amplitude for these sounds to ensure the same source for different specimens.

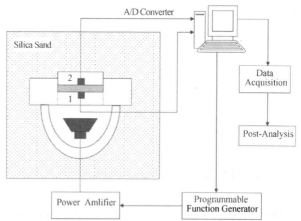

Figure 21 Schematic of the novel technique for measurement of transmission loss

The sample is tested inside the sand bath. The sand bath can prevent the noise disturbance from the environment. The microphones 1 and 2 are placed closely to the sample faces. They have been calibrated for frequency response beforehand and are used to measure the incident and transmitted sound energy. They are connected to an A/D converter so that the analog data can be converted to the digital data. The digitized data is then transferred to the hard disk of the computer. Since the sampling rate of the A/D board is 26.525 kHz that satisfies Nyquist theorem (which states that the sampling rate must be at least 2 times of the maximum frequency of the original signal; otherwise the original signal cannot be recovered completely from the digitized signal), the A/D board is adequate for sequential sampling of a continuous input source within our investigating frequency range. The microphones measure the voltage over a definite period of time. The result is the sine function of voltage against time. Finally we can acquire the data and perform the post-analysis, and finally examine the TL at different frequencies.

1.2.2 Theoretical Background

A strong periodic modulation in either density or sound velocity can create gaps that forbid wave propagation and this has been addressed in many researches (Kushwaha et al 1993, Martines-Sala et al 1995; Kafesaki et al 1995). According to mass area density law to sound shielding/transmission in the absence of absorption, for a given level of sound transmission amplitude, the required mass area density is inversely proportional to the frequency (Liu et al 2000). Usually, the spatial modulation is the same order as the wavelength in the gap. It is thus not practical to shield low frequency acoustic sound by increasing the thickness of a wall or partition because of the limitation of space and structural size.

Liu et al (Liu et al. 2000) developed a subwavelength sonic bandgap material with lattice constants approximately two orders of magnitude smaller than the relevant sonic wavelength to overcome the limitation. The sonic bandgap material utilizes the idea of localized resonant structures. It has proved that composites made from such localized resonant structures behave as a material with effective negative elastic constants and a total wave reflector within certain tunable sonic frequency ranges. This realization relies on the fact that the acoustic wave can't propagate in any direction in the frequency range that the effective elastic modulus starts to be negative.

Indeed, the sonic bandgap material unit is composed of a solid core material, possessing relatively high density and being coated by a layer of elastic soft material. In their experiment, centimeter-sized lead balls are used as the core material, which are coated with a layer of silicone rubber (Figure 22). The coated spheres are then dispersed in an epoxy matrix to form a composite material. The underlying physics can be attributed to the built-in resonance of the coated spheres. Actually, if a sound wave with frequency ω interacts with a medium carrying a localized excitation with frequency ω_0, the linear response function will be proportional to $1/(\omega_0^2 - \omega^2)$. Therefore, resonance occurs when the frequency of the sound matches with that of the local excitation of the coated lead ball. By combining the high density spherical core with an elastically soft coating, there is intuitively a low frequency resonance where the inner core provides the heavy mass and the silicone rubber provides the soft spring. Liu's results (Liu et al. 2000) show that the sonic bandgap material can reduce the sound transmission at low frequency.

1.2.3 Sample Preparation

In additional to a conventional cement-based specimen (sample I, used as reference specimen), a short fiber reinforced cement-based composite panel with incorporation of the newly developed localized resonant unit, lead ball coated with silicone rubber (sample II), was cast. The size of the panel is 290mm × 380mm with a thickness of 30 mm. The diameter of the lead ball is 15mm and the thickness of the silicone rubber coating is 2mm. To cast the sample II, the lead balls were arranged periodically as shown in Figure 23 first. Then cement-based slurry with short glass fibers was poured into the ball bed to form a composite panel. The mix proportion of slurry is listed in Table 6. The panel was then placed inside a curing room for 28 days.

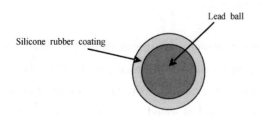

Figure 22   Sketch of sonic bandgap material   Figure 23   Coated lead spheres arrangement
                                                                                 in the cement-based panel

Mix proportion of the cement-based panel   Table 6

Sample	White cement	Fly ash	Slag	Silica fume	PVA powder	PVA fiber (4mm)	w/c	Lead ball
I	40%	15%	40%	5%	1.5%	0.5%	0.4	No
II	40%	15%	40%	5%	1.5%	0.5%	0.4	Yes

1.2.4   Experiment

First, a self-developed computer program is used to control the function generator and then a sound at certain frequency is generated through the speaker. The sound transmits through the panel sample from stereo side to outside. Microphones 1 and 2 are used to record the incident and transmitted signals. These data are digitalized by the A/D converter and saved to the hard disk of the computer. The sampling rate is 26.525kHz, and hence each data is taken at 0.0377ms interval. The post-analysis on the data is performed and the transmission loss at that frequency is determined. Finally the TL plot over a frequency range is obtained. Three tests on the sample I and II are conducted and the average of three results is used as index of TL. In this test, sixteen 1/3 octave frequency bands recommended in ASTM 1991 E90 are adopted and the analysis is done up to 4000 Hz.

1.2.5   Post Analysis

The first step is to use the recorded data to calculate the incident and transmitted energy for every frequency by using Eq. 4.

$$E(watt) = \frac{1}{R}\int_0^t V^2(t)\,dt \tag{4}$$

where V is voltage measured over a period of time t and R is the electrical resistance.

Then the transmission coefficient is estimated by

$$\Gamma = \frac{E_2}{E_1} \quad (5)$$

Where E_1 is incident sound energy; E_2 is transmitted sound energy.

And transmission loss (TL) can be calculated by

$$TL = 10\log_{10}\frac{1}{\Gamma} \quad (6)$$

Finally, the transmission loss at every 1/3 octave frequency can be determined.

The transmission loss against the frequency can be plotted. And the graph can be used to evaluate the acoustical performance at different frequencies

1.2.6 Result and Discussion

Figure 24 shows the incident and transmitted signal detected by the microphones 1 and 2 at 160Hz. It can be observed that the incident wave has a significantly larger magnitude than that of the transmitted wave. The voltage-time graph can be used to calculate the incident and transmitted energy of the sound wave, and hence the tranmission coefficient and transmission loss can be determined. Finally, the transmission loss against the frequency can be plotted to evaluate the acoustic performance. Although the microphones have been calibrated for the frequency response, it is unavoidable to experience about 3% measuring error at some frequencies. Fortunately, this error insignificantly affects the energy calculation.

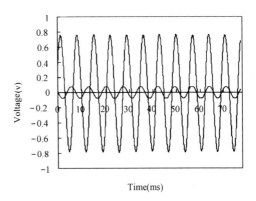

Figure 24 Incident and transmitted signal detected by the micrphpones 1 nd 2 at 160Hz (sample I)

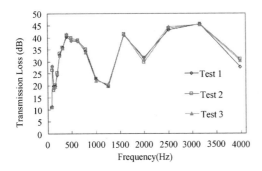

Figure 25 TL plots against 1/3 octave frequency for the sample I (reference)

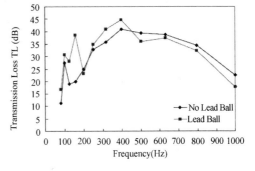

Figure 26 TL plot for the sample I (reference) and the sample II

The TL results of three different tests for the sample I are shown in Figure 25. From the figure, it can be noticed that the results obtained in the three individual tests of the sample I are nearly identical. Similar phenomenon is observed for the sample II. It indicates that the new test method is reliable and has a very good repeatability. Figure 26 compares the sound TL of sample I to that of sample II at frequency range from 70 Hz to 1000 Hz. It can be seen that the transmission loss of the sample II is close to that of the sample I in higher frequency range from 400 Hz to 1000 Hz. The samples' TLs differ by at most 3 dB above 400Hz. It means that the lead ball with the size adopted in this study cannot improve the acoustic insulation in the frequency range from 400 Hz to 1000 Hz. It is consistent to the findings by Sheng et al[3]. However, we can clearly see that the sample with the coated lead spheres shows significant improvement of acoustic insulation as shown in the Figure 26. The sample II has a larger transmission loss from 100Hz to 200Hz, and especially effective at about 150Hz. From 100Hz to 150Hz, the TL of the sample I (reference) decreases from 27dB to 20dB. However, the performance of sample II is totally different. Its TL increases from 30dB to 40dB. The difference in TL between sample I and II is 20dB at about 150Hz which represents 100 times difference in energy transmission. Although the threshold of human's hearing is high at low frequencies below about 100 Hz, the difference in threshold of hearing from 150 Hz to 1000 Hz is not significant. Thus, it is quite important to prevent the people from a noise of low frequency around 150 Hz. In addition, it can be seen from Fig. 26 that lead ball specimen shows a moderate improvement for sound shielding in the frequency range from 200 Hz to 400 Hz. It is clear that the embedment of the coated lead balls can significantly improve the acoustic insulation in the low frequency range. This improvement in the TL can be attributed to localized resonance at the coated lead spheres. It should be indicated that if the reference specimen contains the similar size of aggregate as the lead ball specimen, the transmission losses for the control might be different. However, since the transmission losses at low frequency rang is controlled by mass area density law, such a difference, if any, at the frequency range from 150 Hz to 400 Hz, would be minimal.

1.2.7 Remarks

Concrete using lead balls coated with silicone rubber as aggregate has been developed. The balls serve as local resonant structure. Together with the reference specimen, the samples were tested on their sound transmission loss by using the newly developed novel method. The results show that the new test method is simple, reliable and repeatable. The new concrete is capable of enhancing the acoustic insulation in the frequency range of 100 Hz to 200 Hz. The composite with lead spheres has local resonance that occurs at about 150Hz and improves acoustic insulation whereas the reference sample suffers the reduction of the insulation in that range. Therefore, it can conclude that the coated lead spheres are the effective acoustic insulation units for the cement-based composite at low frequency.

1.3 Controllable Heat Insulation Building Products with Phase Changing Materials

Materials to be used for phase change thermal energy storage must have a large latent heat and thigh thermal conductivity. They should have a melting temperature lying in the practical range of operation, melt congruently with minimum subcooling and be chemically stable, low in cost, nontoxic and non-corrosive. Materials that have been studied during the last 40 years are hydrated salts, paraffin waxes, fatty acids and eutectics of organic and non-organic compounds. Depending on the applications, the PCMs should first be selected based on their melting temperature. Materials that melt below 15 ℃ are used for storing coolness in air conditioning applications, while materials that melt above 90 ℃ are used for absorption refrigeration. All other materials that melt between these two temperatures can be applied in solar heating and for heat load leveling applications. These materials represent the class of materials that has been studied most.

Commercial paraffin waxes are cheap with moderate thermal storage densities and a wide range of melting temperatures. They undergo negligible subcooling and are chemically inert and stable with no phase segregation. However, they have lower thermal conductivity. Fatty acid (capric, lauric, palmitic and stearic acids) and their binary mixtures are attractive candidates for latent thermal energy storage in space heating application (Feldman et al 1989). Neeper (Neeper, 2000) has examined the thermal dynamics of gypsum wall boards impregnated by fatty acids and paraffin waxes as PCM that subjected to the diurnal variation of room temperature. Hydrated salts (for instance, $Na_2SO_4 \cdot 10H_2O$) are attractive materials for use in thermal energy storage due to their high volume storage density, relative high thermal conductivity and moderate cost, with few exceptions. However, the high storage density of hydrated salt materials is difficult to maintain and usually decreases with cycling. This is because most hydrated salts melt congruently with the formation of the lower hydrated salt, making the process irreversible and leading to continuous decline in their storage efficiency. Subcooling is another serious problem associated with all hydrated salts (Farid et al 1994).

Some studies have concentrated on the characterization of thermal properties of cement-based materials. Since the magnitude of the overall thermal conductivity of the composite is a comprehensive index of many parameters, such as the porosity, the thermal conductivities of each phase, their relative proportions, the contact area, and distribution of particles, great effort has been made to characterize the overall thermal conductivity. Intensive works for numerically determining the overall (or effective) thermal conductivity (Deisser et al 1958; Shonnard et al 1989; Saez et al 1991; Tzou et al 1995; Hsu et al 1994; Hus et al 1995). In addition, different techniques have been developed to experimentally measure thermal conductivity (Noazad et al 1985; Olson 1997; Banaszkiewicz et al 1997; Bouguerra et al 1997; Han et al 1998). On the other hand, some studies have been focused on the improvement of thermal insulation properties for cement-based materials. Various kinds of material incorporations in cementitious composites have been investigated in order to reduce the thermal conductivity, k-value. Bourguerra et al (Bourguerra et al 1997) pointed out that the incorporation of wood particles into a cement-based matrix is able to improve thermal performance. Graves et al (Grave et al 1989) attempted to incorporate wax, ranged from 0 to 30% by weight, into gypsum board. His result successfully showed that k-value can be improved by this addition. Foam has been one of the most popular thermal insulation materials because of its extremely small k-value (Burleigh et al 1987; Fricke et al 1989). The incorporation of foam particles is a new approach in cement-based composites. According to Vanderwerf et al (Vanderwerf et al 1997), some systems use a material that combines expanded polystyrene foam beads and Portland cement to form an extremely lightweight insulating concrete.

The selection of phase change materials to meet residential building specifications has received minor attention, although it is one of the most foreseeable applications of PCM. The ability to store thermal energy is important for effective use of solar energy in buildings. Because of the lower thermal mass of lightweight building materials, they tend to high temperature fluctuation, which result in high hearing and cooling demands. It has been demonstrated that paraffins, as mixtures of several linear alkyl hydrocarbons, may be tailored by blending to obtain the desired melting point require for particular application. In this paper, paraffin and expanded polystyrene foam beads have been selected for incorporations in cement-based panels. The advantage of paraffin is its large heat capacity, which is almost three times as high as that of cement-based material. On the other hand, the foam has a very small k-value, which is about one magnitude less than that of cementitious matrix. In addition, the foam incorporation can produce the lightweight composites due to its low density.

Nine material formulae with different amounts of foam, ranging from 0 to 50% by volume, and paraffin have been developed to make cement-based composites. The samples are manufactured by the extrusion technique. It is proved that the extrusion process can enhance the mechanical properties (Shao et al 1995; Shao et al 1996; Li et al 1998; Mu et al 1999).

The samples were tested in a controlled-environment multi-test facility under both steady and transient conditions. Temperature gradient across the samples can be established in the test facility. Different boundary conditions were simulated in the tests. And the thermal performance was evaluated.

1.3.1 Sample preparation

The Portland cement is the major component of the composite. Short PVA fiber is used to improve the toughness. The fiber length is 6mm with average diameter of 14 μm and density of 1.3 g/cm^3. Two governing materials, paraffin and foam plastic, are incorporated to improve the thermal insulation properties of cement-based materials. The paraffin is a waxy crystalline substance that is white, odorless and translucent. It consists mainly of a mixture of saturated straight-chain solid hydrocarbons. Its melting points range from 51 to 53 ℃. The foam plastic is actually expanded polystyrene and in form of beads. The thermal properties of the cement-based matrix, and the paraffin and foam incorporation are shown in Table 7. Nine material formulae (Table 8), that have the same cement-based matrix but different combination of paraffin and foam incorporation, are developed to investigate the thermal resistance.

Thermophysical properties of the cement-based matrix, foam and paraffin. Table 7

Materials	C_p (J/kg℃)	k (W/m℃)	p (kg/m^3)
Cement-based	880	0.7 – 1.2	2400
Paraffin	2364	0.318	860
Foam	1000	0.035	56

Note: cement-based matrix —white cement, slag, silica fume, and silica sand.

Mixing Proportions of the samples Table 8

Type	Sample No.	Paraffin	Foam Plastic	W/C	V_f
I	I-R	—	—	0.28	2%
	I-1	1%	—	0.28	2%
	I-2	2%	—	0.28	2%
II	II-R	—	30%	0.28	2%
	II-1	1%	30%	0.28	2%
	II-2	2%	30%	0.28	2%
III	III-R	—	50%	0.28	2%
	III-1	1%	50%	0.28	2%
	III-2	2%	50%	0.28	2%

Note: i) V_f is PVA fiber volume ratio; ii) W/C is water/cementitious materials ratio; iii) paraffin, foam plastic and PVA fiber are measured by total volume.

1.3.2 Experiment Approaches

Experimental study of the thermal insulation was conducted by using the Controlled-Environment Multi-Test Facility (Dept. of Mechanical Engineering, HKUST). Basically, there are three main parts in the facility, namely indoor room, outdoor room, and control panel. The indoor and outdoor rooms are separated by a wall, which has one opening at the centre. This facility can be used to simulate an indoor and outdoor environment and measure the thermal properties of the materials fixed at the opening. The size of the rooms is approximately 20 m². The temperature range of the outdoor room is from -10 to 60 ℃ while that of the indoor room is from 5 to 40 ℃. The temperature of both rooms can be adjusted with accuracy of 0.1 ℃ through operation of the control panel. In addition, the temperature at some target locations can be measured by thermocouples connected to the control panel.

(1) Heat Conduction Tests

Steady heat conduction test   The target temperature of the indoor and outdoor rooms was set to 15℃ and 45℃ respectively. The room temperatures gradually changed and finally reached a steady state. Therefore, a temperature gradient was correspondingly generated across the two surfaces of the sample. Nine thermocouples were fixed at both surfaces of the specimens to record the temperature at 35 minutes interval.

Transient heat conduction test   The experimental setup on the sample surfaces was similar to that in the steady condition. However, three more thermocouples were placed side by side with 50mm spacing and suspended at 75mm from the cold face (ASTM 1991c; ASTM 1991d), i.e. indoor room side, of the sample. Figure 27 shows this arrangement. The temperature at both rooms was set to 21℃ as an initial condition. Then the temperature of the sample gradually changed to this value. Once the steady condition was attained, the temperature of the indoor room was free to change while that of the outdoor room was set to 60℃. It attempted to model a suddenly changing outdoor environment. This action can show how the outside environment affects the indoor room temperature. In fact, both surface temperatures of the sample were highly uniformly distributed in this case. Hence we can evaluate the thermal insulation capability of the sample by considering the temperature drop across the sample.

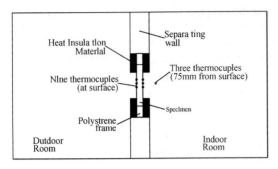
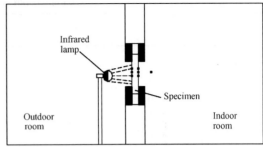

Figure 27　Experimental Setup

Figure 28　Experimental arrangement of thermal capacity test using infrared-lamp as heat source

(2) Thermal Capacity Test

The same initial conditions as the test 2 were set. When the temperature of the rooms and the sample reach the steady state at 21℃, the controls of both rooms were turned off. This means that the temperature of both indoor and outdoor rooms was free to change. Then an infrared lamp was used as a heating device at the outdoor side (Figure 28). The lamp was placed at 300mm from the surface of the sample. The light beams should project on the sample to heat it up. It attempted to stimulate a situation in which a certain heat source directly radiated the sample. Under such condition, the change of the indoor room temperature can be examined. Finally, the thermal insulation performance can be evaluated by considering the change of the air temperature at the indoor side.

1.3.3　Result and Discussion

(1) Steady heat conduction test

In this test, the temperatures at both surfaces of the samples were recorded, and hence the temperature difference across the samples can be determined. Figure 29 shows the surface temperatures of the sample (I-R). It can be observed that the surface temperatures started to rise dramatically at the beginning, and then gradually reached the steady state in approximately two hours. The temperature drop across the hot to the cold surfaces for sample I-R is plotted in Figure 30. Obviously, the temperature drop across sample I-R is about 7℃ under the steady condition.

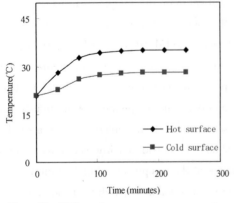
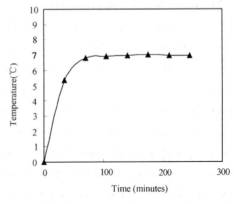

Figure 29　Wall-surface temperatures versus time

Figure 30　Temperature drop across the sample

The effect of foam plastic and paraffin incorporation can be evaluated by comparing the temperature drop across the wall faces. Figure. 31 - Figure33 shows the effect of paraffin for three types of samples. From Figure31, the temperature drop is very close for Type I samples with different paraffin incorporations. The range of the temperature drop is from 6.5 to 7.2 ℃. It indicates that the effect of paraffin is not obvious. If we refer to Figure32, we find that the range of temperature drop is from 12.2 to 13.2 ℃ for Type II samples with different paraffin incorporation, and a slightly increasing effect of paraffin can be noted. The effect of paraffin is significantly increased for Type III sample, as shown in Figure 33. The range of temperature is from 14.9 to 16.5 ℃ for type III samples with different paraffin incorporation. Obviously, the effect of paraffin increases with the foam content. It is because the foam greatly reduces the overall thermal conductivity, the relative paraffin contribution to the insulation becomes larger. This explains why the effect of temperature drop due to paraffin becomes more obvious in the type II and III materials but not in the type I.

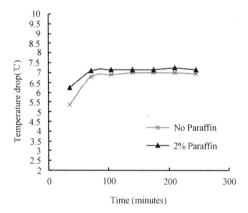

Figure 31 Temperature drop ΔT versus time for the type I samples

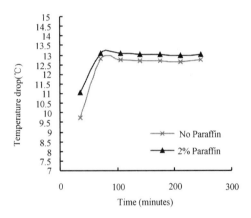

Figure 32 Temperature drop ΔT versus time for the type II samples

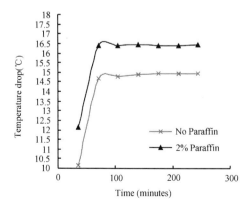

Figure 33 Temperature drop ΔT versus time for the type III samples

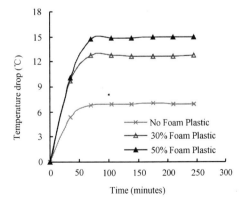

Figure 34 Temperature drop ΔT versus time for Ⅰ-R, Ⅱ-R, and Ⅲ-R.

The previous discussion has pointed out the range that the range of temperature drop increases with the foam plastic content. It implies that the foam plastic content can cause an increment in the temperature gradient, or decrement of thermal conductivity, across the sample. We can compare the three material types in terms of their foam plastic content as shown in Figure 34. The temperature drop increases if the amount of foam plastic content increases. The temperature drop is nearly doubled with 30% foam incorporation. It can achieve a further 25% increment in temperature drop by increasing the foam from 30% to 50%. This proved that the foam plastic bead is an effective incorporation to increase the thermal conduction resistance of the cementitious material by reducing its effective thermal conductivity.

(2) Transient heat conduction test

In this test, the temperature changes at the sample surfaces were measured, and the result is expressed as the temperature difference. We can see the contribution of foam plastic and paraffin incorporation from Figure 35 to Figure 37.

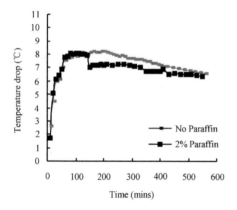

Figure 35 Temperature drop ΔT versus time for type I samples with 0, 1%, 2% paraffin

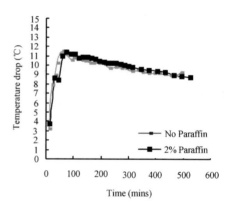

Figure 36 Temperature drop ΔT versus time for type II samples with 0, 1%, 2% paraffin

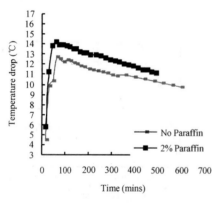

Figure 37 Temperature drop ΔT versus time for type III samples with 0, 1%, 2% paraffin

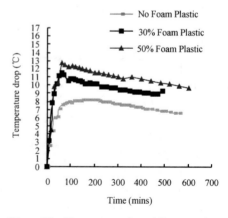

Figure 38 Temperature drop ΔT versus time for the reference sample of type I (no foam), type II (30% foam), and type III (50% foam)

Initially, the temperature rising of the outdoor room caused a temperature increment of the corresponding sample surfaces. At about 100 minutes, the temperature of the outdoor room became steady at 60℃. However, the temperature in the indoor room was still changing since there was no temperature control. It resulted in the decrement of the temperature drop after the outdoor room reached the steady temperature, which is different from the steady heat conduction case. In addition, the temperature drops of type I samples (Figure 35) are very close to each other and hence the enhancement of paraffin is not noticeable. It is same as the results in the steady case and can be explained by the same reason. On the other hand, Figure 36 and Figure 37 illustrate that the temperature difference of the samples with paraffin content is larger than that of the sample without foam. Type III materials (Figure 37) correspondingly showed a greatest improvement from paraffin. This is consistent with the results in the steady state case. Therefore, we can say that the paraffin can increase the temperature drop, or decrease the thermal conductivity, in both steady and transient cases. In addition, this effect becomes more significant with the foam content increasing.

By referring to Figure 38, type I reference sample shows the smallest temperature drop during the transient heat conduction test. However, type II and type III control samples show considerately increased temperature drops. Foam incorporation is capable of increasing thermal insulation. This is consistent with the result obtained in the steady case. It definitely proves that the foam is an effective incorporation, which is able to enhance thermal insulation in both steady and transient heat conduction.

(3) Thermal Capacity Test

In this test, the infrared lamp was used to heat the sample face at the outdoor side. The temperature at the heated surface was not uniformly distributed. It was experienced as high as a 10℃ difference on that surface. However, the temperature difference on the cold face of the sample was greatly reduced, and resulted in at most 6℃ difference. Since this influence occurred approximately the same for every sample, it did not affect the measurement in a relative sense. Since the resolution of the instrument is 0.1℃ and the air temperature has a very large heat capacity, we can confirm the effect from 0.2℃ differences. As the initial condition was 21℃, the air temperature started to rise from this temperature, as shown in Figure 39. By considering the different paraffin contents in type I material, the air temperature rises most rapidly for the type I sample with no paraffin, (I-R). The type I sample with 1% paraffin, I-1, shows decrement of air temperature and that with 2% paraffin content leads to the lowest air temperature among type I materials. This means we can reduce air temperature at the indoor side by increasing the amount of paraffin incorporation. We can also see these enhancements in type II and type III materials, as illustrated in Figure 40 and Figure 41. This successfully proves the effect of paraffin on thermal insulation. Again, we can examine the enhancement of thermal insulation by foam plastic incorporation. From Figure 42, we can see that larger foam plastic incorporation results in lower air temperature. Therefore, the foam plastic can enhance thermal insulation in this test.

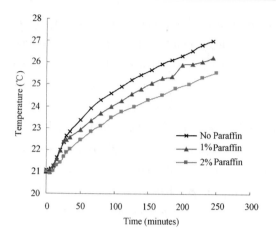

Figure 39　Air temperature at 75mm from the cold surface of the type I samples (no foam) with 0, 1%, 2% paraffin.

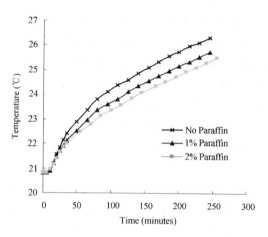

Figure 40　Air temperature at 75mm from the cold surface of the type II samples (30% foam) with 0, 1%, 2% paraffin.

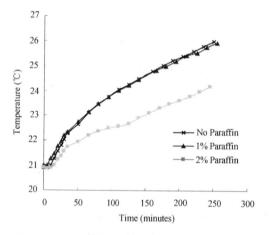

Figure 41　Air temperature at 75mm from the cold surface of the type III samples (50% foam) with 0, 1%, 2% paraffin.

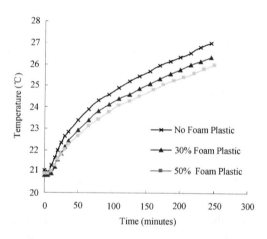

Figure 42　Air temperature at 75mm from the cold surface of the reference sample of the type I (no foam), type II (30% foam), and type III (50% foam)

The test scheme II demonstrates that both paraffin and plastic foam are effective incorporations to enhance the thermal insulation. In particular, paraffin is more functional in this test than in the previous tests, which mainly tests for thermal conductivity. Foam can highly enhance the capability of the thermal insulation for cement-based panel in both transient and steady tests.

1.3.4　Remarks

This study has explored several approaches to evaluate the thermal insulation capability of foam plastic and paraffin incorporation. It has shown that the addition of paraffin provides comparable improvements on the heat capacity even at very low content. However, there is only a slight enhancement for paraffin in the heat conduction tests. In other words, the paraffin can significantly improve the heat capacity, but not the overall thermal conductivity. On the other hand, the foam, despite considerably higher content, 30% to 50%, only provides comparable performance as does the paraffin in the thermal capacity test. However, the foam incorporation largely improves the insulation by reducing the overall thermal conductivity of the cementitious composite materials in both steady and transient heat conduction case. Therefore, foam plastic is a preferable incorporation in reducing thermal conductivity to achieve better thermal insulation.

1.4 Electromagnetic Wave shielding and Absorbing Materials (Guan et al 2006; Du et al 2006)

Electromagnetic interference (EMI) preventing is in increasing demand due to the increasing abundance and sensitivity of electronics, particularly radio frequency devices, which tend to interfere with digital devices. EMI preventing is particularly needed for underground vaults containing transformers and other electronics that are relevant to electric power and telecommunication. It is also needed for deterring any electromagnetic forms of spying. It is in this sense that the cement based building material which is not only a structural material, but also can have some EMI shielding effectiveness and wave absorbing properties through conductive introductions has caused more and more attention.

Cement material which has rich resource and good environmental adaptability is one of the most common structural materials used in engineering constructions. Cement is slightly conducting, but its EMI shielding effectiveness and wave absorbing property are very low, and it is a simple and practical method to increase the cement materials' EMI preventing effectiveness by introducing conductive fillings and loadings.

There are mainly two methods to prevent and attenuate EMI radiation and leakage, which are shielding and absorbing, and the cement materials that are used to prevent EMI can accordingly be divided into EMI shielding materials and wave absorbing materials.

1.4.1 Cement-based EMI shielding materials

So far, extensive researches have been made on shielding and shielding mechanisms, and shielding materials are more extensive and intensive studied compared to absorbing materials. There are, in general, two purposes of a shield namely, to prevent the emissions of the electronics of the product from radiating outside the boundaries of the product and to prevent radiated emissions external to the product's electronics, which may cause electromagnetic interference in the product. So a shield is conceptually a barrier to the transmission of electromagnetic fields. Shielding effectiveness (SE) can be broken into the product of three terms each represents one of the phenomena of reflection loss, absorption loss and multi-reflections. The shielding effectiveness of a shield is defined in decibels (dB). The shielding effectiveness of a cement based material correlates closely to the electric conductivity and the electromagnetic parameters of the composite. In contrast to a polymer matrix, which is electrically insulating, the cement matrix is slightly conductive and its shielding effectiveness is closely pertinent to its conductivity. Generally there are mainly three kinds of conductive fillers used in cement matrix materials, which are conductive polymers, carbon materials and metal materials. The most commonly used fillings are carbon materials (including graphite, carbon black and carbon fiber) and metal materials (mainly metal powder, metal fiber and metal plate).

When the carbon materials are used as the conductive fillers, it is necessary that the fillers be well dispersed, so it often needs to introduce some dispersants. Dispersants are not conductive themselves, but their introduction can obviously improve the dispersion degree of conductive fillers so as to help make more efficient conductive networks. Among the various types of dispersants, styrene butadiene latex and silica fume are the most common for use in cement based composites. Moreover due to the weak strength between the carbon fiber and cement matrix, the introduction of latex, silica fume or methylcellulose can improve the bond between the fiber and matrix, thereby improving the mechanical properties of the cement composites. A surface pretreatment of carbon fiber or treating silica fume with silane can improve the bond strength between carbon fiber and the cement matrix and the dispersion degree of conductive fillers, thereby increase the shielding effectiveness of the composites.

Metal powder has a disadvantage of high density, which makes a lower shielding effectiveness with a small introduction and both an increase of density and a decrease of mechanical strength of the composite material when the loading is increased. By contrast, metal fiber has a high aspect ratio and is liable to be intercalated. So in the same filling with metal powder, metal fiber is more prone to form conductive network; herewith it can provide a better shielding effect with a less loading compared to metal powder. Metal fiber has got a more and more popular application in cement matrix composites as fillers. Only a few studies have been undertaken on metal fiber filled cement matrix shielding materials, and most of the researches are focused on fiber reinforced concretes, which include the effects of concrete structure and the steel bar distribution configurations on the shielding effectiveness of the concrete wall and shielding calculation with finite difference time domain (FDTD) methods. Chung et al. (Shi et al 1995) have studied the magnetic shielding effectiveness of steel paper clips (with a diameter of 0.79 mm and length 31.8 mm) filled cement pastes, which show that paper clips at 5 vol% can provide a magnetic shielding as high as that of a steel mesh with the diameter of 0.6 mm.

1.4.2 Cement based wave absorbing materials

EMI shielding is essentially to form an effective enclosed area with high conductive materials, in which the external electromagnetic radiation cannot penetrate and internal radiation cannot be easily leaked out. But shielding cannot eliminate or weaken EMI radiation, and moreover the reflected wave may interact with the incident wave, which causes disturbance to other units or devices. Only by using electromagnetic wave absorbing materials and transferring the electromagnetic energy to other forms can the EMI radiation be attenuated to the furthest extent.

The electromagnetic absorbing effectiveness of a wave absorbing material is denoted with the reflectivity R, which is expressed as $R = 20 \lg | E_r/E_i |$ (dB), where E_i and E_r refer to the electric field strength of the incident and reflected electromagnetic wave, respectively. A reflectivity of -15 to -20 dB is very good for a plate absorber for civil use and it means that the incident electromagnetic wave has been reduced by 82% - 90%. According to the wave absorption mechanism, traditional wave absorbers can be divided into three types as electric loss, magnetic loss and dielectric loss materials. Carbon materials and conductive polymers are electric attenuation absorbents, which have higher electric loss tangent (tan δe) and the electromagnetic energy is mainly attenuated as a resistor. Ferrites and fine powders are magnetic loss absorbents, which have higher magnetic loss tangent (tan δm), and attenuate and absorb electromagnetic energy by polarization mechanisms such as hysteresis loss and magnetic domain resonance. Metal fiber and many ceramic materials such as barium titanate belong to dielectric loss absorbents, which mainly attenuate electromagnetic energy by electronic and ionic polarization.

As to the cement matrix wave absorbing materials, there are many factors to be considered, such as various physical and chemical reactions between the admixtures and the cement matrix, and the electromagnetic properties of the mixtures, so as to determine the types and contents of the filling admixtures. Taking the economy, practicability and the effects on the other properties of the composite material into account, of all the available wave absorbents, the conductive powders, fibers and magnetic ferrites are suitable to make cement based wave absorbing materials.

Japan has made great efforts on wave absorbing cement materials and has made great progress. In 1992, Nippon Paint Co. Ltd. started the development of a radio wave absorptive material called BMDM (building material to depress multi-path) using ferrite powder and gypsum board. This kind of material has excellent building material characteristics, one of which is that its weight can be handled by an individual worker without difficulty and it also has good fireproof ability. Russian Concrete Research Institute has developed a kind of conductive cement material, which has excellent absorption performance and low reflection coefficient, and moreover it can reduce the EMI defense cost by 99%.

In the field of cement matrix wave absorbing materials, studies on carbon based composites are very few. Most of the absorbing components are metal fibers and ferrites. Ferrite has excellent absorptive abilities at lower frequencies and can widen the frequency band when combined with other absorbents. As for mechanical strength, the ferrite-cement material can keep such a high strength to meet the demands of ordinary building construction as long as the granularity and the weight ratio to cement are controlled. Using the silicon dioxide and nanometre sized carbon black (CB) N234 as the starting materials and with the help of cohesive binder, Guan et al (Guan et al 2006) have developed a kind of wave absorbing material which can be used at the floor of an anechoic chamber. A 200 · 200 mm sample with a thickness of 10 mm and the CB volume fraction of 3.0 vol% to that of the silicon dioxide can give a absorptive performance of 6 - 8 dB at the frequency range 2 - 8 GHz. with the addition of the second wave absorber, the absorbing property increases obviously and there raises two peak values. On the one hand, the adding of the second absorbing component (which has a lower electric conductivity than carbon black) decreases the whole conductivity of the material, therefore the impedance matching between the material surface and the free space is improved, so the reflection decreased and the microwave absorbing ability increased accordingly. On the other hand, when the electromagnetic wave is incident on the surface of the composite, it can penetrate into the material easier and reflected by the backed surface after the introducing of the second component, then the wave reflected from the front and back surface interferes with each other and cause the absorbing peak values.

Most of the metal based absorbents used in cement matrix composites are steel fibers and ferrites, for which Japanese researchers have made many studies. One Japanese institute has successfully introduced ferrite tiles, steel mesh into the fiber reinforced concrete board or other decoration materials to fabricate a kind of wave absorbing material, which can be applied to a building as an absorptive curtain wall. Adding steel fibers to a powdery glass cullet, they have also made a wave absorbing upholstering material, which gives an absorbing effectiveness of 8 dB at the frequency 2.45 GHz and can also depress the multiple reflections in a room. Most of the fibers used in wave absorbing materials have a diameter of several microns and so the cost is usually high. Whereas in building construction, steel fiber with a millimetre diameter has stronger bond strength with cement matrix, and both the fiber and cement can bear loads, which obviously increases the strength of the cement composites. Studies on the wave absorbing property of steel fiber (with the diameter of 0.7 – 1.0 mm and length 2 – 8 cm) reinforced concrete show that a 31 – mm – thick sample can give a peak value of 9.8 dB in the frequency range 2 – 18 GHz and the frequency bandwidth of 15.28 GHz for 4 dB. It also shows that both the peak and the band width increase with the aspect ratio of the steel fiber. But the fiber aspect ratio has a threshold and when it is out of the critic value the absorbing effectiveness will decrease drastically with the increase of the aspect ratio.

Using cement and ferrite or fiber cloth as the materials, Japan has fabricated a kind of curtain wall. It can absorb 90% of the incident electromagnetic waves and give an absorbing effectiveness of 20 – 30 dB in the frequency range 100 – 200 MHz. It also has a light weight and a shockproof ability, and has been successfully used in several buildings in Tokyo and Hiroshima. One Japanese Construction Research Institute has developed a concrete curtain wall material with Mn - Zn and Mg - Zn ferrite, which can give an absorption effectiveness of 5 dB in the frequency 90 – 450 MHz. Using ferrite and aramid fiber as the absorbing components, mixing with steel fiber or metal mesh can also give an absorbing performance of about 14 dB at the frequency of 80 – 220 MHz. Ferrite is a good kind of wave absorbing component, especially in lower frequencies, but its cost is also higher. In view of decreasing cost and using local materials, magnetite sand is also a practicable wave absorbent.

Recently, Du et al (Du et al 2006) investigated wave absorbing characteristics of cement materials composites filled with carbon black-coated expanded polystyrene (EPS) beads was developed. The electromagnetic reflection loss of this composite in the frequency range from 2.6 to 8 GHz was studied experimentally. It was shown that filling with carbon black (CB) -coated EPS beads can improve the reflection loss of plain cement-based materials greatly, and the EPS filler ratio and sample thickness all have remarkable effects on the electromagnetic wave reflection loss of this cement matrix composite material. With a CB-coated EPS filler volume concentration of 80% and sample thickness of 20 mm, the reflection loss can reach -24 dB in the range 2.6 – 8 GHz and the bandwidth for -10 dB reaches 3 GHz. This composite material still has a relatively low bulk density and can be handled easily.

With the rapid development of nano-science, nanostructure wave absorbing materials have caused more and more attention. Nano-sized particles are much less than the incident wavelength, so it can greatly reduce the reflection from the surface, which makes better impedance matching. Moreover, nanometre particles have much large specific surface areas than normal particulates, therefore, when the electromagnetic wave penetrates the material, there will be more particles to be polarized to cause magnetic domain resonance and eddy loss, and thus lead to better wave absorbing performance. After surface treatment, fly ash can be used to preparation of EMI shielding or wave absorbing materials (Shukla et al 2001). Every year, millions of tons of fly ash are generated all over the world and have caused great pollution. Through pretreating, they can be used as admixtures and applied to the cement matrix building materials for providing shielding and wave absorbing functions, which will be a practicable and economical way to dispose of them. There are also many other conductive fillers, such as nickel plated carbon fibers, nickel coated mica, conductive papers and magnetic woods. However, these fillings are not only high in cost, but complicated in processing, so they are not commonly used in cement matrix composites.

2. SUSTAINABLE MATERIALS

2.1 Recycle and Reuse of Industry Waste

American industrial facilities generate and dispose of approximately 7.6 billion tons of industrial solid waste each year. It can be deduced that there are hundreds of billion tons of industrial solid waste in the world each year. Amongst, coal combustion products (CCPs), construction and demolition (C&D) waste, wood waste, metals, glass, plastics, rubber tire et al are one of the main resources of various industry wastes. Due to saving in the use of scarce natural resource, reductions in limited landfill sites and incineration capacity, and environmental consideration, the recycle and reuse of various industrial wastes get an increasing attention from governments of developed countries. Since the recycling of industrial waste is a continual process of ***Collection - Processing - Transportation - Manufacture - Retail - Consumption***, the efficiency of recycling is strongly dependent on the government's support from the aspects of policy, law and regulations. Advantage of recycling of waste: (1) conserves natural resources by reducing the demand for raw materials; (2) conserves energy and water since manufacturing with recycled materials requires less processing than extracting raw materials; (3) reduces air and water pollution since manufacturing from recycled materials is generally a cleaner process and uses less energy; (4) minimizes what is discarded, which maximizes limited landfill capacities and (5) protects our health and the environment when harmful substances, which can be recycled, are removed from the waste stream and processed back into useable products. Recycling facts: (1) Recycling cuts energy consumption and pollution. Paper recycling can reduce air pollutants by 75 percent and water pollution by 67 percent; using scrap steel and iron rather than virgin products results in an 86 percent reduction in air pollution and a 76 percent reduction in water pollution; recycling aluminum saves 95 percent of the energy used to produce it from virgin products. (2) A ton of recycled paper saves 17 trees and three cubic yards of landfill space. (3) Buying recycled products is an essential part of making recycling work (closing the loop). The residential building industry can play a major role in helping to reduce waste and promote recycling by specifying and asking for recycled products.

2.1.1 Recycle of coal combustion products (CCP)

Coal is and will remain in the next future a most important primary energy to produce power and steam all over the world. In coal fired power plants, large quantities of combustion residues and flue gas desulphurization materials are produced. The actual world wide production of ashes from coal and lignite in power plants is estimated to amount to roughly 500 million tones annual (in Figure 43) (von Berg et al, 2000). And it is still increasing rapidly.

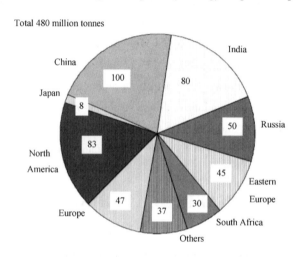

Figure 43 Worldwide Annual Production of Ashes from coal and lignite (von Berg et al 2000)

Due to the mineral components of coal and the combustion technique fly ash (FA), bottom ash (BA), boiler slag (BS) and fluidized bed combustion (FBC) ash as combsution products as well as the products from dry or wet flue gas desulphurization, especially semi dry absorption (SDA) product and flue gas desulphurization (FGD) gypsum are produced.

The major CCPs from coal are:

• Fly Ash (FA) is obtained by electrostatic or mechanical precipitation of dust-like particles from the flue gases of furnaces fired with coal or lignite at 1100 to 1400℃. Fly ash is a fine powder, which is mainly composed of spherical glassy particles. Depending upon the type of boiler and the type of coal siliceous, silico-calcareous and calcereous fly ashes with pozzolanic and/or latent hydraulic properties are produced. *Pulverized Fuel Ash (PFA)* is another term for fly ash in UK and is derived from firing boilers with pulverized coal. Fly ash has pozzolanic properties and consists essentially of SiO_2 and Al_2O_3. Fly ash is obtained by electrostatic or mechanical precipitation of dust-like particles from the flue gases of furnaces fired with pulverized anthracite or bituminous coal. *Cenospheres (floaters)* are recovered from the surface of ash disposal ponds and are of similar chemical composition to fly ash. They float on water and are essentially composed of thin-walled hard-shelled, hollow, minute spheres with a specific gravity (relative density) less than 1.0 kg/dm^3.

• (**Furnace**) **Bottom Ash** (**BA**) is a granular material removed from the bottom of dry boilers, which is much coarser than FA though also formed during the combustion of coal.

• **Boiler Slag** (**BS**) is a vitreous grained material deriving from coal combustion in boilers at temperatures of 1500 to 1700℃, followed by wet ash removal of wet bottom furnaces. The maximum particle diameter of BS is about 8 millimeters.

• **Fluidized Bed Combustion** (**FBC**) **Ash** is produced in fluidized bed combustion boilers. The technique combines coal combustion and flue gas desulphurization in the boiler at temperatures of 800 to 900℃. FBC ash is rich in lime and sulphur.

• **Semi Dry Absorption** (**SDA**) **Product** is a fine grained material resulting from dry flue gas desulphurization with lime acting as the sorbent.

• **Flue Gas Desulphurization** (**FGD**) **Gypsum** is natural gypsum like product which is obtained by wet desulphurization of flue gas and special treatment of the adsorbed products.

Moste of the CCPs produced are used in the building industry, in civil engineering and as construction mateials in underground mining or for restoration of open cast mines, quarries and pits. Each year ECOBA (European Coal Combustion Products Association) prepares a statistic on production and utilization of CCPs in the countries of their members. Since 1993 the statistics cover the data of the EU 15 countries. Since May 1, 2004 the EU has grown by 10 new members. In these countries more than 30 million tons of CCPs are produced in addition. By this, the amount of CCPs produced in Europe (EU 15) totals to more than 95 million tons. Figure 44 display the production and utilization of CCPs in Europe (EU15) in 1999. The evaluation of the data on production and utilization in EU 15 countries in 2003 is shown in Table 9. A graph on the production of CCPs and utilization of fly ash in EU 15 countries is given in Figure 45 (a) and Figure 45(b).

Governments are more and more favouring the use of secondary raw materials. Legislation and rules to increase such use are being improved. In several cases the minerals from coal bring extra quality and higher performance compared to the prime raw materials which are being replaced. Applying minerals from coal adds a geen label to construction due to energy savings and preservation of natural resources. In the European Union (EU 15) approximately 65 million tons of CCPs were produced in 2003. Fly ash, which is obtained by electrostatic or mechanical precipitation of dust like particles from the flue gas, represents the greatest proportion of total CCP production. Within the EU, the utilization for Fly Ash in the construction industry is currently around 47% and for Bottom Ash around 44%, while the utilization rate for Boiler Slag is 100%. In the majority of cases CCPs are used as a replacement for naturally occurring resources and therefore offer environmental benefits by avoiding the need to quarry or mine these resources. CCPs also help to reduce energy demand as well as emissions to atmosphere, for example CO_2, which result from the manufacturing process of the products which are replaced. CCPs are utilized in a wide range of applications in the building and construction industry. Applications for CCPs include their use as an addition in concrete as a cement replacement material and as an aggregate or binder in road construction. They can also be utilized as mineral fillers and as fertilizers. However, the utilization of CCPs is dependent on the government's political intentions and legislation.

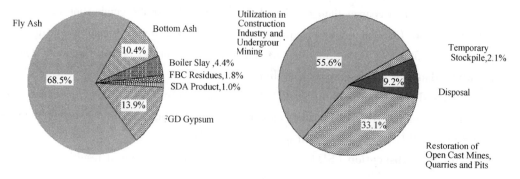

(a) Production of CCPs (Total production 55 million tons)

(b) Utilization and disposal of CCPs

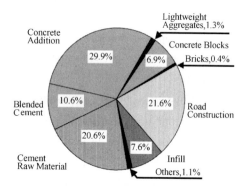

(c) Utilization of Fly Ash (Production 37.6 Million Tons, Utilization: 18.2 Million Tons)

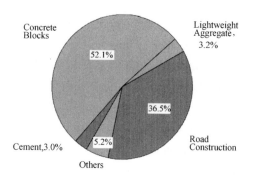

(d) Utilization of Bottom Ash (Production 5.6 Million Tons, Utilization: 2.5 Million Tons)

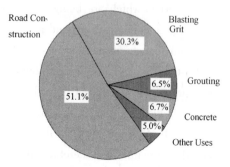

(e) Utilization of Boiler Slag (Production 2.4 Million) Tons, Utilization: 2.4 Million Tons)

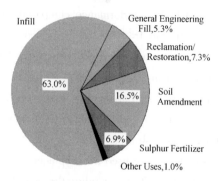

(f) Utilization of SDA product (Production 0.52 Million Tons, Utilization: 0.5 Million Tons)

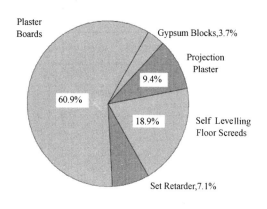
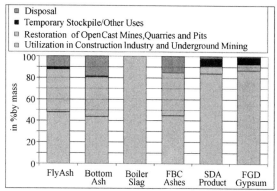

(g)Utilization of FGD gypsum(Production)7.6Million Tons,Utilization:6.6Million Tons)

(h)Utilization and disposal of CCPs

Fiugure 44　Overview of production and utilization of CCPs in EU in 1999（von Berg et al 2000）

**Production and utilization of CCPs in 2003 in Europe
（EU15）［in kilo tones（metric）］（ECOBA 2003）　　Table 9（a）**

		FA	BA	BS	FBC	Other	SDA	FGD		
		1	2	3	4	5	6	7		
CCP Production		44.217	6.045	2.110	1.089	76	490	11.276		
Subtotal（1 - 5）					53.537					
Subtotal（6 - 7）								11.766		
Total（1 - 7）								65.303		
CCP Utilization									Total	%
Cement raw material	1	5.460	151						5.611	8.5
Blended cement	2	2.377	125						2.502	3.8
Concrete addition	3	5.872	98	161					6.131	9.3
Aerated concrete blocks	4	845	16						861	1.3
Non-aerated concrete blocks	5	380	1.264			5			1.649	2.5
Lightweight aggregate	6	93	0						93	0.1
Bricks + ceramics	7	133	23			16			172	0.3
Grouting	8	481		126	84				691	1.0
Asphalt filler	9	158							158	0.2
Subgrade stabilisation	10	184	98		1				283	0.4
Pavement base course	11	387	258	998	72				1.715	2.6
General engineering fill	12	1.777	377		0	35	17		2.206	3.3
Structural fill	13	1.911	145		63				2.119	3.2
Soil amendment	14	37	1		22				60	0.1
Infill	15	689	94		288		15		1.086	1.6
Blasting grit	16	0		660					660	1.0
Plant nutrition	17	3					22		25	0.0

Table 9 (b)

		FA	BA	BS	FBC	Other	SDA	FGD	
		1	2	3	4	5	6	7	
Set retarder for cement	18							642	642	1.0
Projection plaster	19							760	760	1.2
Plaster boards	20							4.897	4.897	7.4
Gypsum blocks	21							261	261	0.4
Self levelling floor screeds	22						1.401	1.401	2.1
Other uses	23	329	6	165	27	9	41	5	582	0.9
Reclamation, Restoration	24	18.964	2.686	0	178		180	1.645	23.653	35.9
Temporary stockpile	25	3.507	128	0	40			1.581	5.256	8.0
Disposal	26	1.207	613	0	314	11	215	84	2.444	3.7
Total utilization 1 – 23	27	21.116	2.656	2.110	557	65	95	7.966	34.565	52.4
Utilization rate in %	28	47	44	100	51	86	19	71		
Average utilization rate in %	29							52		
Total utilization 1 – 24	30	40.080	5.342	2.110	735	65	275	9.611	58.218	88.3
Utilization rate in %	31	89	88	100	67	86	56	85		
Average utilization rate in %	32							89		
Reuse of stockpiled CCPs	33	577	38	0	0	0	0	0	615	0.9
Total production 1 – 26	34	44.794	6.083	2.110	1.089	76	490	11.276	65.918	100.0

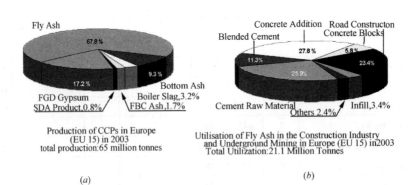

(a)   (b)

Figure 45 Production of CCPs and utilization of Fly Ash in EU 15 countries (ECOBA 2003)

2.1.2 Recycle of construction and demolition waste

Construction and demolition (C&D) debris consists of the materials generated during the construction, renovation, and demolition of buildings, roads, and bridges. C&D debris often contains bulky, heavy materials that include: concrete, wood (from buildings), asphalt (from roads and roofing shingles), gypsum (the main component of drywall), metals, bricks, glass, plastics, salvaged building components (doors, windows, and plumbing fixtures), and trees, stumps, earth, and rock from clearing sites. Figure 46 displays the fraction of various wastes in overall waste stream in California in 2003 (California Integrated Waste Management Board, 2004). It can be seen that construction and demolition (C&D) waste account for almost 22 percent of the waste stream. It was estimated that 136 million tons of building-related C&D debris was generated in the United States in 1996. And C&D waste for the EU was estimated at 221 to 334 million tons in 1995, which is about twice the amount of municipal solid waste generated. The composition of C&D debris varies significantly, depending on the type of project from which it is being generated. For example, debris from older buildings is likely to contain plaster and lead piping, while new construction debris may contain significant amounts of drywall, laminates, and plastics. For building debris, U. S. Environmental Protection Agency (EPA) estimates the overall percentage of materials in C&D debris, it was shown in Table 10.

Composition of various materials in C&D debris Table 10

Type	Concrete and mixed rubble	Wood	Drywall	Asphalt roofing	Metals	Bricks	Plastics
Percentage	40% - 50%	20% - 30%	5% - 15%	1% - 10%	1% - 5%	1% - 5%	1% - 5%

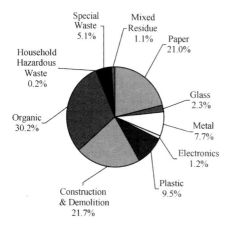

Figure 46 Material Classes in California's Overall Disposed Waste Stream, 2003 (California Integrated Waste Management Board, 2004)

Figure 47 wood particle based magnesium oxychloride cement products generated by extruding technology

Many of these materials can be reused or recycled, thus prolonging our supply of natural resources and potentially saving money in the process. Reducing and recycling C&D debris conserves landfill space, reduces the environmental impact of producing new materials, creates jobs, and can reduce overall building project expenses through avoided purchase/disposal costs. For instance, concrete waste, crushed bricks and blocks, crushed ceramic tile and glasses debris can be used as recycling aggregate to produce pre-case concrete products (such as kerbs), paving blocks, terrazzo tiles (Poon 2000). It is estimated that 80% - 90% of all C&D waste is of the brick/concrete fraction, which is recyclable and may be used as a substitute for natural stone and gravel resources. Most paper can be recycled unless it has been tainted with food or coated with wax. Building and construction materials that utilize recycled paper products include cellulose for insulation, cellulose fiberboard and gypsum board sheeting material. Wood waste can be used produce particleboard, fiberboard, and flakeboard. In addition, wood waste (such wood fiber or particles) can be combined with waste plastics to produce wood/plastic composites. If it is mixed with inorganic binder, such as Portland cement, magnesia cement, and gypsum etc, various wood-based inorganic composite is generated. HKUST also developed an extruding technology to reuse wood waste. Figure 47 displays product generated by such technology. Waste gypsum board can be recycled at gypsum factories. In Japan, 10% of recycled materials can be included in raw materials used in the production of gypsum board (Senbu et al 2004). The building and construction materials industry utilizes recycled aluminum in flashing material and window components. Recycled steel is used for framing connectors, nails and structural framing. In Netherlands, over 95% of C&D waste can be recycled (Hendriks et al 2000; Hendriks 2000) due to the limited landfill space and restriction by law. There is a long way to go for China.

2.1.3 Recycling of plastics

There are many different types of plastics from various resources. Even a small amount of a different type can make the entire batch unusable. So, the separating process is very important. What follows lists the type of plastics and their potential recycling routes (AMI 2006; Hulse 2000).

(1) PETE or PET (Polyethylene Terephthalate) Used for soda, liquor and juice bottles and peanut butter jars and some jars for oils. This plastic can be recycled into new construction products including fabrics and carpet fibers.

(2) HDPE (High-Density Polyethylene) Used for milk, juice, detergent, bleach and motor oil containers. When recycled, this plastic is used for lumber substitutes, trash and compost containers among other products.

(3) PVC (Vinyl/Polyvinyl Chloride) Used for windows, doors, shower curtains, and similar products. This plastic can be recycled into fencing, sewer pipes and garden hoses.

(4) LDPE (Low-Density Polyethylene) Used for cellophane wrap, stretch wrap and squeeze bottles. This plastic is recycled to make similar products.

(5) PP (Polypropylene) Used for food containers and long underwear. This plastic is recycled into furniture, carpet and auto parts.

(6) PS (Polystyrene) Also know as Styrofoam. This plastic is recycled into plastic wood, packing peanuts, office and desk accessories.

(7) Other Plastics. This designation is for all other plastics that are difficult to recycle.

2.1.4 Recycling of glass cullet (Dhir 2004)

Glass cullet is recycled container glass (previously used for bottles, jars and other similar glass vessels) prior to processing. The material is typically collected via bottle banks, kerbside collection schemes and from premises handling large quantities of containers, with the primary aim of processing it for returning to the glassmaking process to manufacture glass containers or other products. The term 'cullet' also refers, however, to waste glass produced as a result of breakage and rejection on quality control grounds during the manufacturing process. Container glass is manufactured by combining quantities of minerals (and usually glass cullet) to obtain a feedstock rich in silica (SiO_2), soda (Na_2O) and lime (CaO). By the time cullet has been recovered for recycling, it is typically broken into smaller fragments and may contain other substances which have been introduced and carried with it during its use and disposal. These substances can include paper, adhesives, aluminium foil, plastics, and residues of the substances originally held in the containers. The quantity of foreign matter present with cullet varies depending on the method of collection, storage conditions and any pre-processing carried out on the material.

Glass cullet is a hard, granular material and this has led many engineers to consider using it as a construction aggregate. In recent years, glass cullet has been used in the UK as an aggregate in bituminous highway materials. The extent of the material's use as an aggregate in concrete is currently small, although the production of precast concrete containing glass aggregate is being carried out on a small-scale in the UK at present. The use of powdered glass in concrete has been rare in the UK, but it has been used routinely for a number of years in Sweden as filler.

In terms of its physical and mechanical properties, crushed glass cullet behaves in a very similar way to sand, being a hard, granular material with a similar particle density. In many circumstances there may be economic benefits in using glass cullet in place of natural sand. However, the material possesses other beneficial properties which may be exploited. Concrete containing glass cullet as aggregate typically displays higher resistance to abrasion than equivalent materials made using quartz sand or similar materials as fine aggregate. Reduced drying shrinkage is also often observed for mixes containing glass cullet aggregate. However, the potential alkaline-aggregate reaction should be checked out before using glass cullet as aggregate. Powdered glass cullet is pozzolanic, meaning that it will react with lime to form products that contribute towards strength development. As with all pozzolanas, the contribution to strength development is less than that of Portland cement, meaning that cement combinations comprising these two materials will produce lower compressive strengths than Portland cement alone. However, there are clear economic benefits of using cullet as a cement component. Abrasion resistance of concrete containing powdered cullet is also improved.

2.1.5 Recycling of tires (Ohio Department of Natural Resources, 2005)

The U.S. Environmental Protection Agency estimates that 250 million scrap tires are generated in the United States each year, not counting another 45 million scrap tires that are used to make 34.5 million automobile and truck tire retreads every year. Scrap tires not only waste landfill space, they can damage the linings put in place to keep groundwater and surface water from mixing with landfilled contaminants. Tires discarded illegally - individually as litter or collectively in clandestine tire dumps - are an eyesore and a drag on surrounding property values. They also pose threats to public health and safety. Tire dumps provide excellent breeding grounds for mosquitoes, and elevated incidents of mosquito-borne diseases have been noted near large tire piles.

Tire pile fires have been an even greater environmental problem. Tire pile fires can burn for months, sending up an acrid black plume that can be seen for dozens of miles. That plume contains toxic chemicals and air pollutants, just as toxic chemicals are released into surrounding water supplies by oily runoff from tire fires. Fighting a tire pile fire is not only futile in some cases, it can actually make the pollution problem worse. In 1989, only 10 percent of the scrap tires generated in the United States were reclaimed through recycling or other uses. Today, more than 80 percent of scrap tires are pulled from the waste stream and reused in some way.

The single biggest use of scrap tires in the United States and the European Union is as fuel. Environmentally or even economically speaking, it is not the best method of reclaiming scrap tires, but it beats the alternative of disposing or dumping unused tires. In 2000, approximately 47 percent of the 273 million scrap tires generated in the United States were burned for fuel. The best management strategy for scrap tires that are worn out beyond hope for reuse or retreading is recycling. Chips of shredded tire rubber can be used as replacement of conventional construction material such as road fill, gravel, crushed rock or sand in a variety of construction and civil engineering projects. Tire chips are successfully used as fill for embankments and retaining walls, drainage material and daily cover at solid waste landfills and as an insulating layer beneath roads and behind retaining walls. More finely chipped and screened tire rubber is used as playground and landscaping mulch. Crumb rubber is used to make better asphalt, while rubber mixed with urethane is used to make athletic track surfaces and a variety of molded products.

2.1.6 Reuse of cement kiln dust

Cement kiln dust (CKD) is a by-product of the cement manufacturing process. The accumulated industrial waste from this process (CKD) is a source of concern both to the cement industry and environmental agencies, creating an urgent need to find cheap, eco-friendly uses for it. During the cement manufacturing process, particulates are produced. About three quarters of these particulates are returned to the process, while the remaining one quarter captured in air pollution control equipment are discarded. This discarded material-cement kiln dust (CKD) -contains extremely alkaline materials, such as sodium and potassium, and may also include heavy metals, such as lead and cadmium, making it difficult to dispose of and it can't be reused in the cement manufacturing process or stored or dumped. And because CKD is unstable (heavy metals are encapsulated but not fixed in CKD's molecular structure), it can't be easily handled or treated.

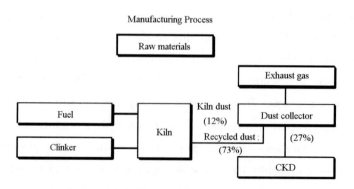

Figure 48　Cement manufacturing process (Raki 2001)

CKD has some positive attributes, however. It is similar to ordinary Portland cement, both in terms of its chemical composition and physical properties. This similarity may be the key to its successful reuse in durable and eco-efficient products, such as blended cements containing CKD along with suitable additives to immobilize alkali and heavy metals.

Raki (Raki 2001) and Al-Harthy et al (Al-Harthy et al 2003) have already produced cement containing significant amounts of CKD that demonstrates excellent strength characteristics. Sreekrishnavilasam et al (Sreekrishnavilasam et al 2007) employed CKD for soil treatment. But before considering its possible uses, the level, mobility and leachability of the trace metals in the CKD have to be assessed. The current research is taking another approach, which focuses on the development of new structures to encapsulate and chemically bind the trace elements. With this approach, concerns about the leaching of heavy metals can be eliminated.

2.1.7 Blast furnace slag, phosphorus slag and steel slag

In the production of iron, iron ore, iron scrap and fluxes (limestone and/or dolomite) are changed into a **blast furnace slag** along with coke for fuel. The composition of blast furnace slag varies with the type of iron made and the type of ore used, which can be represented in a $CaO - SiO_2 - Al_2O_3 - MgO$ quaternary diagram. A slow cooling of slag leads to lower or no cementitious property. The molten slag must be quickly cooled in order to enhance its cementitious property. It is well-known that the ground granulated blast furnace slag is widely used as part of binder in various ways. So, we do not talk much about it.

Phosphate slag (also called phosphorus slag) is produced as a by-product of white phosphorus. Every ton of elemental phosphorus roughly produce 4 ~ 10 tons of slag (Corbrideg 1995, Shi et al, 2006a). Phosphate mineral always contain some impurities, such as fluorapatite, phosphatized limestone, sand and clay. Those compounds come out in the slag. Phosphorus slag is composed mainly of SiO_2 and CaO. The minor components in phosphorus slag depend on the nature of phosphate ores used. Table 11 lists the chemical composition of several phosphorus slags produced in the world.

Chemical composition of phosphorus slag from several countries in the world (% by mass) (modified from Shi et all 2006a) Table 11

Origin	SiO_2	Al_2O_3	Fe_2O_3	CaO	MgO	P_2O_5	F	SiO_2
Nanjing, China	38.2	3.9	0.4	49.0	1.6	1.6	2.9	38.2
Yunnan, China	41.1	4.1	2.5	44.8	2.8	2.4	2.7	41.1
Germany	42.9	2.1	0.2	47.2	2.0	1.8	2.5	42.9
Former USS	43.0	3.4	3.2	45.0	—	3.0	2.7	43.0

Molten phosphorus slag can be cooled either in air or through water-quenching. If phosphorus slag is air-cooled, it consists mainly of crystalline compounds such as $CaO \cdot SiO_2$ and $3CaO \cdot 2SiO_2 \cdot CaF_2$ can also be detected if CaO/SiO_2 ratio of the slag is at the high end. If it is water-quenched, it mainly consists of vitreous structure with a glass content of up to 98% due to the high viscosity of the molten slag. Air-cooled phosphorus slag does not exhibit any cementitious property and can only be crushed for uses as ballast or aggregate. Granulated phosphorus slag (GPS) has a vitreous structure similar to that of granulated blast furnace slag (GGBFS). GPS is a latent cementitious material but less reactive than GGBFS at early age due to its lower Al_2O_3 content and the presence of P_2O_5 and F. However, if a portion of siliceous materials used as flux is replaced with bauxite to give Al_2O_3 content similar to that in GGBFS, the hydraulic reactivity of the GPS will be comparable with that of BBGFS. Lebedeva et al used phosphate slag to prepare porous glasses. (Lebedeva et al, 2000). Shi et al (Shi et al 1989) used phosphorus slag to develop alkali-slag cement. This cement was further used for the purpose of stabilization/solidification of hazardous and radioactive wastes (Shi et al 2006b).

Steel slag is a by-product from either the conversion of iron to steel in a basic oxygen furnace (BOF), or the melting of scrap to make steel in an electric arc furnace (EAF). There are several methods for cooling molten steel slag: natural air-cooling, water-spray, water-quenching, air quenching and shallow box chilling. Obviously the reactivity of steel slag by different cooling methods is different. Chemical composition of steel slag is highly variable and changes from batch to batch even in one plant depending on raw materials, type of steel made, furnace condition, etc. Table 12 lists the chemical composition range from different type of steel slag.

Chemical composition range of steel slags (%) (Shi 2004)   Table 12

Components	BOF	EAF (carbon steel)	EAF (alloy/stainless)	Ladle
SiO_2	8–20	9–20	24–32	2–35
Al_2O_3	1–6	2–9	3.0–7.5	5–35
FeO	10–35	15–30	1–6	0.1–15
CaO	30–55	35–60	39–45	30–60
MgO	5–15	5–15	8–15	1–10
MnO	2–8	3–8	0.4–2	0–5
TiO_2	0.4–2	—	—	—
S	0.05–0.15	0.08–0.2	0.1–0.3	0.1–1
P	0.2–2	0.01–0.25	0.01–0.07	0.1–0.4
Cr	0.1–0.5	0.1–1	0.1–20	0–0.5

Since the chemical composition of steel slag varies significantly from source to source, it can be expected that the mineralogical composition of steel slag can be very different from source to source. Reported mineral in steel slag include olivine, merwinite, C_3S, $\beta-C_2S$, $\gamma-C_2S$, C_4AF, C_2F, RO phase (CaO–FeO–MnO–MgO solid solution), free-CaO and free-MgO. Table 13 summarizes the relationship between basicity, main mineral phase and hydraulic reactivity of steel slag. It can be seen that reactivity of steel slag increases with its basicity. However, free-CaO content also increases with the increase of the basicity of steel slag. The C_3S content in steel slag is much lower than that in Portland cement. Thus, steel slag can be regarded as a weak Portland cement clinker (Tang 1973.).

reactivity, basicity and mineral compositions of steel slag (Tang 1973)　　Table 13

Hydraulic reactivity	Types of steel slag	Basicity		Major mineral phases
		CaO/SiO_2	$CaO/(SiO_2+P_2O_5)$	
Low	Olivine	0.9 – 1.5	0.9 – 1.4	Olivine, RO phase and Merwinite
	Merwinite		1.4 – 1.6	Merwinite, C_2S and RO phase
Medium	Dicalcium silicate	1.5 – 2.7	1.6 – 2.4	C_2S and RO phase
High	Tricalcium silicate	>2.7	>2.4	C_3S, C_2S, C_4AF, C_2F and RO phase

The lower reactive steel slag is particularly useful in areas where good-quality aggregate is scarce. By comparing with concrete with crushed limestone aggregate, Maslehuddin et al (Maslehuddin et al 2003) have ever investigate the mechanical properties and durability of steel slag aggregate concrete. And Wu et al (Wu et al 2006 Xue et al 2006) used steel slag as aggregate in asphalt mixture. For the relative high hydraulic reactive steel slag, they can be used as part of binder (Shi 2006b). A speciality cement, steel slag cement, which is composed mainly of steel furnace slag, GGBFS, cement clinker and gypsum has been commercially marketed in China for more than 20 years (Sun 1983, Wang and Lin 1983).

2.1.8 Remarks

Since the recycling of industrial waste is a continual process of **Collection - Processing - Transportation - Manufacture - Retail - Consumption**, the efficiency of recycling is strongly dependent on the government's support from the aspects of policy, law and regulations. Recycling of waste is helpful to conserves natural resources, to reduces air and water pollution and to minimize landfill. However, with the increase in reuse and recycling, researchers should consider some new questions, such as: are reused materials and products more hazardous to the environment and to our health? Does the processing require use up a disproportionate amount of energy? What about the maintenance and lifespan?

2.2　New Binders-Energy Efficient and with Less CO_2

Portland cement (PC) concrete is the most popular and widely used building materials, due to its availability of the raw materials over the world, its ease for preparing and fabricating in all sorts of conceivable shapes. The applications of concrete in the realms of infrastructure, habitation, and transportation have greatly prompted the development of civilization, economic progress, stability and of the quality of life (Mehta 1999). Nowadays, with the occurrence of high performance concrete (HPC), the durability and strength of concrete have been improved largely. However, due to the restriction of the manufacturing process and the raw materials, some inherent disadvantages of Portland cement are still difficult to overcome. There are two major drawbacks with respect to sustainability. (1) About 1.5 tons of raw materials is needed in the production of every ton of PC, at the same time, about one ton of carbon dioxide (CO_2) is released into the environment during the production. The world's cement production is expected to increase from 1.4 billion tons in 1995 to almost 2 billion tons by the year 2010 (Nehdi et al 2001). Therefore, the production of PC is extremely resource and energy intensive process. (2) Concrete made of PC deteriorates when exposed to the severe environments, either under the normal or severe conditions. Cracking and corrosion have significant influence on its service behavior, design life and safety.

Here, two different cementitious materials will be discussed. One is geopolymer and the other is magnesium phosphate cement (MPC). Compared with Portland cement, the above two cements possess some common and individual characters, respectively. Their properties are very favorable to the sustainable development of our modern society.

2.2.1 Advantages and application of geopolymer

Compared with Portland cement, geopolymers possess the following characteristics:

• Abundant raw materials resources: any pozzolanic compound or source of silicates or alumino-silcates that is readily dissolved in alkaline solution will suffice as a source of the production of geopolymer.

• Energy saving and environment protection: geopolymers don not require large energy consumption. Thermal processing of natural alumino-silicates at relative low temperature (600℃ to 800℃) provides suitable geopolymeric raw materials, resulting in 3/5 less energy assumption than Portland cement. In addition, a little CO_2 is emitted.

• Simple preparation technique: Geopolymer can be synthesized simply by mixing alumino-silicate reactive materials and strongly alkaline solutions, then curing at room temperature. In a short period, a reasonable strength will be gained. It is very similar to the preparation of Portland cement concrete.

• Good volume stability: geopolymers have 4/5 lower shrinkage than Portland cement.

• Reasonable strength gain in a short time: geopolymer can obtain 70% of the final compressive strength in the first 4 hours of setting.

• Ultra-excellent durability: geopolymer concrete or mortar can withdraw thousands of years weathering attack without too much function loss.

• High fire resistance and low thermal conductivity: geopolymer can withdraw 1000℃ to 1200℃ without losing functions. The heat conductivity of geopolymer varies form 0.24w/m·k to 0.3w/m·k, compared well with lightweight refractory bricks (0.3 w/m·k to 0.438 w/m·k).

Geopolymer has properties such as abundant raw resource, little CO_2 emission, less energy consumption, low production cost, high early strength, fast setting. These properties make geopolymer find great applications in many fields of industry such as civil engineering, automotive and aerospace industries, non-ferrous foundries and metallurgy, plastics industries, waste management, art and decoration, and retrofit of buildings.

(1) Toxic waste treatment. Immobilization of toxic waste may be one of the major areas where geopolymer can impact significantly on the statues quo. The molecular structure of geopolymer is similar to those of zeolites or feldspathoids, which are known for their excellent abilities to adsorb and solidify toxic chemical wastes such as heavy metal ions and nuclear residues. It is the structures that make geopolymer a strong candidate for immobilizing hazardous elemental wastes. Hazardous elements present in waste materials mixed with geopolymer compounds are tightly locked

into the 3-D network of the geopolymer bulk matrix.

(2) Civil engineering. Geopolymer binders behave similarly to Portland cement. It can set and harden at room temperature, and can gain reasonable strength in a short period. Some proportions of geopolymer binders have been tested and proved to be successful in the fields of construction, transportation and infrastructure applications. They yield synthetic mineral products with such properties as high mechanical performance, hard surface (on the Mohs Scale) (Davidovits 1994a; Lyon et al 1997), thermal stability, excellent durability, and high acid resistance. Any current building component such as bricks, ceramic tiles and cement could be replaced by geopolymer.

(3) Global warming and energy saving. It is well known that a great amount of CO_2 is emitted during the production of Portland cement, which is one of the main reasons for the global warming. Studies have shown that one ton of carbon dioxide gas is released into the atmosphere for every ton of Portland cement which is made anywhere in the world. In contrast, geopolymer cement is manufactured in a different way than that of Portland cement. It does not require extreme high temperature treatment of limestone. Only low temperature processing of naturally occurring or directly man-made alumino-silicates (kaoline or fly ash) provides suitable geopolymeric raw materials. These lead to the significant reduce in the energy consumption and the CO_2 emission. It is reported by Davidovits (Davidovits 1994b) that about less 3/5 energy was required and 80%-90% less CO_2 is generated for the production of geopolymer than that of Portland cement. Thus it is of great significance in environmental protection for the development and application of geopolymer cement.

(4) High temperature and fire resistance. Geopolymer cement possesses excellent high temperature resistance up to 1200℃ and endures $50kW/m^2$ fire exposure without sudden properties degradation. In addition, no smoke is released after extended heat flux. The merits make geopolymer show great advantages in automotive and aerospace industries. At present, some geopolymer products have been used in aircraft to eliminate cabin fire in aircraft accidents.

(5) Archaeological analogues. It is proved that the micro-structure of hardened geopolymer materials is quite similar to that of ancient constructs such as Egyptian pyramid, Roman amphitheater. Consequently, many experts suspended that these ancient constructs might be cast in place through geopolymerization, rather than made of natural stones. To confirm the viewpoint, many scientists make much attempt to explain the unsolved enigma for some ancient long-term constructs by means of geopolymer theories in recent years.

2.2.2 Advantages and applications of MPCs

MPCs are artificial stone made from acid-base reaction of magnesia and phosphates. They possess some properties that Portland cements do not possess according to the previous studies. Therefore, they can be utilized in the field in which Portland cements are not suitable (Kingery 1950; Yoshizake etl 1989; Seehra et al 1993; Yang et al 2002; Singh et al 1997; Wagh et al 1999). (1) Very quick setting, high early strength. (2) Recycling a lot of non-contaminated industrial waste to building material. (3) Recycling organic waste to building materials. (4) Stabilization of toxic and radioactive waste. (5) Very good durability, including chemical attack, resistance, deicer scaling resistance, permeation resistance.

The applications of MPCs include following aspects:

(1) Due to its rapid setting and high early strength, magnesium phosphate cement (MPC) has been utilized in raped repair of concrete structure, such as highway, airport runway, and bridge decks for many years. It can save a lot of waiting time and cost caused by long disrupting time by use other materials. If the interrupt period is too long for the busy highway, airport runway, and bridges, etc., it will cause lose of millions dollars. By using MPC materials, the interrupt time of transportation can be greatly shortened. Therefore, the valuable time and resource can be saved.

(2) MPC can incorporate with lot of non toxic industrial waste, such as Class F fly ash (FA) and translate it into useful construction materials. The addition FA in MPC can be over 40% by mass of MPC, about two times comparing with PC. In addition, MPC can combine the FA that is not suitable incorporated in PC because of its high carbon content and other impurities. Besides FA, even acid blast furnace slag, red mud (the residua of alumina industry), even tails of gold mine can also be utilized in MPC at large amount. These wastes are difficult to use in PC concrete in a considerable amount.

(3) Due to the high alkali environment of PC (pH over 12.5), when they are use as reinforcement, some components natural fibers, notably lignin, and hemicelluose will be susceptible to degradation. However, the lower alkalinity of MPC matrices (pH value 10 to 11) makes them potentially better suited to vegetable fiber reinforcement. Furthermore, the sugar in some natural fibers, such and sugarcane and corn stalk can prohibit the setting of PC, and weakens the bonding between Portland cement and fiber. But, the set of MPC is not influenced by sugar.

(4) Management and stabilization of toxic and radioactive wastes, including solids and liquids. The wastes can be micro and/or micro-encapsulated and chemically bonded by MPC, form a strong, dense and durable matrix that stores the hazardous and radioactive contaminants as insoluble phosphates and microencapsulates insoluble radioactive components. The waste forms are not only stable in groundwater environments, but also are nonignitable and hence safe for storage and transportation.

(5) Very suitable for repairing of the deteriorated concrete pavements in the cold areas. MPC can develop strength at low temperature due to its exothermic hydration and low water to binder ratio. At the same time, MPCs possess a higher deicer scaling property than Portland cement.

(6) The raw material of MPC is hard burnt magnesia. In fact, it is a refractory material. Therefore, MPC can be designed to be fire proof and/or as cold setting refractory according to the practical need.

2.2.3　Summary on geopolymer development

2.2.3.1　Work done by others

Since France scientist Davidovits invented geopolymer materials in 1978 (Davidovits 1993), great concerns on the development of geopolymer have been received across the world. More than 28 international scientific institutions and companies have presented updated research and published their results in public journals. These works mainly focus on the following aspects:

(1) Solidification of toxic waste and nuclear residues Davidovits et al. (Davidovits 1994c) firstly began to investigate the possibilities of heavy metal immobilization by commercial geopolymeric products in the early 1990s. The leachate results for geopolymerization on various mine tailings showed that over 90% of heavy metal ions included in the tailings can be tightly solidified in 3D framework of geopolymer. In the middle of 1990s, J. G. S. Van Jaarsveld and J. G. S. Van Deventer et al. (Davidovits et al 1990; van Jaarsveld et al 1997, 1998, 1999a) also set out to study the solidification effectiveness of geopolymer manufactured from fly ash. The bond mechanism between heavy metal ions and geopolymer matrix is also simply explained on the basis of the XRD, IR, MAS-NMR and leaching results. Recently, the European research project GEOCISTEM (van Jaarsveld et al 1999b) successfully tested geopolymerization technology in the context of the East-German mining and milling remediation project, carried out by WISMUT. Another research project into the solidification of radioactive residues was jointly carried out by Cordi-Geopolymer and Comrie Consulting Ltd., and was documented in reference (European 1997).

(2) Fire resistance Recently The Federal Aviation Administration (FAA), USA, and the Geopolymer Institute of Cordi-Geopolymere SA, France (Comrie 1988), have jointly initiated a research program to develop low-cost, environmentally-friendly, fire resistant matrix materials for use in aircraft composites and cabin interior applications. The flammability requirement for new materials is that they withstand a 50 kW/m^2 incident heat flux characteristic of a fully developed aviation fuel fire penetrating a cabin opening, without propagating the fire into the cabin compartment. The goal of the program is to eliminate cabin fire as cause of death in aircraft accidents. As with this program, the fire resistance properties of geopolymer reinforced by various types of fiber such as carbon fiber, glass fiber, SiC fiber etc. were tested and the fire-proof mechanics were also analyzed. In addition, the comparisons were made among geopolymer composite and carbon-reinforced polyester, vinyl, epoxy, bismaleinide, cyanate ester, polyimide, phenolic, and engineering thermoplastic laminates. The test results showed that these organic large molecular polymers ignited readily and released appreciable heat and smoke, while carbon-fiber reinforced Geopolymer composites did not ignite, burn, or release any smoke even after extended heat flux exposure. On the basis of these fireproof studies, some non-flammable geopolymer composites for aircraft cabin and cargo interiors were produced and introduced on November 18, 1998, in Atlantic City, NJ, USA.

(3) Archeological research In the 1970s Professor J. Davidovits proposed a controversial theory that documented in a book (Lyon 1994) and has since gained widespread support and acceptance. He postulated that the great pyramids of Egypt were not built by natural stones, but that the blocks were cast in place and allowed to set, creating an artificial zeolitic rock with geopolymerization technology. He collected a great amount of evidences which come from old ancient Egyptian literatures and samples in sites to confirm his geopolymerization theory. From then on, many experts began to focus their concerns on geopolymer studies. Some related papers [(Davidovits 1987; Morris 1991; Campbell et al 1991; Folk et al 1992; Mckinney 1993)] and patents were also published.

2.2.3.2 Work done in HKUST

(1) Reaction mechanism

Much attempt on formation mechanism has been made since the invention of geopolymer. However, only one described formation mechanism was proposed by Davidovtis. He believed that the synthesis of geopolymer consist of three steps. The first is dissolution of alumino-silicate under strong alkali solution. The second is reorientation of free ion clusters. The last is polycondensation. But each step includes many pathways. Taking dissolution step for example, it includes 8 pathways according to the thermodynamics. Different pathway can create different ion clusters that directly determine the final properties of geopolymer. Thus it is very important to understand the actual pathway for producing geopolymer in order to gain insight into the mechanism of geopolymerization. However, until now, these studies are not still done. It is because that the forming rate of geopolymer is very rapid, as a result, these three steps take place almost at the same time, which make the kinetics of these three steps inter-dependent. Thus it is impossible to separate these steps in experimental studies. This leads to the use of molecular simulation to solve these problems.

In our studies, two 6-membered-rings molecular structural models to represent the chemical structure of metakaolinite (main raw material for synthesizing geopolymer) were established in order to quantitatively analysis the formation process of geopolymer, as shown in Figure 49 (a), and Figure 49 (b). Based on these two 6-memberedrings models, all possible dissolution pathways of metakaoline under strongly alkali environment were numerically simulated using quantum mechanics, quantum chemistry, computation chemistry and thermodynamics theories. All possible pathways (Eq. 7 to Eq. 14) involved in the formation process of geopolymer were analyzed, and the enthalpies of each possible pathway were also calculated (Table 14). As a result, the optimum pathways in theory, that is the actually occurring pathways in the geopolymerization process, were determined. During molecular simulation, some interesting phenomena were found, and were explained by experimental results.

$$(Si(OH)_2O)_6 + 3NaOH \Rightarrow (OH)_3Si-(Si(OH)_2O)_3-Si(OH)_3 + HO-Si\equiv(ONa)_3 + H_2O \quad \Delta E1 \quad (7)$$

$$(Si(OH)_2O)_6 + 3KOH \Rightarrow (OH)_3Si-(Si(OH)_2O)_3-Si(OH)_3 + HO-Si\equiv(OK)_3 + H_2O \quad \Delta E2 \quad (8)$$

$$(Si(OH)_2O)_6 + 4NaOH \Rightarrow (OH)_3Si-(Si(OH)_2O)_3-Si(OH)_2-ONa + HO-Si\equiv(ONa)_3 + 2H_2O \quad \Delta E3 \quad (9)$$

$$(Si(OH)_2O)_6 + 4KOH \Rightarrow (OH)_3Si-(Si(OH)_2O)_3-Si(OH)_2-OK + HO-Si\equiv(OK)_3 + 2H_2O \quad \Delta E4 \quad (10)$$

$$(Al^-(OH)_2O)_6 + 3NaOH \Rightarrow (OH)_3Al^--(Al^-(OH)_2O)_3-Al^-(OH)_3 + HO-Al^-\equiv(ONa)_3 + H_2O \quad \Delta E5 \quad (11)$$

$$(Al^-(OH)_2O)_6 + 3KOH \Rightarrow (OH)_3Al^--(Al^-(OH)_2O)_3-Al^-(OH)_3 + HO-Al^-\equiv(OK)_3 + H_2O \quad \Delta E6 \quad (12)$$

$$(Al^-(OH)_2O)_6 + 4NaOH \Rightarrow (OH)_3Al^--(Al^-(OH)_2O)_3-Al^-(OH)_2-ONa + HO-Al^-\equiv(ONa)_3 + 2H_2O \quad \Delta E7 \quad (13)$$

$$(Al^-(OH)_2O)_6 + 4KOH \Rightarrow (OH)_3Al^--(Al^-(OH)_2O)_3-Al^-(OH)_2-OK + HO-Al^-\equiv(OK)_3 + 2H_2O \quad \Delta E8 \quad (14)$$

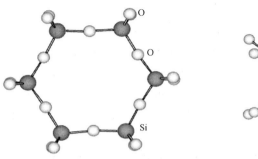 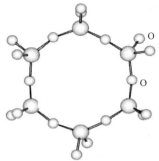

(a) 6-member rings structure cluster of SiO₄ terrahedron   (b) 6-member rings structure cluster of AlO₄ tetrahedron

Figure 49   Molecular structure representing model of metakaolinite

Reaction heat of single 6-member rings structure model under strongly alkaline solution   Table 14

a. Single 6-member rings of SiO_4 tetrahedra

The molecular structural unit	Formation enthalpy (a. u.)	Reaction enthalpy (kJ/mol)			
		$\Delta E1$	$\Delta E2$	$\Delta E3$	$\Delta E4$
$(Si(OH)_2O)_6$	−1491.44763	−5.48908	12.12039	−36.22681	−19.22697
$(OH)_3Si-(Si(OH)_2O)_3-Si(OH)_3$	−1294.64502				
$(OH)_3Si-(Si(OH)_2O)_3-Si(OH)_2-ONa$	−1385.43022				
$(OH)_3Si-(Si(OH)_2O)_3-Si(OH)_2-OK$	−1370.88091				
$HO-Si\alpha(ONa)_3$	−500.93548				
$HO-Si\alpha(OK)_3$	−437.84919				
NaOH	−119.29816				
KOH	−104.13922				
H_2O	−59.25069				

b. Single 6-member rings of AlO_4 tetrahedra

The molecular structural unit	Formation enthalpy (a. u.)	Reaction enthalpy (kJ/mol)			
		$\Delta E1$	$\Delta E2$	$\Delta E3$	$\Delta E4$
$(Al^-(OH)_2O)_6$	−619.66576	−299.7974	−245.78806	−302.69174	−235.0346
$(OH)_3Al^--(Al^-(OH)_2O)_3-Al^-(OH)_3$	−776.45042				
$(OH)_3Al^--(Al^-(OH)_2O)_3-Al^-(OH)_2-ONa$	−839.39222				
$(OH)_3Al^--(Al^-(OH)_2O)_3-Al^-(OH)_2-OK$	−810.58557				
$HO-Al^-\equiv(ONa)_3$	−441.65654				
$HO-Al^-\equiv(OK)_3$	−342.17037				
NaOH	−119.29816				
KOH	−104.13922				
H_2O	−59.25069				

(2) Microstructure characterization

The structure characteristics of products directly determine the final mechanical and durability properties. The case is also true for geopolymer. Many researchers have investigated its microstructure using different advanced techniques. But because geopolymer is a type of amorphous 3D materials with complex composition, It is very difficult to quantitatively measure the exact arrangement and chemical atmosphere of different atomic in geopolymer. If we want to solve this difficulty, we should have to turn to statistical theories for establishing its molecular model. But unfortunately, until now, these studies are not still been done. Therefore, the structural nature of geopolymer is not yet understood thoroughly.

In our studies, many microstructure techniques, such as XRD, IR, XPS, MAS-NMR, ESEM-EDXA and TEM were used to investigate the structural characterization in atomic, molecular, nanometer, micrometer and centimeter scales. The relationship between geopolymers and the corresponding zeolites were also investigated. The inter-transformation between geopolymers and zeolites can be realized under specified conditions. On basis of these results, the micro-structure of geopolymers can be clearly characterized: geopolymer is an amorphous 3D alumino-silicate material, which is composed of AlO_4 and SiO_4 tetrahedra lined alternatively by sharing all oxygen atoms. Positive ions (Na^+, K^+) are present in the framework cavities to balance the negative charge of Al^{3+} in four-fold coordination. In addition, 3D statistical models (Figure 50) were also simulated according to the decomposition results of MAS-NMR spectra.

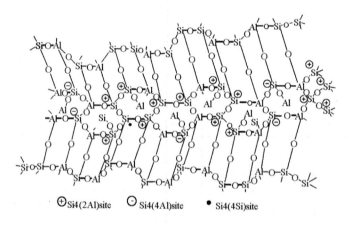

Figure 50 Statistical structure model of K-Geopolymer

(3) Mechanical properties

More concerns have been received on the solidification of heavy metal ion and nuclear waste and fire resistance since 1990, but at present, less experimental data is available for the systematical investigation on mechanical properties and durability. Up to now, more than 100 geopolymer concrete specimens were prepared to study mechanical behaviors such as compressive, flexural, splitting tensile, shear strength and their stress-strain responds. PSS geopolymer concrete has the highest mechanical performance among various geopolymer concretes, next to PSDS, and PS has the lowest strength. Another 147 geopolymer concrete or mortar specimens were also produced to investigate the durability properties such as chloride ions permeability, resistance to freezing and thawing cycles, resistance to chemical attack including HCl, H_2SO_4 and Na_2SO_4 attack, long-term volume stability, and alkali aggregate reaction (AAR). At present, the durability tests are still under way.

2.2.4 The Development of MPCs

2.2.4.1 Work done by others

The phosphate bonding has been known for about a century, since the advent of dental cement formulations. In refractory industry, the properties of cold-setting and heat-setting compositions were used as chemically bonded refractory. According to the comprehensive studies of Kingery in 1950, The phosphate bonding can be classified as (1) zinc-phosphate bond, (2). Silicate-phosphoric acid bond, (3) oxidephosphoric acid bond, (4) acid phosphate bond, and (5) metaphosphate/polyphosphate bond (Kingery 1950). The oxides such as magnesium, aluminum, zirconium, will react with phosphoric acid or acid phosphate at room temperature, forming a coherent mass, setting quickly and giving high early strength. The hydration system based on magnesia and ammonia phosphate (Kingery 1950; Yoshizake et al 1989; Seehra et al 1993; Yang et al 2002; Singh et al 1997; Wagh et al 1999) had drawn most of the attention in the past years.

From 1970s, many patents using the reaction of magnesia and acid ammonia phosphate have been granted for rapid repair of concrete. The variation in patents arises from the use of different raw materials, inert materials to reduce cost, and retarders to control the reaction rate. Most claims are supported by a few examples cited in the patents without systematic scientific approach. From the middle of 1980s, systematic studies about the system of magnesia and ammonia phosphate were made by researchers (Kingery 1950; Yoshizake et al 1989; Seehra et al 1993; Yang et al 2002). The hydration products, setting process, and strength development were the main content among those previous investigations. Very few papers focused on the durability of the system (Yoshizake et al 1989; Seehra et al 1993; Yang et al 2002). Entering the middle of 1990s, it was found that MPC can incorporate with lot of industrial waste and solidify toxic waste (Singh et al 1997; Wagh et al 1999; Davidovits 1993). Therefore, MPC became a forceful candidate for sustainable development. The benefits in environment may be obtained from two aspects, (1) the non-toxic industrial waste can be recycled to useful building materials, and (2) many toxic and radioactive wastes treated difficultly with traditional process can be treated by MPC easily. This function endues MPC more promising use in the future, especially to the sustainable development of the modern society.

About the durability of MPCs, research work had been done by other investigators mainly includes, superior durability such as freezing-thawing and scaling resistance, protection steel from corrosion, better bond properties with waste organic materials, transfer non-contaminated industrial wastes into useful construction materials, and stabilization of toxic or radioactive wastes.

The deterioration of concrete pavements is mainly cause by frost action in cold areas. It is severely amplified by the use of deicer chemicals. The repair material must possess high frost/deicer-frost resistance. The result shown that MPC have very high deicer-frost resistance (Yoshizake et al 1989; Yang et al 2002). The scaling does not occur on the surfaces of MPC materials until 40 freeze-thaw cycles. The regime of freeze-thaw cycling was achieved with cooling rate of about 0.5℃/min. for 4 hours at -20 ± 2℃ and then thawed for 4 hours at 20 ± 5℃. A 3% NaCl solution was used as the deicer solution. The studies showed that the freezing thawing resistance of MPCs was basically equal to the well air-entrained PC concrete in general.

Steel corrosion in PC concrete was a very serious problem. However, MPC is inhibitor of corrosion of steel, forming an iron phosphate film at the surface of the steel. The pH of hardened MPC mortar is 10 to 11, this may be considered as contributing to inhibition of reinforcing steel corrosion. In addition, the ratio of permeability of MPC to PC concrete is 47.3%, or more than double in resistance to permeation (Yoshizake et al 1989). Abrasion resistance test shown that MPC mortar possesses approximately double the abrasion resistance compared with slab-on-grade floor concrete and to be nearly equal to that of pavement concrete (Yoshizake et al 1989; Seehra et al 1993;). With respect to chemical corrosion resistance, in the case of continuous immersion of specimens in sulphate solution and potable water, results indicate that MPC mortar patches will practically remain durable under sulphate and moist conditions.

A wide range of waste particle sizes can be utilized when producing structural products using the MPC. Styrofoam materials are the candidate for optimal results. The styrofoam particles can be completely coated with a thin, impermeable layer of the MPC. The uniform coating of the styrofoam particles not only provides structural stability but also confer resistance to fire, chemical attack, humidity and other weathering conditions. The styrofoam insulation material provides superior R values. Furthermore, wood waste (suitable size range from 1 to 5 mm long, 1 mm thick and 2 to 3 mm wide) can be bonded with MPC to produce particleboard having flexural strength. For example, samples containing 50wt% of wood and 50wt% of binder display approximately 10.4 MPa in flexural strength. Samples containing 60wt% and 70wt% of wood exhibit flexural strength of 2.8 and 2.1 MPa, respectively. Once the wood and binder is thoroughly mixed, the samples are subjected to pressurized molding on the order of approximately 18.3 MPa, and for approximately 30 to 90 minutes.

With the progress of modern civilization, the living conditions had been greatly improved; at the same time, however, a large amount of industrial waste (including toxic and nontoxic) had been produced. MPC can bind lot of non-toxic industrial waste to useful construction materials. If the wastes were toxic, MPC can solidify and stabilize them. There is a significance to recycle and/or stabilize the waste, especially under the condition of natural resources becoming more and more deficient. The waste is in various forms in aqueous liquids, inorganic sludge, particles, heterogeneous debris, soils, and organic liquids. However, there was only a few part of the total waste can been recycled, such as fly ash, red mud was manufactured blended Portland cement or concrete. Most of the wastes need to be solidified and stabilized. Because of the divers nature of the physical and chemical composition of these wastes, no single solidification and solidification technology can be used successfully treat and dispose of these wastes. For example, the low-level wastes contain both hazardous chemical and low-level radioactive species (Singh et al 1997). To stabilize them requires that contaminants of two kinds be immobilized effectively. Generally, the contaminants are volatile compounds and hence cannot be treated effectively by high-temperature processes.

In a conventional vitrification or plasma hearth process, such contaminants may be captured in secondary waste stream or off-gas particulates that need further low temperature treatment for stabilization. Also some of these waste streams may contain pyrophorics that will ignite spontaneously during thermal treatment and thus cause hot spots that may require expensive control system and equipment with demanding structure integrity on. Therefore, there is a critical need for a low temperature treatment and stabilization technology that will effectively treat the secondary wastes generated by high-temperature treatment process and waste that are not amenable to thermal treatment. Now, those wastes can be successfully solidified by magnesia phosphate cement, or chemically bonded phosphate ceramics (CBPC) (Singh et al 1997). Other forms of waste, such as ashes, liquids, sludge and salts can be also solidified by MPC.

MPC is very extremely insoluble in ground water and this will protect ground water from contamination by the contained wasted. The long-term leaching tests conducted on magnesium phosphate systems shown that these phosphates are insoluble in water and brine. The radiation stability of MPC is excellent (Wagh et al 1999). Changes in the mechanical integrity of the materials were not detected after gamma irradiation to cumulative dosage of 10^8 rads.

2.2.4.2 Work done in HKUST

From the late of 2001, we started the project of new MPC system based on potassium phosphate. The main advantages of new system are binding lot of industrial waste and no ammonium gas was emitted. Up to now, the mechanical and chemical properties, hydration process and mechanism, durability and binding properties with old PC concrete had been investigated. Here the mechanical property and durability will be mainly introduced.

(1) Mechanical properties

Strength development of MPC made of different hard burnt magnesia Two kinds of hard burnt magnesia and a Class F fly ash (FA). The magnesia contains 89.6% magnesium oxide was named M9, whose average size of particle was 30.6 μm. The other contains 71.6% magnesium oxide was named M7, whose average size of particle was 59.8 μm.

Compressive strength versus fly ash content for MPC mortars at 3, 7, and 28 days is presented in Figure 51 for M7 and M9 series. From the figures, it can be seen that for the two series, the MPC mortars with 30%-50% fly ash exhibit higher strength than the sample without fly ash, and the highest strength occurred at the samples with 40% fly ash. To the mortars made from M9, from 10% -40% of FA, the strength gradually increases with the addition of fly ash at all ages (except M9F1 at 28 days has lower strength than that of M9F0). When the fly ash content surpasses 40%, the strength decreases. But, the strength of sample with 50% fly ash is still comparable to that of sample with 30% fly ash.

The modulus of elasticity of MPC mortar M9F0 and M9F4 was determined at age of 7 days. The elastic modulus of M9F0 and M9F4 is 27.47 and 31.85 GPa, respectively.

The compressive strength of MPC mortar at 1, 4, 7, and 24 hours under room temperature is

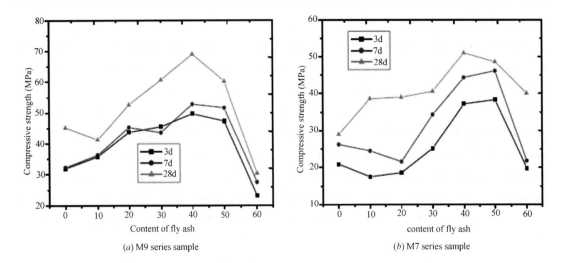

Figure 51 Strength development of MPC mortar sample

shown in Figure 52 (a). The specimens containing 40% FA had very fast development of strength than the specimens not containing FA, And Figure 52 (b) was the strength development of MPC (AF content was 40%) mortar at 1, 3, 7, and 24 hours under negative temperature (After the specimens were formed they were put into the environmental chamber immediately together with molds. And they were demolded after one hour.) The test results show that FA has the effect of reinforcement to strength, even if MPC mortar were cured under very low temperatures.

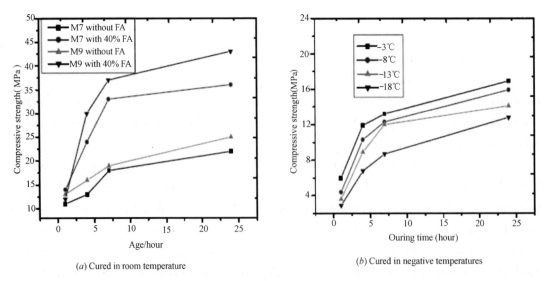

Figure 52 Early strength development of MPC under different temperatures

(2) Durability

(a) Deicer frost scaling resistance

The deicer used here is calcium chloride ($CaCl_2$) and the concentration in water is 4% by weight of water. The MPC mortar sample and PC mortar samples together immersed completely in $CaCl_2$ solution in a plastic box, which has no cover. Then the box was placed inside the environmental room, KATO for freezing and thawing. After these samples were frost ($-18°C$) for 16 hours, they were removed from the environmental room and placed in laboratory air at normal condition for 8 hours, which is a freezing-thawing cycle. Add water each cycle as necessary to maintain the proper depth of solution. Repeat the cycle daily. The surfaces of samples were flushed off thoroughly at the end each 5 cycles. Compressive strength was determined of MPSC mortar after following every curing stage: (I) The sample of MPSC mortar formed after 3 days, and 7 days for PC mortar; (II) These samples were suffered 30 freezing-thawing (FT) cycles; (III) The same above samples were cured 30 days under normal conditions; (IV) After that, the samples were tested after aging 60 days under normal condition.

The compressive strength of MPC mortar was 54.1 MPa after hydration 3 days, and strength of PC mortar was 59.8 MPa for after hydration 7 days. They had the comparable strength when they were suffered FT cycles at same time. After 30 FT cycles, the surface of PC mortar samples were severe scaled and cannot be used to determine compressive strength (due to the very rough surfaces). However, the surface of MPC mortars is intact, smooth as the surfaces before FT cycles. This indicates that MPSC mortar has a superior deicer scaling resistance to PC mortar. The compressive strength test result, Figure 53, showed that the strength of MPC sample increased a little after 30 FT cycles, comparing to the 3-day strength. Furthermore, the strength can increase continually when MPSC samples were set in normal condition after the FT cycles. This shows that the microstructure of MPC mortar was not damaged also after 30 FT cycles.

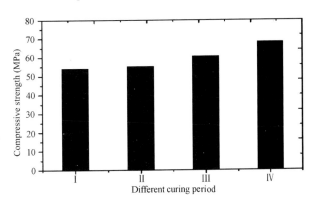

Figure 53 Strength of MPC after FT salt scaling cycles

The resistance of concrete to freezing and thawing mainly depends on its degree of saturation and the pore system of the hardened cement paste. If concrete is never going to be saturated, there is no danger of damage from freezing and thawing. Even in a water cured specimen, not all residual space is water-filled and indeed this is why such a specimen does not fail on first freezing. Space available for expelled water must by close enough to the cavity in which ice is being formed, and this is the basis of air entrainment: if the hardened cement paste is subdivided into sufficiently thin layers by air bubbles, it has no critical saturation.

When the dilating pressure in the concrete exceeds its tensile strength, damage occurs. The extent of the damage varies from surface scaling to complete disintegration as ice is formed, starting at the exposed surface of the concrete and progressing through its depth. Each cycle of freezing causes a migration of water to locations where it can freeze. These locations include fine cracks, which become enlarged by the pressure of the ice and remain enlarged during thawing when they become filled with water. Subsequent freezing repeats the development of pressure and its consequences. When salts are used for deicing road or bridge surface, some of these salts become absorbed by the upper part of the concrete. This produces a high osmotic pressure, with a consequent movement of water toward the coldest zone where freezing takes place, which aggravates the scaling condition of concrete.

The reason of MPSC mortar possesses higher deicer scaling resistance than PC mortar can be attributed two aspects. First is less water inside the former than in the latter. Usually, the water to binder ratio of MPSC mortar was around 0.20, but for Portland cement mortar it was around 0.44. Therefore, the former has denser microstructure than the latter. MIP test result indicates that the total porosity of MPC paste is about 9 percent by volume, while the total porosity of PC paste is about 20 percent by volume. The second reason is that there are many closed pores inside the MPC paste, very like the entrained PC concrete. These closed pores can prohibit water permeates into the inner of MPC matrix. The specimens were far from saturation of water.

(b) Wet-dry cycles in fresh water and natural sea water

The compressive strengths were determined at the end of each following curing stages: (I) After the MPC mortar samples were formed 3 days; (II) They were immersed in fresh water (FW) and sea water (SW) respectively, under room temperature. One wet-dry cycle kept 24 hours, including 12 hours in air and immersing in water 12 hours. The samples were put in water and taken out manually every day during wet-dry cycle; (III) Then those samples were set in lab air for another 30 days and test strength; (IV) After that, the samples were immersed in FW and SW for another 60 days, respectively.

Figure 54 shows the strength development after wet-dry cycles in FA and SW. After 30 wet-dry cycles in FW and SW, the strength of MPC samples even increased a little. After then, the strengths of MPSC mortars recovered and continued increasing when set in lab air for another 30 days. However, when the MPSC samples were immersed in FA and SW for 60 days again, the strength reduced. The result shows that there is no inverse effect under the wet-dry cycle in FW or SW. However, the strength reduced some when they were immersed in water for a long time, though the deduction of strength was not larger than 17.0%. MPCs were suitable utilized in the environments that are dry or wet-dry alternatively.

(c) Sulphate attack resistance test

The compressive strength was determined after each following stage: (I) After 3 days formed for MPC, 7 days for the PC mortar samples after they were molded, respectively; (II) The MPC mortars were immersed

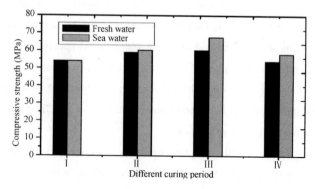

Figure 54 Strength of MPC after wet-dry cycles in fresh water and sea water

completely in solution of sodium sulphate (NS) and magnesium sulphate (MS) respectively, their concentration is 5 wt%. MPC mortars were immersed 30 days in the two solutions; (III) The same samples set in normal condition for another 30 days; (IV) Afterwards, all the mortars immersed in the corrosive solutions for 60 days; (V) At last, those specimens were set in lab air for 90 days.

After immersing 30 days in the NS solutions, comparing with the strength at 3 days, the strength of MPC sample increased. But, after 30 days immersed in solution of MS, strength of CON decreased 7.2%. However, the strength loss of PC mortar is 29.4%; see Figure 55.

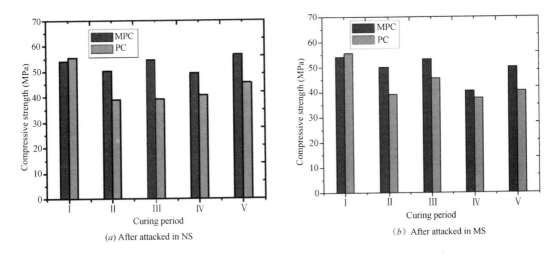

Figure 55 Strength development after attacked by sulphate solutions

After then, the corroded samples set in normal condition, the strengths of MPC mortar increase continually, and surpass their strengths at 3d. However, the strength of PC sample basically did not recover anymore. This indicated that the microstructure of MPSC can recover when separated from the attacking agents; however, the microstructure of PC had been damaged in the attacking agents. Then, these samples were immersed the sulphate solution again for another 30 day, respectively. The strength of MPC and PC decreased once more. However, after the specimens were put in lab air for another 90 days, the strength of MPC recovered much more (even catch up with the un-eroded specimens), the strength of PC mortar only recovered a little. From the results, it can be deduced that MPC sample has more resistance to NS attach than MS attack. In spite of which type of sulphate solution, MPC possesses high salt attack resistance than PC mortar in the present research.

2.2.5 Remarks

Geopolymer is a type of amorphous alumino-silicate cementitious material. Geopolymer can be synthesized by polycondensation reaction of geopolymeric precursor, and alkali polysilicates. Comparing to Portland cement, the production of geopolymers consume less energy and almost no CO_2 emission. Geopolymers are not only energy efficient and environment friendly, but also have a relative higher strength, excellent volume stability, better durability, good fire resistance, and easy manufacture process. Thus geopolymer will become one of the prospective sustainable cementitious materials in 21st century.

As new sustainable cementitious materials, MPCs have much beneficial advantages in environments. Not only non-toxic wastes can be transferred into useful building materials, but also the toxic and/or radioactive waste can be solidified and stabilized safely with MPCs. Furthermore, MPC can incorporate with natural organic fibers to form composites, light weight or insulation materials. These natural organic fibers are not suitable bonding with Portland cement. This is very meaning to the recycling of agricultural organic fibers in larger degree.

MPCs are high early strength and quick setting, very suitable to repair highways, airport runways, and bridges that are busy for transportation. The short waiting time for repairing means that saving lot of costs. In addition, MPCs have very good durability. Such as higher freezing-thawing and scaling resistance, low permeability, higher abrasion resistance, higher ability of sulphate attack resistance. MPCs are very suitable utilized in severe environments, such as frosty areas and corrosive conditions.

3. MULTI DISCIPLINARY DEVELOPMENT

3.1 Nanotechnology in Construction (Sobolev et al 2005a, 2005b)

Nanotechnology deals with the production and application of physical, chemical and biological systems at scales ranging from a few nanometers to submicron dimensions. It also deals with the integration of the resulting nanostructures into larger systems and investigation of matter to individual atoms. When the dimensiona of a material are decreased from macrosize to nanosize, significant changes in electronic conductivity, optical absorption, chemical reactivity and mechanical properties occur. With decrease in size, more atoms are located on the surface of the particle. Noanpowers have a remarkable surface area (Figure 56). The surface area imparts a serious change of surface energy and surface morphology. All these factors alter the basic properties and the chemical reactivity of the nanomaterials. The change in properties causes improved catalytic ability, tunable wavelength-sensing ability and better-desinged pigments and pains with self-cleaning and self-healing features. Nanosized particles have been used to enhance the mechanical performance of plastics and rubbers. They make cutting tools harder and ceramic materials more ductile. For example, ductile behavior has been reported for nanophase ceramics, such as titania and alumina, processed by consolidation of ceramic nanoparticles. New nanomaterials based on metal and oxides of silicon and germanium have demonstrated superplastic behavior, undergoing 100% ~ 1000% elongation before failure.

Nanotechnology has changed and will continue to change our vision, expectations and abilities to control the material world. These developments will definitely affect construction and also the field of construction materials. The major achievements in this domain include: the ability to observe the structure at its atomic level and measure the strength and hardness of micro- and nanoscopic phases of composite materials; discovery of a highly ordered crystal nanostructure of "amorphous" C-S-H gel; development of paints and finishing materials with self-cleaning properties, discoloration resistance, anti-graffiti protection, high scratch and wear resistance; self-cleaning materials based on photocatalyst technology; nanometer-thin coatings protecting carbon steel against corrosion and enhancing thermal insulation of window glass; smart stress-sensing composites; and others. Among new nano-engineered polymers are highly efficient superplasticizers for concrete and high strength fibers with exceptional energy absorbing capacity. Nanoparticles, such as silicon dioxide, were found to be a very effective additive to polymers and also concrete, a development recently realized in high-performance and self-compacting concrete with improved workability and strength.

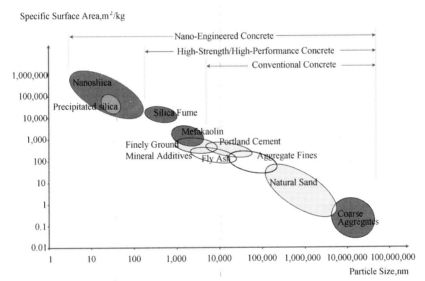

Figure 56 The particle size and specific surface area scale related to concrete materials show the general trend to use finer materials (Sobolev et al 2005a, 2005b)

Portland cement, one of the largest commodities consumed by mankind, is obviously the product with great- but at the same time- not completely explored potential. Better understanding and precise engineering of an extremely complex structure of cement based materials at the nanolevel will apparently result in a new generation of concrete, stronger and more durable, with desired stress-strain behavior and possibly possessing the range of newly introduced "smart" properties such as electrical conductivity, temperature-, moisture-, stress-sensing abilities. At the same time this new concrete should be sustainable, cost and energy effective-in essence exhibiting the qualities modern society demands. Nano-binders or nano-engineered cement based materials with nano-sized cementitious component or other nano-sized particles are the next ground-breaking development.

3.1.1 Impact on construction materials

Nano-chemistry with its "bottom-up" possibilities offers new products that can be effectively applied in concrete technology. One example is related to the development of new superplasticizers for concrete, such as polycarboxylic ether polymer based PCE Sky. This product was recently nanodesigned by Degussa targeting the extended slump retention of concrete mixtures. It was proposed that, when nanoparticles are incorporated into the conventional building materials, such materials can possess advanced or smart properties required for the construction of high-rise, long-span or intelligent civil and infrastructure systems. For example, silicon dioxide nanoparticles (nanosilica) can be used as an additive for high-performance and self-compacting concrete improving workability and strength of concrete. The combination of carbon nanotubes and conventional polymer based fibers and films is another challenge. For example, the incorporation of 10% SWNTs (single walled nanotube) into the strongest man-made fiber Zylon, resulted in the new material with 50% greater strength. Resent research reported by A. B. Dalton et al. from University of Texas at Dallas introduces a further breakthrough related to SWNT-reinforced fibers. The composite fiber comprising 60% of nanotubes and 40% of polyvinyl alcohol (PVA) produced by continuous spinning with modified coagulation method, achieved the strength of 1.8 GPa and higher. Clearly, application of these new super-fibers in cement composites is very promising.

3.1.2 Concrete with nanoparticles

Mechanical properties of cement based materials with nano-SiO_2, TiO_2, Fe_2O_3 and carbon nano-tubes were recently studied. Experimental results demonstrated an increase in compressive and flexural strength of mortars containing nano-particles. Based on the available data, the beneficial action of the nano-particles on the microstructure and performance of cement based materials can be explained by the following factors:

- Well-dispersed nano-particles increase the viscosity of the liquid phase helping to suspend the cement grains and aggregates, improving the segregation resistance and workability of the system;
- Nano-particles fill the voids between cement grains, resulting in the immobilization of "free" water ("filler" effect);
- Well-dispersed nano-particles act as centers of crystallization of cement hydrates, therefore accelerating the hydration;
- Nano-particles favor the formation of small-sized crystals (such as $Ca(OH)_2$ and AF_m) and small-sized uniform clusters of C-S-H;
- Nano-SiO_2 participates in the pozzolanic reactions, resulting in the consumption of $Ca(OH)_2$ and formation of extra C-S-H;
- Nano-particles improve the structure of the aggregates' contact zone, resulting in a better bond between aggregates and cement paste;

- Crack arrest and interlocking effects between the slip planes provided by nano-particles improve the toughness, shear, tensile and flexural strength of cement based materials.

3.1.3 Expected developments

Vast progress in concrete science is to be expected in coming years by the adaptation of new knowledge generated by a quickly growing field of nanotechnology. The development of the following concrete-related nanoproducts can be anticipated:

- Catalysts for the low-temperature synthesis of clinker and accelerated hydration of conventional cements;
- Grinding aids for superfine grinding and mechano-chemical activation of cements;
- Binders reinforced with nano-particles, nano-rods, nano-tubes (including SWNTs), nano-dampers, nano-nets, or nano-springs;
- Binders with enhanced/nanoengineered internal bond between the hydration products;
- Binders modified by nano-sized polymer particles, their emulsions or polymeric nano-films;
- Bio-materials (including those imitating the structure and behavior of mollusk shells);
- Cement based composites reinforced with new fibers containing nanotubes, as well as with fibers covered by nano-layers (to enhance the bond, corrosion resistance, or introducing the new properties, like electrical conductivity etc.);
- Next generation of superplasticizers for "total workability control" and supreme water reduction;
- Cement based materials with supreme tensile and flexural strength, ductility and toughness;
- Binders with controlled internal moisture supply to avoid/reduce micro-cracking;
- Cement based materials with engineered nano- and micro-structures exhibiting supreme durability;
- Eco-binders modified by nanoparticles and produced with substantially reduced volume of portland cement component (down to 10% – 15%) or binders based on the alternative systems (MgO, phosphate, geopolymers, gypsum);
- Self-healing materials and repair technologies utilizing nano-tubes and chemical admixtures;
- Materials with self-cleaning/air-purifying features based on photocatalyst technology;
- Materials with controlled electrical conductivity, deformative properties, non-shrinking and low thermal expansion;
- Smart materials, such as temperature-, moisture-, stress-sensing or responding materials.

3.1.4 Emerging research

Incorporation of nano-tubes into the cement matrix would result in a ductile and energy absorbing concrete. The performance of such concrete can be further enhanced by the addition of polymers and nano-structured materials, such as nano-rods, nano-dampers, nano-nets, nano-springs or nano-engineered fibers.

Nano-binder can be proposed as a logical extension of the two concepts: Densified System with Ultra Fine Particles (DSP) and Modified Multi-Component Binder (MMCB) extended to the nano-level. In these systems the densification of binder is achieved with the help of ultra-fine particles: silica fume (SF) dispersed with superplasticizer (SP) in DSP and finely ground mineral additives (FGMA) and SF modified by SP in MMCB; these particles fill the gaps between cement grains. In these systems portland cement component is used at its "standard" dispersion to provide the integrity of composition. In contrast to DSP and MMCB, the nano-binder can be designed with a nano-dispersed cement component applied to fill the gaps between the particles of mineral additives (including FGMA).

In nano-binder, the mineral additives (optionally, finely ground), acting as the main component, would provide the structural stability of the system and the micro- or nano-sized cementitious component (which can also contain the nano-sized particles other than portland cement) would act as a glue to bind less reactive particles of mineral additives together. Such nano-sized cementitous component can be obtained by the colloidal milling of a conventional (or especially sintered/high C_2S) portland cement clinker (the top-down approach) or by the self-assembly using mechano-chemically induced topo-chemical reactions (the bottom-up approach). Development of nano-binders can lead to more than 50% reduction of the cement consumption, capable to offset the demands for future development and, at the same time, combat global warming.

In addition to nano-binders, the mechano-chemistry and nano-catalysts could change the face of modern cement and concrete industry by the great reduction of clinkering temperature and even realizing the possibility of cold sintering of clinker minerals in mechano-chemical reactors.

3.1.5 Nano-devices in construction (McCoy et al 2005)

3.1.5.1 Workability monitoring

Clearly, good workability is essential for quality concrete construction. Although the slump and flow tests are somewhat successful tools for characterizing concrete consistency, they provide only partial information regarding workability. The workability of concrete is more accurately evaluated by measuring its viscosity, which is usually measured using rheometers. However, these devices are quite expensive and primarily used only in the laboratory; in-field measurement of concrete viscosity therefore is not possible. Research at TIS2 is currently exploring the use of nano-technology-based devices for in-situ monitoring of concrete density and viscosity. Figure 57 shows a wireless densitometer/viscometer which was originally developed by Sandia Laboratory to measure the density and viscosity of oil. TIS2 is miniaturizing and modifying the device for use in fresh concrete. As shown in Figure 57, the device is composed of smooth and textured nano-resonators, a micro resistance temperature detector (RTD) and a power-source-free radio frequency (RF) communication system. With proper packaging and sealing, the device could be embedded into concrete allowing measurements to be made in-situ and online during concrete mixing and placement. Each resonator is composed of a silicon substrate coated with a piezoelectric thin film and patterned metal NEMS transducers. The textured resonator includes 80 nm wide metal gratings, so the response from the rough resonator is different from that of the smooth resonator. When a material (in this case, concrete) comes into contact with the substrate of the resonators, it interacts mechanically and perturbs the resonant frequency and damping of the substrate, with the changes being proportional to the material's viscosity and density. Thus, given calibration data, concrete viscosity and density can be measured. As designed, a separate, external interrogator emits a RF signal which is received by the antenna of the device. The NEMS transducer connected to the sensor's antenna transforms the received signal into a bulk acoustic wave (BAW), which propagates on the substrate and is partially reflected by the gratings. The NEMS transducer then converts the resulting BAW into a RF signal that is transmitted, along with data from the RTD, to the interrogator unit's antenna. The RF signal is processed to extract the response frequency and amplitude data for the device's resonators, using the temperature data from the RTD to compensate for temperature effects.

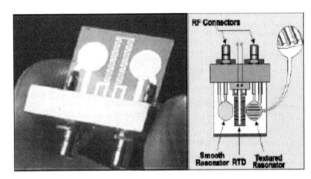

Figure 57　Typical NEMS device for concrete density and viscosity measuring (McCoy et al 2005)

3.1.5.2　Monitoring early-age concrete properties

Certainly, accurate field measurements of early-age concrete properties could lead to enhanced construction efficiency and quality control. It is well established that, by tracking temperature and moisture data, in-place concrete strength can be accurately estimated. These estimates can, in turn, be used for establishing timing of stripping and prestressing operations as well as opening of highways to traffic. Further, research has shown that rapid loss of moisture, including through the hydration process itself, can lead to insufficient strength development and to cracking. Because widespread in-situ measurement of measuring early-age temperature, moisture content could be useful not only for monitoring projects, but for improving the composition of concrete as well, we are developing low-cost wireless devices suitable for embedding into concrete. Figure 58 shows a recently developed sensor for measuring internal humidity and temperature. The sensor comprises four micromachined NEMS cantilever beams (120 μm × 380 μm). A 75 nm thick PolyViol G 2810 film, designed to expand and contract during exposure to water, is bonded to the top of each cantilever beam. In addition, a micro-strain gauge is embedded in each cantilever beam so that strains can be measured using the piezoresistive effect. The four strain gauges are connected in a Wheatstone Bridge configuration directly on the sensor chip. An onchip semiconductor (thermistor) temperature sensor is also bonded to the top of the chip. During sensor operation, water vapor molecules are absorbed into the sensing film surface, producing shear stresses at the film/beam interface and therefore causing the cantilever beams to deflect. This deflection is measured as a proportional resistance change in the embedded strain gauges, effectively measuring the water concentration. Thermistor data are used to measure concrete temperatures and for temperature compensation of strain data.

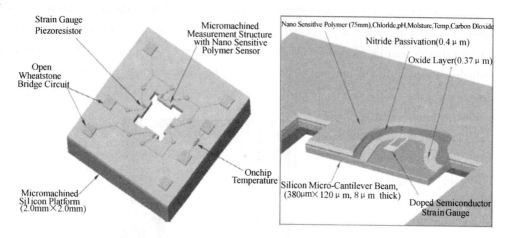

Figure 58　Typical layout of NEMS for moisture and temperature, and other environmental parameters sensing ((McCoy et al 2005))

3.1.5.3　Detecting environmental damage

For concrete structures, the mechanisms for corrosion, carbonation, freeze/thaw damage, and alkali-silica reaction (ASR) are coupled and therefore create a highly complex system. For example, the corrosion of steel in concrete is affected by chloride (Cl^-) ions, temperature, acidity (pH), carbon dioxide (CO_2) and relative humidity (RH), requiring a minimum RH of approximately 50%. ASR and freeze-thaw mechanisms affect local pH levels, are functions of temperature, and also require a minimum RH of 50%. The monitoring of pH, Cl^-, CO_2, temperature, strain, and RH could therefore potentially allow tracking the onset and progress of depassivation and corrosion of steel, the expansion of ASR gel, and the pore pressures generated during freezing and thawing cycles. The sensor shown in Figure 58 could be used to measure many of the above environmental parameters using different sensitive nano-polymers. The pH sensitive nano-polymer consists of poly (methacrylic acid) with poly (ethylene glycol) dimethacrylate; the chloride sensitive nano-polymer is a trimethincyanin, a family of J-aggregation cyanines dye; and the CO_2 sensitive nano-polymer is a polyethyleneimine.

3.1.5.4　Wireless acquisition of NEMS data

McCoy et al (2005) evaluated a system for wireless acquisition of data from humidity/temperature NEMS sensors. Figure 59 (a) shows an embeddable sensing node containing a NEMS humidity/temperature sensor, a rectifier, a 2 bit A/D converter with a sampling rate of 80 Hz, a

110 dB common-mode rejection ratio amplifier, a 16 bit memory and an antenna, all packaged in the form of a 25 mm diameter ×6 mm thick aggregate particle (Figure 59 (b)). The interrogator consists of oscillator, demodulator/level shifter, data logger and antenna (Figure 59 (c)). The smart aggregates were embedded in 300 mm × 300 mm × 200 mm concrete (w/cm = 0.6) blocks. As the blocks cured at approximately 21 ℃ and 50% RH, the sensing node outputs were recorded. In these tests, the interrogation antenna was placed 100 mm above the surface and the sensing node outputs were captured by the data acquisition unit and presented in the form of a graph showing the evolution of these outputs over time.

(a) NEMS sensing node layout

(b) packaged sensing node

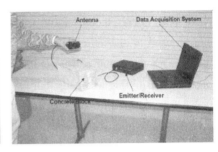
(c) wireless concrete curing monitoring system

Figure 59 Wireless sensing for concrete moisture/temperature measuring (McCoy et al 2005)

The typical calibrated humidity and temperature response of the embedded sensor is shown in Figure 60 for the first 20 days of curing. As can be seen, the measured temperature and humidity reflect the known behavior of concrete, with the temperature reaching a maximum value of 37 ℃ after 15 hours of curing, then decreasing and stabilizing at about room temperature. Also, the internal humidity gradually decreased after 2 days of curing.

Till now three workshops about nanotechnology in construction were held, in Paisley (in June 2003), in Bilbao (Nov, 2005) and Gainesville (Aug. 2006), respectively. According Scrivener mentioned (Partl et al 2006), 21 Academic partner and 12 industrial partners from 11 European countries join the core research program of nanocement. It can be predicted that nanotechnology may bring new opportunity for the development of construction and building materials.

Actually, nanotechnology can be used not only in construction materials,

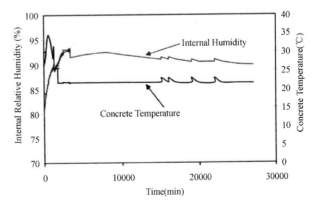

Figure 60 Typical measured moisture and temperature using wireless NEMS sensing node (McCoy et al 2005)

but also in other building materials, such as self-cleaning window and self-cleaning coating. The self-cleaning window, tile coating employs the film of nano titanium dioxide coving the surface of glass to realize the cleaning function.

3.2 Dynamic Shading Window System (DSWS)

A team of researchers at Rensselaer Polytechnic Institute has developed the first of its kind solar-powered, integrated window system that could significantly reduce dependency on the same energy grid that caused the biggest power outage in U.S. history. Designed to function as a shading system, the Dynamic Shading Window System (DSWS) (in Figure 61) uses a newly developed solar-energy technology to convert the sun's light and diverted heat into storable energy that can be used to also efficiently heat, cool, and artificially light the same office building. It acts in three ways: as a solar cell for collecting energy through photovoltaic cells (active solar), by heating and distributing water through clear plastic tubes (passivesolar), and a "dynamic" louver system for regulating the passage of light into a building.

The DSWS system is made of clear plastic panels that fit in between two panes of glass. On each panel are dozens of small, pyramid-shaped units, or "modules," made from semi-translucent focusing plastic lenses, that track the motion of the sun. Sensors, embedded in the walls or the roof, ensure that the units are always facing the sun to capture all incoming rays while at the same time deflecting harsh, unwanted rays from a building's interior. Each unit holds a miniaturized photovoltaic (PV), or solar-cell, device used to collect light and heat that is then transferred into useable energy to run the motors, also embedded in the building's interior walls. The remaining energy is used for heat, air conditioning, and artificial lighting. The surplus energy can be directly and automatically distributed through wires inside a building's walls, or can be stored in a group of batteries, for later use. This solar-powered technology will provide the typical business office the most superior lighting available-natural daylight. It will allow for better views outside your window that are no longer hidden by a standard shade or obscured by penetrating glare.

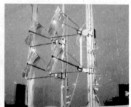

Figure 61　Dynamic shading window systems (from http://isites.harvard.edu/fs/docs/icb.topic31837.files/anna_dyson.pdf)

3.3 Self Compacting Concrete (Ouchi et al 2006; Grunewald, 2004; Wallevik et al 2003)

The application of concrete without vibration in highway bridge construction is not new. For examples, placement of seal concrete underwater is done by the use of a tremie without vibration, mass concrete has been placed without vibration, and shaft concrete can be successfully placed without vibration. These seal, mass and shaft concretes are generally of lower strength, less than 34.5 MPa and difficult to attain consistent quality. Modern application of self-compacting concrete (SCC) is focused on high performance-better and more reliable quality, dense and uniform surface texture, improved durability, high strength, and faster construction.

Recognizing the lack of uniformity and complete compaction of concrete by vibration, researchers at the University of Tokyo, Japan, started out in late 1980's to develop SCC. By the early 1990's, Japan has developed and used SCC that does not require vibration to achieve full compaction. More and more applications of SCC in construction have been reported in Japan as shown in Figure 62. As of the year 2000, the amount of SCC used for prefabricated products (precast members) and ready-mixed concrete (cast-in-place) in Japan was about 400,000 m^3. And various type of fiber reinforced SCC was also developed (Grunewald, 2004).

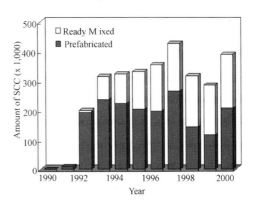

Figure 62 Amount of SCC Placement in Japan (Ouchi et al 2003)

SCC offers many advantages for the precast, prestressed concrete industry and for cast-in-place construction: (1) Low noise-level in the plants and construction sites; (2) Eliminated problems associated with vibration; (3) Less labor involved; (4) Faster construction; (5) Improved quality and durability; (6) Higher strength. SCC mixes must meet three key properties: (1) Ability to flow into and completely fill intricate and complex forms under its own weight; (2) Ability to pass through and bond to congested reinforcement under its own weight; (3) High resistance to aggregate segregation. Therefore, SCC can not survive without development of superplastizer and proper mixing design method.

Several European countries were interested in exploring the significance and potentials of SCC developed in Japan. These European countries formed a large consortium in 1996 to embark on a project aimed at developing SCC for practical applications in Europe. The title of the project is "Rational Production and Improved Working Environment through using Self-compacting Concrete." In the last six years, a number of SCC bridges, walls and tunnel linings have been constructed in Europe. According to Takada's statistical data in 2004, the ratio of SCC to total amount concrete in the Netherlands exceeded that in Japan (Takada, 2004).

In the United States, SCC is beginning to gain interest, especially by the precast concrete industry and admixture manufacturers. The precast concrete industry is beginning to apply the technology to commercial projects when specifications permit. The applications range from architectural concrete to complex private bridges.

3.4 Fiber reinforced Composites

Civil structures made of steel reinforced concrete normally suffer from corrosion of the steel by the salt, which results in the failure of those structures. Constant maintenance and repairing is needed to enhance the life cycle of those civil structures. There are many ways to minimize the failure of the concrete structures made of steel reinforce concrete. The custom approach is to adhesively bond fiber polymer composites onto the structure. This also helps to increase the toughness and tensile strength and improve the cracking and deformation characteristics of the resultant composite. But this method adds another layer, which is prone to degradation. These fiber polymer composites have been shown to suffer from degradation when exposed to marine environment due to surface blistering. As a result, the adhesive bond strength is reduced, which results in the delamination of the composite. Another approach is to replace the bars in the steel with fibers to produce a fiber reinforced concrete and this is termed as FRC. FRC is Portland cement concrete reinforced with more or less randomly distributed fibers. In FRC, thousands of small fibers are dispersed and distributed randomly in the concrete during mixing, and thus improve concrete properties in all directions. Fibers help to improve the post peak ductility performance, pre-crack tensile strength, fatigue strength, impact strength and eliminate temperature and shrinkage cracks.

Several different types of fibers have been used to reinforce the cement-based matrices. The choice of fibers varies from synthetic organic materials such as polypropylene or carbon, synthetic inorganic such as steel or glass, natural organic such as cellulose or sisal to natural inorganic asbestos. Currently the commercial products are reinforced with steel, glass, polyester and polypropylene fibers. The selection of the type of fibers is guided by the properties of the fibers such as diameter, specific gravity, young's modulus, tensile strength etc and the extent these fibers affect the properties of the cement matrix.

Recently, with the development of superplastizer, various new type of FRC is widely used, such as fiber reinforced self compacting concrete (Grunewald, 2004), reactive power concrete (RPC) (Dauriac, 1997), engineered cementitious composites (ECC) (Li, 2005).

Reactive Powder Concrete is an ultra high-strength and high ductility composite material with advanced mechanical properties. Developed in the 1990s by Bouygues' laboratory in France. It consists of a special concrete where its microstructure is optimized by precise gradation of all particles in the mix to yield maximum density. It uses extensively the pozzolanic properties of highly refined silica fume and optimization of the Portland cement chemistry to produce the highest strength hydrates. RPC represents a new class of Portland cement-based material with compressive strengths in excess of 200 MPa range. By introducing fine steel fibers, RPC can achieve remarkable flexural strength up to 50 MPa. The material exhibits high ductility with typical values for energy absorption approaching those reserved for metals. The advantages of RPC: (1) RPC is a better alternative to High Performance Concrete and has the potential to structurally compete with steel. (2) Its superior strength combined with higher shear capacity results in significant dead load reduction and limitless structural member shape. (3) With its ductile tension failure mechanism, RPC can be used to resist all but direct primary tensile stresses. This eliminates the need for supplemental shear and other auxiliary reinforcing steel. (4) RPC provides improve seismic performance by reducing inertia loads with lighter members, allowing larger deflections with reduced cross sections, and providing higher energy absorption. (5) Its low and non-interconnected porosity diminishes mass transfer making penetration of liquid/gas or radioactive elements nearly non-existent. Cesium diffusion is non-existent and Tritium diffusion is 45 times lower than conventional containment materials. An application of RPC can be seen in the Pedestrian Bridge in the city of Sherbrooke, Quebec, Canada (in Figure 63). RPC has also been used for isolation and containment of nuclear waste of several projects in Europe.

(a) Overview Sherbrooke bridge　　(b) Design of the Pedestrian Bridge　　(c) in construction

Figure 63　Application of RPC in Sherbrook Bridge (http://www.new-technologies.org/ECT/Civil/reactive.htm)

Engineered cementitious composites (ECC) is a fiber reinforced cement based composite material systematically engineered to achieve high ductility under tensile and shear loading. By employing micromechanics-based material design, maximum ductility in excess of 3% under uniaxial tensile loading can be attained with only 2% fiber content by volume. This moderate amount of short discontinuous fibers allows flexibility in construction execution, including self-consolidation casting and shotcreting. Structural products have been manufactured by extrusion of ECC. Recent research indicates that ECC holds promise in enhancing the safety, durability, and sustainability of infrastructure. Field application of ECC include composites steel/ECC deck for a cable-stayed bridge in Hokkaido, Japan, repair of the Mitaka Dam in the Hiroshima-Prefectrure in 2003 and an impending bridge deck retrofit with ECC link-slabs in Michigan, in the US in 2005 (Li, 2005).

Besides FRC, Fiber reinforced Polymer (FRP) is an unneglectable category of fiber reinforced composites. In the field of construction material, the common-used FRP is carbon fiber reinforced polymer (CFRP) and glass fiber reinforced polymer (GFRP). CFRP bar can be used as replacement of steel bar. Fiber mat reinforced polymer can be used as repairing materials.

Recently, instead of polymer as matrix, our group in HKUST also investigated the possibility of using carbon fiber mat (or basalt fiber mat) reinforced magnesia cement repairing material (Chan 2006).

3.5 Integration of Inorganic and Organic Materials

Cement concrete is one of the most important construction materials. It is strong market position is due to its high compressive strength, stiffness, ability to shape products easily and lower prices. However, it also shows a number of deficiencies, such as lower tensile strength, lower adhesion strength to most substrates and a relative lower resistance to a number of degradation mechanisms, such as steel corrosion. Most polymer often show just the opposite characteristics, high tensile strength but relatively low compressive strength, low stiffness, high cost, good adhesion and high resistance to physical (i.e. abrasion, erosion, impact) and chemical attack. Combinations of these two materials can exploit the useful properties of both and yield composites with excellent strength and durability properties. For instance, FRP bar can be used as a replacement of steel bar to make FRP reinforced concrete, it avoids the influence of steel corrosion on safety of concrete structure. In addition, FRP bar is lighter than steel bar. Fiber mat reinforced polymer can be used as a repairing materials. Engineering may also employer organic fiber to product FRC.

There are three fundamentally different ways in which organic polymers may be combined with inorganic material: (1) the polymer may be employed to fill the pore space between inorganic matrix or as the reinforcement of inorganic matrix; (2) The inorganic is used as the filler or reinforcement in polymer matrix; (3) a new material may be created in a chemical interaction between the polymer and constituent of the inorganic materials.

3.5.1 MDF cement (Odler 2000, Drábik et al 2002)

MDF cement, first reported by the research group of Birchall at Imperial Chemical Industries (ICI, United Kindom) (Drábik et al 2004), is not a cement but a cementitious polymer-modified and specially processed materials. The designation "MDF" stands for "macro defect-free", and it originated at a time when it was believed that the absence of macrodefets was solely responsible for the high strengths of this materials.

The starting constituents in making an MDF materials are an inorganic cement, an organic polymer, and mixing water. Small amounts of gylcerine have been ground to improve the processing, and may also be added. Two cement/polymer combinations have been found to be particularly suitable, and are employed nowadays almost exclusively: (a) calcium aluminate cement in combination with a polyvinyl alcohol-polyvinyl acetate co-polymer; and (2) Portland cement in combination with polyacrylamide.

The calcium aluminate cement used should preferably have a high Al_2O_3 content. A product with the brand name Secar 71 by Lafarge, with an Al_2O_3 content of around 70%, has been found to be particularly suitable, but other calcium aluminate cements may also be used. The polyvinyl alcohol-polyvinyl acetate co-polymer is the product of an incomplete hydrolysis of polyvinyl acetate. The selection of an appropriate product is critical. The simultaneous presence of hydroxyl and acetyl group within the polymer appears to be essential. It has been suggested, however, that both the strength and the water resistance of the resultant MDF cement will increase with decreasing degree of hydrolysis of the polymer and with decreasing molecular weight (Santos et al 1999). The polymer droplets in the dispersion should be small enough (that is, below 100μm) to ensure rapid and effective distribution of the polymer within the MDF mix during processing. A PVA-PVA1 product widely and successfully used in the producing MDF materials is one with the brand Gohsenol KH-172 by Nippon Gohsel.

There do not seem to be any particular requirements for the composition of the Portland cement emplyed. Ordinary Portland cement performs better than slag-modified Portland cement. The polyacrylamide polymer should also be available in a well-dispersed form, to facilitate processing. In addition to polyvinyl alcohol/acetate and polyacrylamide, some other polymer have also been employed as the organic constitutents of MDF cement, including polypropylene glycol and hydroxypropyl-methyl cellulose. As well as Portland and calcium aluminate cements, sulfoaluminate-ferrite-belite cement (in combination with hydroxypropyl-methy cellulose) has also been employed as constituent of an MDF materials.

The following ratios of the individual constituents may be considered typical for MDF materials: 75 wt%-85 wt% inorganic cement, 5 wt%-10 wt% organic polymer, 8 wt%-15wt% water and 0.3 wt%-0.6wt% glycerine. It is also possible to add various inorganic fillers to the mix, such as finely ground silica, fly ash, metallic powders to increase electrical and thermal conductivity, or silicon carbide to increase abrasion resistance. They may be introduced in amounts of up to 50wt% without any dramatic affect on the final strength of the resulting MDF material. MDF materials have also been developed that are reinforced with small amounts of organic fiber (or whiskers), steel fiber.

The production of MDF materials consists of a series of steps: premixing, high-shear mixing, forming, and curing.

Premixing: The components are premixed in a conventional low-shear blender.

High-shear-mixing: This procedure is usually performed in a two-roll mill, a device commonly used in the plastics and rubber industry. The apparatus consists of two rolls that rotate counter to each other. The gap between them is adjustable, and may vary between about 0.5 and 2mm. Owing to a difference in their speeds of rotation, shear forces are generated in the material that is allowed to pass through the gap between the rolls. In this way shear rates exceeding 1000/s may be realized. High-shear mixing is an important step in making MDF cement. Originally it was believed that high-shear mixing just helps to eliminate macrodefects (represented by large pores and air voids) from the paste, and the polymer acts as a processing aid that helps to achieve this goal. This belief was based on the well-known fact that the flexural strength of porous bodies is a function not only of overall porosity but also of the critical flaw size, which in cement pastes is characterized by the size of the largest pores present. In a further development it was recognized, however, that high-shear mixing, in addition eliminating large pores, also helps to induce chemical interactions between polymer and the inorganic cement, and it was recognized that much of the mechanical strength of MDF materials is due to these mechanically induced chemical changes. The way in which the material is high-shear-mixed largely determines its ultimate physico-mechanical properties. Within a reasonable range the strength will increase with increasing shearing rate and increasing mixing time. However, the shear rate cannot be increased with impunity, because with an increasing rate of mixing, the amount of heat generated by viscous damping also increase, and increased temperature of the mix may cause the material to stiffen before the processing is completed. An additional increase of temperature also takes place as a consequence of exothermic chemical reactions taking place within the materials. A cooling of the rolls of the two-roll mill may extend the time available for high-shear mixing and thus affect positively the quality of the MDF material produces. During high-shear mixing the consistency of the mix is gradually altered. The material ultimately attains a rubbery consistency, with vis-coelastic properties.

Forming Different methods may be used to obtain the desired shape of the materials. Sheets with thicknesses between about 0.5 and 2mm may be produced by calendering. These may be densified further by pressing between two plates, which may be heated. Other possible processing methods include extrusion and injection molding or pressing.

Curing After forming, the material may be cured in humid air at ambient or elevate temperature. A final drying at about 80℃ increases the ultimate strength significantly. The hardening and strength development of MDF materials is the result of chemical reactions taking place in the materials during processing and curing. These include both an interaction between the inorganic cement and the polymer, and cross-linking reactions within the polymer itself. Van der Wall forces may also be involved.

In the system comprising calcium aluminate cement and polyvinyl alcohol/acetate the cement hydrate is limited, as the polymer slows down the hydrate rate. Non-hydrated particles of the cement become constituents of the developed structure, held together by a polymeric 0 inorganic matrix. The phase C_2AH_8 is formed as the main hydration products of the cement. In the system comprising Portland cement and polyarcylamide more hydrated material is formed, but again a significant part of the cement remains non-hydrated. The cement hydration products are similar to those formed in the absence of the polymer, but they exhibit a denser packing. Because of the high pH of the liquid phase brought about by the hydration of the cement, the polymer also undergo chemical reactions. In the polyvinyl alcohol/acetate co-polymer ahydrolysis of the acetate groups take place. The liberated acetate groups react with cations in the liquid phases, and in particular with Ca^{2+} ions, and calcium acetate is precipitated. On the micrometer scale the structure of MDF materials consists of densely packed residual unreacted cement grains embedded in a polymeric matrix. An important role in the development of strength is played by an interphase region that surrounds the unreacted grains and is responsible for the high degree of bonding between cement particles and the polymeric matrix. In this region crystallites of C_2AH_8 (5 – 8mm in size) reside in an amorphous matrix.

MDF materials exhibit a relatively small but distinct porosity. In generally the porosity found in Portland cement based MDF materials are larger than those of calcium aluminate cement based products. From mercury porosimetry value of up to 20 vol. % have been reported for the former and below 5 vol. % for the latter. Value found by nitrogen adsorption is usually very low. The maximum flaw sizes in MDF materials are reduced typically to 10 ~ 100μm.

Hardened MDF pastes exhibit strengths that-if compared with conventional cement materials-must be considered extremely high. Out of the two main types of MDF materials, those based on calcium aluminate cement are about twice as strong as those based on Portland cement. Flexural strength exceeding 100MPa may be achieved in the Portland cement/polyacrylamide system and the strength exceeding 200Mpa in a calcium aluminate cement/polyvinyl alcohol/acetate system. Such high strengths can be attributed to several factors, including a high degree of compaction, a low overall porosity, an absence of large pores acting as macro-defects, and the extremely favorable intrinsic strength properties of the cement/polymer composite material, in which an important role is played by the interphase region. The typical physico-mechiancal properties of MDF cement may be summarized as follow: flexural strength of 150 ~ 200MPa; compressive strength of 300MPa; Young's modulus 50GPa; critical stress intensity factor $3MPa \cdot m^{1/2}$; Poisson ratio of 0.2; density of $2500kg/m^3$; coefficient of thermal expansion of 9.7×10^{-6} m/mK. For further improvement of mechanical properties MDF material may be reinforced with fibers. Carbon and alumina fibers gave the best result. The optimum fiber amount was 10 % ~ 15% by volume.

Similar to DSP cement, a serious drawback of MDF materials is their sensitivity to water. Upon immersion in water they exhibit a significant uptake of water and an expansion, associated with a loss of strength, mainly within the first two weeks. The residual strength may be as low as 20% of the original strength of the oven-dried materials, and the linear expansion may exceed 20 mm/m. This loss of strength must be attributed to a swelling and softening of the organic polymer phase constituting the MDF material. It has also been suggested that the reduction of strength seen under moist conditions is due to a weakening of the bonding between the polymer and cement matrix by van der Waals forces and a base/acid interaction between the polymer and water. If kept under water, the material may lose weight, in spite of water uptake and swelling. This loss is due to a partial dissolution of the cement and the polymer in the surrounding water. Upon redrying the material retains a significant fraction, but usually not all, of its original dry strength.

Just as in samples kept under water, an uptake of water and expansion of the material associated with a loss of strength may also be observed in samples kept in humid air. Here the amount of absorbed water and the extent of expansion increase with increasing relative humidity. Weight increases of up to 10% and linear expansions up to 0.8% may be observed under these conditions. The actual extent of water uptake and expansion will also depend on the processing and curing conditions. Samples hot-pressure at 90℃ have been reported to show a lower expansion that those processed at only 40℃. The loss of strength may exceed 50%, and will also increase with relative humidity and curing time. Along with an uptake of water, swelling and loss of strength, MDF cements also exhibit a distinct increase of creep deformation upon expose to water. This effect increases with increasing relative humidity. Under the same experimental conditions all these moisture effects generally increase with increasing polymer content in the MDF cement. Efforts have been made to eliminate or at least reduce the adverse effect of moisture on MDF materials. A significant improvement was achieved by the use of organosilane or titanate coupling agents or by incorporating an isocyanate compound to cross-link the polyvinyl alcohol chains through urethane bonding.

It has also been suggested that the water resistance of MDF materials might be improved by an *in situ* reticulation of the polymer. This would entail combining with Portland cement an aqueous solution of partially hydrolyzed polyvinyl alcohol and sodium silicate, which would act as the reticulation agent. Drabik et al achieved a reduction of water susceptibility of an MDF system consisting of a combination of a sulfoaluminoferrite belitic clinker and hydroxypropylmethyl cellulose by adding sodium polyphosphate to the original mix. This improvement was attributed to the formation of Al(Fe)-O-C(P) cross-links in the hardened materials. The resistance of MDF cement to elevated temperature is limited by the presence of the organic constituent. At around 500℃ the polymer is burnt out, and along with this the strength of the material declines distinctly. Effects have also been made to produce MDF-like materials in which the role of the organic polymer is taken over by an inorganic polymer, specially sodium polyphosphate. In this way flexural strength of up to about 30MPa were achieved in combination with calcium aluminate cement. The resulting phosphate-containing reaction products were found to be amorphous.

Following the initial invention and basic findings of Birchall and co-workers, there are an increasing number of reports that the polymer plays a more active role than simply being a rheological aid. These reports include: (i) series of flexural vs. compressive strength values, the ratio of these values being much higher than typical of hydraulic cements; (ii) the composition of cement as well as polymer is crucial for material performance; (iii) removal of polymer reduces the strength by 90% and, if the porosity created by that removal is filled by subsequent hydration, the strength returns to only one third of the original value of MDF cement; (iv) dramatic loss of strength occurs on prolonged exposure to water. It has been concluded that MDF cements have a microstructure with close packed unreacted cement particles (consequent on a low w/s ratio) acting as a filler within the binding matrix. The matrix itself consists of two interpenetrating phases, a cross-linked polymer and a nanocomposite interphase region around the cement particles. Several models for the local atomic-level structures are reviewed in a significant aspect of the models essentially correspond to grafting of functional polymers to solid surfaces as presented in.

The monitoring of the synthetic sequence by a Banbury-type rheometer (which mimics a twin roll mill) suggests that the formation of microstructure in MDF materials is a mechanochemical process, and both chemical reactions and mechanical effects are involved to create a highly filled, cross-linked viscoelastic polymer system. The mechanical effects are postulated to be due to fission of polymer chains during high shear mixing, creating fragments which subsequently react to form cross-links. The chemical effects arise from release of ionic species from the hydrating cement, which participate in controlled cross-linking reactions that do not proceed as rapidly as the mechanical mixing. The time dependence of the increase in the cross-link density results in a characteristic plateau in torque vs. mixing time plot, referred to as the window of processability. Crucial aspects of the microstructure are the binding of the matrix with cement particles (formed due to the mechanochemical process) and the grafting of cross-linked polymer chains to the surface of cement grains. The control and improvement of the moisture resistance is a strategic target for MDF materials. Moisture enters an MDF material by diffusion through the polymer-containing phase, the residual cement particles hydrate destroying the interphase region and ultimately degrading the original binding matrix. The extent of moisture absorption of high aluminium MDF cement test pieces at ambient humidity and on immersion in water has been reported, with the consequent effects on the performance and phase composition. Since the polymer is the conduit for moisture uptake, resistance should be enhanced by increasing the hydrophobicity of the polymer matrix itself. Studies of the control/improvement of moisture resistance rely on up to two of the following methods: (i) removal of unreacted cement (reduction of the concentration of active clinker); (ii) improved cross-linking by increasing the amount of water soluble polymer in cross linked component; (iii) substitution of the water soluble polymer by a hydrophobic polymer.

A few potential applications of MDF materials have been examined, with a view to meeting specific market needs. Thermal and sound insulators for specific applications, reinforcement for ordinary cement pastes or mortars, and formed body shells for specific applications are promising. The technology utilized in making a cement shell for a solar powered car can be easily applied to form armours also. Loudspeaker cabinets were also considered by Alford and Birchall; generally a material higher in damping ($\tan^{TM}$), Young's modulus and density has good acoustic properties. These applications are by no means a comprehensive list, but illustrate the range of potential uses that exist for high flexural strength polymer-cement composites or, at least, provide guidelines for application-oriented experiments. A comprehensive overview of potential market applications for MDF materials (Table 15) has been given by the Maeta Techno-Research Inc. in a market survey of Japanese and European manufacturing industries (Drábik et al 2002).

Information on market needs, adapted from (Drábik et al 2002) Table 15 (a)

Field	Intended applications	Required Performance
Construction	Foofing (tiles), permanent Formwork, fire resistant doors Cover for underfloor cable pits, OA floor, Sewage pipe, Airport bridge, monuments, electric poles in coastal area, partition panel, door or window shutters	Light weight, durability, fire proofing, impact resistance, smaller deflection, productivity, possible size, cost, workability, maximum angle sheet folded without damage, nailing

Table 15 (b)

Field	Intended applications	Required Performance
Steel, machinery	Replacement for angle steel, replacement for steel towers moulds for plastics or leather industry, printing rollers for newspapers, repairing material for moulds, thermal insulators for injection molds, tubes, exhaust, handgrips for iron pots	Thermal resistance, acoustic damping, incombustiblity, insulating properties, high strength, hardness, cost, machineability, ability to apply colours, dimensional accuracy, ability to process by injection moulding
Chemical engineering	Corrosion-resistant tanks, oil tank	Corrosion resistance, setting shrinkage, cost
Electrical engineering	Cover for cable dusts, globe insulators, propellers in electrical generators, pipe, housing for rotators, electrical parts	Lightweight, thermal resistance, dimensional stability, shaping, machineability, ability to apply colours, surface roughness, adhesion
transportation	Tyre wheel, yacht, airplane, frame for motorcycle, boat deck, interior panel for train, thermal insulation, stiffening plate for outer shell of cars, brake lining, pallets	Thermal and chemical resistance, outdoor stability, acoustic damping, stiffness, dimensional accuracy, shaping, mixing with powders, from waste FRP products
miscellaneous	Toy, swing, playground slide, model plane, model house, signboard, pipe, fireproof safety box, cooler box, artificial teeth, handicraft material for children	Lowe thermal conductivity, outdoor stability, non-toxicity, cost, smaller in deflection, productivity, strength equal to steel

3.5.2 Polymer modified concrete, polymer-impregnated concrete and polymer concrete (Su 1995, Older 2000; Chandra and Ohama 1994; Kardon 1997; Ohama 1997)

Polymer-modified concrete (PMC), or polymer Portland cement concrete (PPCC or PCC), is normal Portland cement concrete with a polymer admixture. The polymer and the cement hydration products comingle and create two interpenetrating matrices, which work together, resulting in an improvement in the material properties of PCC alone. PMC is the term for such concrete with lower dosages of polymers, typically 5% or less, and PPCC or PCC generally is the term for composites with more than 5% polymer by weight of concrete (Chandra et al 1994).

Polymers are materials with long chain molecules made up of many individual monomers connected end to end. Polymers made up of one type of monomer are called homopolymers, and those made up of more than one monomer type are called copolymers. Polymers are classified as elastomers, thermoset-ting, or thermoplastics, each characterized by the type and relative amount of crosslinking of the polymer chains. Polymers most often used in PMC are the thermosetting and thermoplastic type (Su 1995). The effectiveness and density of the crosslinking influences the physical characteristics of the polymer. Elastomers have weaker crosslinking of the molecular chains, thermosetting polymers have strong crosslinking, and thermoplastics have crosslinking that is less effective than thermosets and more effective than elastomers. The physical characteristics of the polymers used in PMC affect the properties of the hardened PMC. A characteristic physical property of the polymers used in PMC is their glass transition temperature range (T_g). At temperatures above T_s, the material acts more rubbery, undergoing plastic deformation when loaded. Below T_g, the material behaves more in a glassy manner, deforming elastically and being susceptible to brittle failure (Su 1995). Thermosetting polymers have relatively high T_g values, elastomers have low T_g values, and thermoplastics have intermediate T_g values. The significance of the polymer's T_g in PMC is the influence of the polymer's plastic or elastic behavior on the properties of the composite. Another characteristic of a polymer is its minimum film temperature (MFT). This is the temperature below which polymerization will not form a continuous film, but instead forms discrete particles. It is desirable for polymers used in PMC to have MFT values lower than the temperatures to which the material is exposed during curing (Su 1995). The atomic composition of the monomers, and of the polymer side chains, affect the amount and type of crosslinking and the chemical properties of the polymer, and influence the properties of the fresh PMC. Polymerization can occur with little energy input or only in the presence of a chemical catalyst, depending on the chemical nature of the monomers. Polymers can be ionic or nonionic, depending on the atomic charge of the side chains. Ionic polymers tend to orient themselves around other particles, such as cement grains, in the same manner as water-reducing admixtures. They act as surfactants and tend to disperse the clinker grains, reducing water demand and facilitating hydration.

Polymers are incorporated into PMC in several forms: as a latex, or suspension of monomers or polymers in water; as a redispersible powder; or as a resin, which is a monomer or polymer in liquid form. Polymerization of the monomers can take place prior to combination with Portland cement mortar or concrete, or can be initiated after mixing. Polymerization can take place by combination with a hardener or activator, by thermal catalysis, or by drying. When polymers are incorporated in PMC as latex, the composite is referred to as latex-modified concrete (LMC). LMC is the most common type of PMC because latex is relatively simple to incorporate in mortar or concrete. It is added with the other ingredients during the mixing of fresh concrete. The water portion of the latex must be considered in the overall mix design of the concrete. Elastomeric polymers used in LMC include bitumen, natural rubber, and styrene-butadiene (SB). Thermoplastics used in LMC include vinyl acetate-ethylene (VAE), polyacrylic ester (PAE), styrene-acrylic (SA), and others. Polyvinylidene chloride (PVDC) was used during the popularization of LMC in the early 1970s, but the chlorine resulted in corrosion of reinforcing steel and the use of PVDC was discontinued. Poly-vinyl acetates (PVAC) hydrolyze in moist environments, so their use in LMC is limited (Ramakrishnan 1994). Polymers can be incorporated in LMC as a powder. They are referred to as redispersible powders since they are manufactured by spray-drying polymer latexes, and must be returned to the latex form in order to disperse within the cement paste and to coalesce into a film (Chandra et al 1994). An advantage of redispersible powders is that they are easier to store and handle than latexes, and can be supplied in cheaper packaging. Most importantly, they can be prepackaged with cement and sold as a one-part product that needs only water.

Current applications of PMC are primarily as overlays on roadways and bridges, both as new construction and as repair of existing deteriorated structures. PMC is also being used in flooring, water tanks, swimming pools, septic tanks, silos, drains, pipes, and ship decks (Su 1995). Two very promising, relatively new applications of PMC are its use in combination with fiber reinforcing (Chen et al 1996; Zaya et al 1996), and its use as a pneumatically applied material or shotcrete. Possible future applications of PMC mentioned in the literature include: in roller-compacted concrete (RCC) for air strips, roadways, and parking lots; and in ductile concrete foundation and shear wall construction, as well as marine and offshore structures (Ramakrishnan 1994). PMC use has also been predicted in concrete structures wherever there is a need for its tensile strength, resistance to cracking, and impermeability (Kardon 1997).

Polymer-impregnated concrete (PIC) is produced by infusing a monomer into the cracks and voids of already hardened concrete. The monomers are polymerized after they enter the voids by the action of a chemical hardener or the application of heat. Since the polymer ideally fills the voids and binds with the cement matrix and the aggregates, there is no need to have high-quality concrete for PIC. Impregnation depth is limited by the porosity of the concrete, the viscosity and volatility of the monomer and hardener, the setup time of the polymer, and the pressure applied. PIC strength is dependant on the type and amount of polymer used, and the degree of polymerization achieved. Applications of PIC are typically limited to precast thin panels and to the repair of highway surfaces due to the limit on impregnation depth and the difficulties in applying the material on anything but the top of a horizontal surface.

Polymer concrete (PC) is a material made of aggregate and a polymer binder. There is no Portland cement in polymer concrete. The polymer matrix binds very well to the aggregate particles with no transition zone, unlike Portland cement concrete. And various fiber reinforced PC were also developed (Garas et al 2003). To improve the bonding strength between polymer and inorganic materials, coupling agent (such as Methacryloxy propyltrimethoxy silane, MPS) is sometimes also used. Since polymer materials are more expensive than Portland cement, and can generate heat and undergo shrinkage during curing, PC is made with evenly graded aggregates to achieve close packing, minimizing the space between the aggregates to be filled with polymer. Uses for PC include cast-in-place PC connections for precast concrete construction, precast PC elements, and overlays for concrete repairs. A possible problem with some PC is its sensitivity to high temperature and to cyclical temperature changes (O'Connor et al 1993). An interesting type of PC (Rebeiz et al. 1994) involves the use of recycled polyethylene terephthalate (PET). Plastics represent a large portion of the garbage stream, and significant environmental benefits can be realized if good-quality construction materials can be made from some plastics. Considering the sustainability, recycled aggregate (such as concrete waste, brink waste and wood waste) can also be used to produce polymer concrete.

In addition, concrete or mortar with recycled rubber aggregates, thermal insulated cement-based materials incorporated with expanded polystyrene beads (EPS) or EPS waste from packing material can also be regarded as inorganic/organic integrated system. They may be used for special purpose in construction.

4 REMARKS

Building materials development is largely depending on demand and application requirements. They have to be designed and produced according to the end user's needs. In 21st century, due to the limitation of natural resources and energy supply, the change of environment and the strong demand on high-quality working and living conditions, building materials will be developed forward the direction of sustainable, functional and multi-disciplinary integration.

REFERENCES

Aizawa, S., Kakizawa, T. and Higasino, M. (1998). "Case studies of smart materials for civil structures". *Smart Materials and Structures*, **7** (5), 617-626.

Ali-Harthy, A. S., Taha, R. and Al-Maamary, F. (2003). "Effect of cement kiln dust (CKD) on mortar and concrete mixtures", *Construction and Buidling Materials*, **17** (5), 353-360.

AMI's Guide (2006). *The Plastic Recycling Industry in Europe*. http://www.amiplastics.com/ami/Assets/press_releases/newsitem.aspx?item=88.

ASTM (1991a). "Standard Test Method for Laboratory Measurement of Airborne Sound Transmission Loss of Building Partitions" 1991 Annual Book of ASTM Standards, Vol. 04.06, ASTM E-90-90, Philadelphia: American Society for Testing and Materials, 678-687.

ASTM (1991b). "Standard Test Method for Measurement of Airborne Sound Insulation in Building" 1991 Annual Book of ASTM Standards, Vol. 04.06, ASTM E-336-90, Philadelphia: American Society for Testing and Materials, 696-705.

ASTM (1991c). "Test Method for Thermal Transmission Properties of Nonhomogeneous Insulation Panels Installed vertically", 1991 Annual Book of ASTM Standards, Vol. 04.06, ASTM C-1061-86, Philadelphia: American Society for Testing and Materials, 1991, 581-591.

ASTM (1991d). "Standard Test Method for Steady-State Heat Flux Measurements and Thermal Transmission Properties by Means of the Guard Hot Plate Apparatus" 1991 Annual Book of ASTM Standards, Vol. 04.06, ASTM C-177-85, Philadelphia: American society for Testing and Materials, 1991, 20-31.

Banaszkiewicz, M., Seiferlin, K., and Spohn, T. (1997). "A new method for the determination of thermal conductivity and thermal diffusivity from linear heat source measurements", *Review of Scientific Instruments*, **68** (11), 4184-4190.

Banks, H. T., Smith, R. C. and Wang, Y. (1996). *Smart Material Structures: Modeling, Estimation, and Control*, Wiley, New York, USA.

Banno, H. (1988). "Recent developments of piezoelectric composites in Japan", *Advanced Ceramics*, S. Saito, ed. , Oxford University Press, Oxford, U. K. , 8-26.

Von Berg, W. and Feuerborn, H. J. (2000). "CCPs in Europe", http://www.energiaskor.se/rapporter/ECOBA_paper.pdf.

Bouguerra, A. , Laurent, J. P. , Goual, M. S. , and Queneudec, M. , (1997) "The measurement of the thermal conductivity of solid aggregates using the transient plane source technique", *Journal of Physics D: Applied Physics*, **30** (20), 2900-2904.

Brennan, M. J. , Garcia-Bonito, J. , Elliott, S. J. , David, A. and Pinnington, R. J. (1999). "Experimental investigation of different actuator technologies for active vibration control", *Smart Materials and Structure*, **8** (1), 145-153.

Burleigh, D. D. , and Gagliani, J. (1987). "Thermal conductivity of PVC foams for spacecraft applications", *Thermal Conductivity*, **20**, 41-45.

California Integrated Waste Management Board (2004). *Statewide Waste Characterization Study*, http://www.ciwmb.ca.gov/Publications/default.asp?pubid=1097.

Campbell D. H. , and Folk R. L. (1991). "The ancient pyramids – concrete or rock", *Concrete International: Design & Construction*, **13** (8), 28-39.

Chan, J. (2006). *The development of Magnesium Oxychloride Cement as Repairing Materials*. MSc Thesis, The Hong Kong University of Science and Technology, HongKong.

Chandra, S. and Ohama, Y. (1994). *Polymers in Concrete*. CRC Press, Boca Raton, USA.

Chang, F. K. (1999). *Structural Health Monitoring 2000*, Technomic Publishing, Lancester, USA.

Chaurand, P. Rose, J. and Briois, B. et al. (2006). "Environmental impacts of steel slag reused in road construction: A crystallographic and molecular (XANES) approach", *Journal of Hazardous Materials*, in press.

Chen, P. W. and Chung, D. D. L. (1996). "A comparative study of concretes reinforced with carbon, polyethylene, and steel fibers and their improvement with latex addition", *ACI Materials Journal*, **93** (20), 129-133.

Comrie Consulting Ltd (1988), *Comrie Preliminary Examination of the Potential of Geopolymers for Use in Mine-Tailings Management*, D. Comrie Consulting Ltd. , Mississauga, Ontario, Canada.

Corbridde, D. E. C. (1995). *Phosphorus. An Outline of Its Chemistry, Biochemistry and Technology*, Elsevier, Amsterdam, the Netherlands.

Dauriac, C. (1997). *Special Concrete may Give Steel Still Competition*. http://www.djc.com/special/concrete97/10024304.htm

Davidovits, J. (1987) Ancient and modern concretes: what is the real difference? *Concrete International: Design & Construction*, **9** (12), 23-35.

Davidovits, J. , Comrie, D. C. , Paterson, J. H. and Ritcey, D. J. (1990). "Geopolymeric concretes for environmental protection", *Concrete International: Design & Construction*, **12** (7), 30-40.

Davidovits, J. (1993). "Geopolymer cement to minimize carbon-dioxide greenhouse warming". *Ceramic Transactions*, **37**, 165-182.

Davidovits, J. (1994a). "Recent progresses in concretes for nuclear waste and uranium waste containment", *Concrete International*, **16** (12), 53-58.

Davidovits, J. (1994b). "Properties of geopolymer cements", *Alkaline Cements and Concretes*, KIEV Ukraine, page 9.

Davidovits, J. (1994 c). "Geopolymers: Inorganic polymeric new materials", *Journal of. Materials Education*, **16**, 91-139.

Deisser, R. G. , and Boegli, J. S. , (1958) "An investigation of effective thermal conductivities of powders in various gases", Transactions of ASME, 80 (10), 1417-1425.

Dhir, R. K. (2004). "Realising a high-value sustainable recycling solution to the glass cullet surplus". Guidance for the Use of Crushed or Powdered Glass in Concrete. http://www.azobuild.com/details.asp?ArticleID = 7686.

Dias, C. J. and Das-Gupta, D. K. (1996), "Inorganic ceramic/polymer ferroelectric composites electrets", IEEE Transactions on Dielectrics and Electrical Insulation, 3 (5), 706-734.

Dong, B. Q. (2005). Cement-Based Piezoelectric Ceramic Composites as Sensor Applications for Civil Engineering, PhD Thesis, The Hong Kong University of Science and Technology, Hong Kong.

Drábik, M., Mojumdar, S. C. and Slade, R. C. T. (2002). "Prospects of novel macro-defect-free cements for the new millennium", *Ceramics*, **46** (2), 68-73.

Drábik, M. Gáliková, L., and Varshney, K. G. et al (2004). "MDF cements: synergy of the humidity and temperature effects", *Journal of Thermal Analysis and Calorimetry*, **76** (1): 91-96

Du, J., Liu, S. and Guan, H. (2006). "Research on the absorbing characteristics of cement matrix composites filled with carbon black-caoted expanded polystyrene beads", *Advances in Cement Research*, **18** (4), 161-164.

ECOBA (2003), Information about Coal Combustion Products (CCPs) and their applications in Europe, European Coal Combustion Products Association (Ecoba), http://www.energiaskor.se/

European R&D project BRITE-EURAM BE-7355-93 (1997). *Cost-Effective Geopolymeric Cement for Innocuous Stabilization of Toxic Elements (GEOCISTEM)*. Final Report, April 1997.

Farid, M. M. Khalaf, A. N. (1994). "Performance of direct contact latent heat storage units with two hydrated salts", *Solar Energy*, **52**, 179-189.

Feldman, D. and Shapiro, M. M. (1989). "Fatty acids and their mixtures as phase-change materials for thermal energy storage", *Solar Energy Mater*, 18, 201-216.

Folk R. L., and Campbell D. H. (1992). "Are the pyramids built of poured concrete blocks?", *Journal of Geological Education*, **40**, 25-34, and 344.

Fricke, J., and Arduini-Schuster, M. C. (1989). "Opaque silica aerogel insulation as substitutes for polyurethane (PU) foams", *Thermal Conductivity*, **21**, 235-245.

Furukawa, T. Ishida, K. and Fukada, E. (1979). "Piezoelectric properties in the composites systems of polymer and PZT ceramics", *Journal of Applied Physics.*, **50** (7), 4904-4912.

Furukawa, T. (1989). "Piezoelectricity and pyroelectricity in polymers", *IEEE Transactions on Electrical Insulation*, **24** (3), 375-394.

Garas, V. Y. and Vipulanandan, C. (2003) *Review of Polyester Polymer Concrete Properties.* http://www2.egr.uh.edu/~civeb1/CIGMAT/03_poster/11.pdf

George, E. P., Gotthardt, R., Otsuka, K., Trolier-McKinstry, S. and Wun-Fogle, M. (1997). *Materials for Smart Systems II*, Symposium Proceedings of Materials Research Society, Vol. 459. Materials Research Society, Pittsburgh, USA.

Grave, R. S., McElory, D. L., Yarbrough, D. W., Fine, H. A. (1989). "The thermophysical properties of gypsum boards containing wax", *Thermal Conductivity*, **21**, 343-357.

Green, D. W. and Sherry, C. W. (1982). "Sound Transmission Loss of Gypsum wallboard Partitions", *The Journal of the Acoustical Society of America*, **71** (4), 90-96.

Grunewald, S. (2005). *Performance-Based Design of Self-Compacting Fibre Reinforced Concrete*. Ph.D.

Thesis, Delft University Press, Delft.

Guan, H. T., Liu, S. H. and Duan, Y. P et al. (2006). "Cement based electromagnetic shielding and absorbing building materials", Cement & Concrete Composites, **28** (5), 468-474.

Han, J. H., Cho, K. W., and Lee, K. H., (1998) "Transient one dimensional heat flow technique applied to porous reactive medium", Review of Scientific Instruments, **69** (8), 3079-3080.

Harrell, J. A., and Penrod, B. E. (1993). "The great pyramid debate - Evidence from the Lauer sample", Journal of Geological Education, **41**, 358-363.

Hendriks, C. F., Nijkerk, A. A. and van Koppen, A. E. (2000). The Building Cycle, Aeneas, Best, the Netherlands.

Hendriks, C. F. (2000). Durable and Sustainable Construction Materials, Aeneas, Best, The Netherlans.

Huang, S. F., Chang, J. and Xu, R. H. et al. (2004). "Piezoelectric properties of 0-3 PZT/sulfoaluminate cement composites", Smart Materials and Structures, **13** (2), 270-274.

Huang, S. F. (2005). Fabrication and Properties of Cement-Based Piezoelectric Composites, Ph. D. Thesis, Wuhan University of Technology, Wuhan.

Hsu, C. T., Cheng, P., and Wong, K. W. (1994). "Modified Zehner-Schlunder models for stagnant thermal conductivity of porous media", International Journal of Heat and Mass Transfer, **37** (17), 2751-2759.

Hsu, C. T., Cheng, P., and Wong, K. W., (1995) "A lumped-parameter model for stagnant thermal conductivity of spatially periodic porous media", ASME Journal of Heat Transfer, **117**, 264-269.

Hulse, S. (2000). Plastic Products Recycling: Technology and Market Trends. Rapra Market Report.

Van Jaarsveld, J. G. S. and van Deventer, J. S. J. (1997). "The potential use of geopolymeric materials to immobilize toxic metals: Part I. Theory and Applications", Minerals Engineering, **10** (7), 659-669.

Van Jaarsveld, J. G. S., van Deventer, J. S. J. and Lorenzen, L. (1998). "Factors affecting the immobilization of metals in geopolymerized fly ash", Metallurgical and Materials Transactions B, **29B** (1), 283-291.

Van Jaarsveld, J. G. S. and van Deventer, J. S. J. (1999a). "The effect of metal contaminants on the formation and properties of waste-based geopolymers", Cement and Concrete Research, **29** (12), 1189-1200.

Van Jaarsveld, J. G. S., van Deventer, J. S. J. and Schwartzman, A. (1999b). "The potential use of geopolymeric materials to immobilize toxic metals: Part II. Material and Leaching Characteristics", Minerals Engineering, **12** (1), 75-91.

Kafesaki, M., and Economou, E. (1995), "Interpretation of the band-structure results for elastic and acoustic waves by analogy with the LCAO approach", Physics Review B, **52** (18), 13317-13331.

Kanda, T. and Li, V. C. (1999). "A new micromechanics design theory for pseudo strain hardening cementitious composite", ASCE Journal of Engineering Mechanics, **125** (4), 373-381.

Kardon, J. B. (1997). "Polymer-modified concrete: review", ASCE Journal of Materials in Civil Engineering, **9** (2), 85-92.

Kingery, W. D. (1950). "Fundamental study of phosphate bonding in refractories (I): literature review", Journal of the American ceramic society, **33** (8), 239-250.

Kushwaha, M., Halevi, P., Dobrzynski, L. and Djafari-Rouhani, B. (1993), "Acoustic band structure of periodic elastic composites", Physical Review Letter, **71** (13), 2022-2025.

Lam, K. H. and Chan, H. L. W. (2005). "Piezoelectric cement-base 1 – 3 composites", Applied Physics A: Materials Science & Processing, **81** (7), 1451-1454.

Latour, M., Jolivet, S., Rahmoune, M., Lagarrigue, O. and Roure, A. (1994). "Piezopolymer transducers in the active control of vibrations", *Proceedings of* 8th *International Symposium on Electrets*. IEEE, New York, USA, 985-990.

Lebedeva, O. E., Dubovichenko, A. E. and Kotsubinskaya, O. I., et al. (2000). "Preparation of porous glasses from phosphorus slag", *Journal of Non-Crystalline Solids*, **277** (1), 10-14.

Levassort, F., Lethiecq, M., Gomez, T. and de Espinosa, F. M. (1996). "Modeling the effective properties of highly loaded 0-3 piezocomposites", *Proceeding of IEEE 1996 Ultrasonics Symposium*, IEEE, New York, USA, 463-466.

Li, Z., and Mu, B., (1998) "Application of extrusion for manufacture of short fiber reinforced cementitious composite", *Journal of Materials in Civil Engineering ASCE*, **10** (1), 2-4.

Li, Z. J., Zhang, D. and Wu, K. R. (2001). "Cement matrix 2-2 piezoelectric composites. Part I. sensory effect", *Materials and Structures*, **13** (242), 506-512.

Li, Z. J., Zhang, D. and Wu, K. R. (2002). "Cement-based 0-3 piezoelectric composites", *Journal of the American Ceramic Society*, **85** (2), 305-313.

Li, Z. J., Dong B. Q. and Zhang, D. (2005). "Influence of polarization on properties of 0-3 cement-based PZT composites", *Cement and Concrete Composites*, 27 (1), 27-32.

Li, V. C. (2005). "Engineered Cementitious composites", *Proceedings of ConMat'05* (CD-ROM), Vancouver, Canada, Aug 2005.

Litvin, A. and Belliston, H. W. (1978) "Sound Transmission Loss through Concrete and Concrete Masonry Walls", *ACI Journal*, **75** (12), 641-646.

Liu, Z., Zhang, X., Mao, Y., Zhu, Y. Y., Yang, Z., Chan, C. T., and Sheng, P., (2000). "Locally resonant sonic materials", *Science*, **289** (5485), 1734-1736.

Lyon, R. E. (1994). Technical Report DOT/FAA/CT-94/60.

Lyon, R. E., Foden, A., Balaguru, P. N., Davidovits, M. and Davidovits, J. (1997). "Fire resistant alumino-silicate composites", *Journal Fire and Materials*, **21** (1), 67-73.

Martínez-Sala, R., Sancho, J. Sánchez, J. V. Gómez, V. LLinares, J. and Meseguer, F. (1995), "Sound attenuation by sculpture", *Nature*, **378** (6554), 241.

Maslehuddin, M. Sharif, A. M., Shameem, M., Ibrahim, M. and Barry, M. S. (2003), "Comparison of properties of steel slag and crushed limestone aggregate concretes", *Construction and Building Materials*, **17** (2), 105-112.

Mazur, K. (1995). "Polymer-ferroelectric ceramic composites", *Ferroelectric Polymers: Chemistry, Physics, and Applications*, H. S. Nalwa, ed., Marcel Dekker, New York, USA, 539-610.

McCoy, M., Bett, J. and Norris, A. et al (2005). "Nanotechnology in construction: nano materials and devices offer macro improvements in concrete materials", *Proceedings of the 2nd Internation Symposium on Nanotechnology in Construction* (CD-ROM), Bilbao, Spain.

Mckinney, R. G. (1993). "Comments on the work of Harrell and Penrod", *Journal of Geological Education*, 41, p369.

Mehta, P. K. (1999). "Advanced cements in concrete technology", *Concrete International*, **21** (6), 69-76.

Morris, M. (1991). "The cast-in-place theory of pyramid construction", *Concrete International: Design & Construction*, **13** (8), 29, 39-44.

Mu, B., Li, Z., Chui, N. C., and Peng, J., (1999) "Cementitious composite manufactured by extrusion technique", *Cement and Concrete Research*, 29 (2), 237-240.

Neeper, D. A. (2000). "Thermal dynamics of wallboard with latent heat storage", *Solar Energy*, **68**, 393-403.

Nehdi M. and Mindess, S. (2001). "Microfiller partial substitution for cement", *Materials Science of Concrete VI*, S. Mindess and J. Skalny, eds., The American Ceramic Society, Westerville, Ohio, USA.

Noazad, S., Carbonell, R. G., and Whitaker, S. (1985). "Heat conduction in multiphase systems. I: theory and experiments for Two-Phase Systems", *Chemical Engineering Science*, **40** (5), 843-855.

O. Connor, D. N. and Saiidi, M. (1993). "Polyester concrete for bridge deck overlays", *Concrete International* **15**, 36-39.

Odler I. (2000) *Special Inorganic Cements*. E&FN Spon, London, New York.

Ohama, Y. (1997) "*Recent progress in concrete-polymer composites*", *Advanced Cement Based Materials*, **5** (2), 31-40.

Ohio Department of Natural Resources (2005). *Recycling Tires*, http://exchange.dnr.state.oh.us/recycling/awareness/facts/tires/

Okazaki, K. (1985). "Developments in fabrication of piezoelectric ceramics", *Piezoelectricity*, G. W. Taylor, J. J. Gagnepain, T. R. Meeker, T. Nakamura, and L. A. Shuvalov, eds. Gordon and Breach, Basel, Switzerland, 131-150.

Olson, J. R., (1997) "Thermal conductivity of fibrous insulating materials", *American Ceramic Society Bulletin*, **76**, 81-84.

Olszowy, M. (1996). "Piezoelectricity and dielectric properties of PVDF/$BaTiO_3$ composites", *Dielectric and Related Phenomena: Materials Physico-Chemistry, Spectrometric Investigations and Applications*, A. Wlochowicz, and E. Targosz-Wrona, eds., Szczyrk, Poland, 69-72.

Ouchi, M., Nakamura, S. Osterberg, T., Hallverg, S. E. and Lwin, M. (2006). Applications of Self-Compacting concrete in Japan, Europe and the Unite States. http://www.fhwa.dot.gov/BRIDGE/scc.htm

Partl, M. N., Scrivener, K. and Bartos, P. J. M. (2006) "From nanocem to nanobit: Perspective on nanotechnology in construction materials with a focus on asphaltic materials", *NSF Workshop on Nanomodifications of Cementitious Materials*, Gainesville, Florida, Aug, 2006.

Poon, C. S. (2000). "Use of recycled materials in construction products". http://www.cse.polyu.edu.hk/~cecspoon/eco.html

Raki, L. (2001). "Eco-efficient materials incorporate hazardous waste", *Construction Innovation*, **6** (2).

Ramakrishnan, V. (1994). "Properties and applications of latex-modified concrete", *Advances in Concrete Technology*, 2nd Ed. V. M. Malhotra, ed. Canada Ctr. For Mineral and Energy Technology, 839-890.

Rebeiz, K. S., Yang, S. and Fowler, D. W. (1994). "Polymer mortar composites made with recycled plastcs", *ACI Materials Journal*, **91** (1), 313-319.

Reed, D. M., Srinivasan, T. T., Xu, Q. C. and Newnham, R. E. (1990). "Effect of particle size on the dielectric and piezoelectric properties of $PbTiO_3$-polymer composites", *Proceedings of IEEE 7th International Symposium on Applications of Ferroelectrics*, IEEE, New York, USA, 324-327.

Saez, A. E., Perfetto, J. C., and Rusinek, I. (1991). "Prediction of effective diffusivities in porous media using spatially periodic models", *Transport in Porous Media*, **6** (2), 143-157.

Safari, A., Sa-gong, G., Giniewicz, J. and Newnham, R. E. (1992). "Composites piezoelectric sensors", *Piezoelectricity*, C. Z. Rosen, B. V. Hiremath, and R. Newnham, eds., American Institute of Physics, College Park (MD), USA., 195-204.

Seehra, S. S., Gupta, S. and Kumar, S. (1993). "Rapid setting magnesium phosphate cement for quick repair of concrete pavements-characterization and durability aspects", *Cement and Concrete Research*, **23** (2), 254-266.

Senbu, O. Ochi, M. and Tomosawa, F. (2004). "Current status of building materials for recycling in Japan and study of their evaluation methods", *Proceedings of Conference on the Use of Recycled Materials in Building and Structures*, Barcelona, Spain, http://congress.cimne.upc.es/rilem04/frontal/Papers.htm

Shao, Y., Marikunte, S., and Shah, S. P. (1995). "Extruded fiber-reinforced composites", *Concrete International*, **17** (4), 48-52.

Shao, Y., and Shah, S. P. (1996). "High performance fiber-cement composites by extrusion process", *Materials for the New Millennium*, Proceedings of the 4th Materials Engineering Conference, Vol. 2, K. P. Chong, ed., Washington, D. C., 251-260.

Shen, B., Yang, X. M. and Li, Z. J. (2006). "A cement-based piezoelectric sensor for civil engineering structure", *Materials and Structures*, **39** (1), 33-37.

Shi, C. J. and Li Y. Y. (1989). "Investigation on some factor affecting the characteristics of alkali-phosphorus slag cement", *Cement and Concrete Research*, **19** (4), 527-533.

Shi, C. J. (2004). "Steel slag-its production, processing, characteristics and cementitious properties", *Jounral of Materials in Civil Engineering*, **16** (3), 230-236.

Shi, C. J., Krivenko, P. V. and Roy, D. (2006). *Alkali-Activated Cements and Concretes.* Taylor & Francis, London.

Shi, C. J. and Fernández-Jiménez, A. (2006b). "Stabilization/solidification of hazardous and radioactive wastes with alkali-activated cements", *Journal of Hazardous Materials*, in press.

Shi, Z. and Chung, D. D. L. (1995). "Concrete for magnetic shielding", *Cement and Concrete Research*, **25** (5), 939-944.

Shonnard, D. R., and Whitaker, S. (1989). "The effective thermal conductivity for a point-contact porous medium: an experimental study", *International Journal of Heat and Mass Transfer*, **32** (3), 503-512.

Shukla S, Seal S, Akesson J, et al. (2001). "Study of mechanism of electroless copper coating of fly-ash cenosphere particles", *Applied Surface Science*, **181** (1/2), 35-50.

Singh, D., Wagh, A., Cunnane, J., and Mayberry, J. (1997). "Chemically bonded phosphate ceramics for low-level mixed-waste stabilization", *J. Environ. Sci. Health*, **A32** (2), 527-541.

Sobolev, K. and Ferrada-Gutiérrez, M. (2005a). "How nanotechnology can change the concrete world", *American Ceramic Society Bulletin*, **84** (10), 14-17.

Sobolev, K. and Ferrada-Gutiérrez, M. (2005b). "How nanotechnology can change the concrete world", *American Ceramic Society Bulletin*, **84** (11), 16-19.

Sreekrishnavilasam, A., Rahardja, S., Kmetz, R. and Santagata, M. (2007). "Soil treatment using fresh and landfilled cement kiln dust", *Construction and Building Materials*, **21** (2), 318-327.

Sun, S. (1983). "Investigations on steel slag cements", *Collection of Achievements on the Treatment and Applications of Metallurgical Industrial Wastes*, Sun .S. ed., Chinese Metallurgical Industry Press, Beijing, 1-71.

Su, Z. (1995). *Microstructure of Polymer Cement Concrete*. PhD Thesis, Delft University Press, Delft.

Takada, K. (2004). *Influence of Admixtures and Mixing Efficiency on the Properties of Self Compacting Concrete*. PhD Thesis. Delft University Press, Delft.

Tang, M. S. (1973). *Investigation of Mineral Compositions of Steel Slags for Cement Production*, Research Report, Nanjing Institute of Chemical Technology, Nanjing, China.

Tao, B. (1997). *Smart/Intelligent Materials and Structures* (In Chinese), Defense Industry Press, Beijing, China.

Tzou, H. S. and Guran, A. (1998). *Structronics Systems: Smart Structures, Devices and Systems (Parts I & II)*, World Scientific, Singapore.

Tzou, D. Y. and Li, J. (1995) "Some scaling rules for the overall thermal conductivity in porous materials", *Journal of Composite Materials*, **29** (5), 634-652.

Vanderwerf, P. A. and Feige, S. J. (1997). *Insulating Concrete Forms for Residential Design and Construction*, McGraw-Hill, New York, USA.

Wagh, A., Strain, R., Jeong, S., Reed, D., Krouse T., and Singh, D. (1999). Stabilization of rocky flats Pu-contaminated ash within chemically bonded phosphate ceramics, *J. Nucl. Mat.*, **265** (3), 295-307.

Wallevik, O. and Nielsson, I. ed. (2003). *Self-Compacting Concrete*. RILEM Pro33, RILEM, Cachan.

Wang, S., Han, J. and Du, S. (1999). "Development of research on fabrication and properties of piezoelectric ceramic/polymer composites (in Chinese)", *Functional Materials*, **30** (2): 113-117.

Wang, Y. and Lin, D. (1983). "The steel slag blended cement", *Silicates Industrielles*, **6**, 121-136.

Wu, S. P., Xue, Y. J. and Ye, Q. S. (2006). "Utilization of steel slag as aggregates for stone mastic asphalt (SMA) mixture", *Building and Environment*, in press.

Xu, Y. (1991). *Ferroelectric Materials and Their Applications*, North-Holland, Amsterdam, the Netherlands.

Xue, Y. J., Wu S. P. and Hou H. B. et al (2006). "Experimental investigation of basic oxygen furnace slag used as aggregate in asphalt mixture", *Journal of Hazardous Materials*, in press.

Yang, Q. B., Zhang, S. Q and Wu X. L. (2002). "Deicer-scaling resistance of phosphate cement-based binder for rapid repair of concrete", *Cement and Concrete Research*, **32** (1), 165-168.

Yoshizake, Y., Ikeda, K; Yoshida, S; and Yoshizumi, A. (1989). "Physicochemical study of magnesium-phosphate cement", *MRS Int'l. Mtg. On Adv.* **13**, 27-38.

Zayat, K. and Bayasi, Z. (1996). "Effect of latex on the mechanical properties of carbon fiber reinforced cement", *ACI Materials Journal*, **93** (2), 178-181.

Zheng, Z., Qu, Y., Ma, W. and Hou, F. (1998). "Electric properties and applications of ceramic-polymer composites (in Chinese)", *Acta Materiae Compositae Sinica*, **15** (4), 14-19.

第六章 | **Chapter 6**

高强高性能工程结构材料与现代工程结构及其设计理论的发展

叶列平，陆新征，冯鹏，Asad Ullah Qazi，汪训流，林旭川

清华大学土木工程系

北京，100084 E-mail：ylp@mail.tsinghua.edu.cn

摘　要：介绍了近年来高强高性能工程结构材料的发展现状，从结构体系和结构功能的需求论述了高性能工程结构材料在现代工程结构中的合理应用原理和方法。重点研究了高强钢筋对提高混凝土框架结构抗震性能和减轻结构的地震损伤程度的积极作用。通过静力弹塑性推覆分析和弹塑性动力时程分析方法，对在框架柱中分别使用高强钢筋和普通钢筋的混凝土框架结构的抗震性能进行了对比分析研究。分析结果表明高强钢筋配筋结构具有更优越的抗震性能，主要表现在高强钢筋配筋混凝土框架结构底层柱端塑性铰的出现显著推迟，在大震作用下可以形成更合理的屈服破坏机制，地震动力响应与普通钢筋混凝土结构基本相同，且震后残余位移小，有利于震后结构修复。最后，针对高强高性能材料工程结构与传统材料工程结构在受力性能上的差异，以及高强高性能材料结构体系的受力性能特征，提出了工程结构安全储备理论和结构设计概念及理论的发展。

关键词：高强高性能混凝土，高强高性能钢材，工程结构，安全储备，设计理论，意外事件，结构抗震

HIGH STRENGTH/PERFORMANCE STRUCTURAL MATERIALS AND THE DEVELOPMENTS OF MODERN ENGINEERING STRUCTURES AND THE DESIGN THEORY

L. P. Ye, X. Z. Lu, P. Feng, Asad Ullah Qazi, X. L. Wang, X. C. Lin

Department of Civil Engineering, Tsinghua University, Beijing, 100084, P. R. C

Abstract: This paper firstly presents the latest development of high strength/performance structural materials in recent years. The rational applications and examples of high strength/performance structural materials in the structural systems to obtain high performance are discussed and presented. The positive functions of high strength reinforcement used in reinforced concrete frame structures, that can enhance the structural performance against earthquake and reduce the structural

---

国家自然科学重点基金项目（编号50238030）；教育部科技创新工程重大项目培育资金项目（编号704003）；高等学校博士学科点专项科研基金项目（编号20040003095）资助。

seismic damage, are investigated in detail with nonlinear pushover analysis and dynamic analysis. The results show that the high seismic performance of the reinforced concrete frames, including a delayed appearance of plastic hinges at the bottom story columns feet, to form a rational failure mechanism under strong earthquake, and a small residual displacement after earthquake that cause an easy retrofitting after earthquake, can be obtained by replacing normal strength reinforcement with high strength reinforcement in the columns. Finally, the development of the safety and design theory for the structures using high strength/performance materials are discussed.

Keywords: High strength/performance concrete, High strength/performance steel, Engineering Strucrues, Structure safety, Design theory, Accidental event, Earthquake resistance.

1. 引言

工程结构材料的发展是工程结构技术和理论发展的基础。在目前和未来相当长时期内，土木工程结构的主要材料将仍然是混凝土和钢材。因此，高性能混凝土和高性能钢材是工程结构材料发展主要方向。

自从 1824 年英国人阿斯普丁（J. Aspdin）发明硅酸盐水泥，到 1872 年美国纽约建造第一座钢筋混凝土房屋，混凝土结构作为现代工程结构的主要形式距今仅 130 多年的历史。

虽然铁的出现可上溯到古代，但 1859 年贝塞麦转炉炼钢法出现至今也不到 150 年。1883 年美国的 W. B. Jenney 在芝加哥建造了 11 层的钢结构住宅保险大楼和 1889 年法国巴黎的 300m 高艾菲尔铁塔，成为现代钢结构的标志。

众所周知，长期以来工程结构中使用的混凝土材料和钢材强度较低，并具有显著的弹塑性受力特征。基于这种工程材料的力学性能，在过去的 100 多年中，随着人们对工程结构材料弹塑性受力性能和结构弹塑性承载能力认识的不断深入，工程结构设计理论经历了从弹性容许应力法到目前的考虑弹塑性极限承载能力的基于可靠度理论的极限状态设计法的发展历程。应该说，目前的结构分析理论和设计方法已发展得十分成熟和完善。然而应该指出的是，目前的结构弹塑性分析理论和设计方法的发展是基于现有材料技术所能提供工程应用的低强度弹塑性材料性能基础上的，更应该注意到的是，结构材料进入塑性，意味着产生某种程度的损伤。事实上，这种因材料达到强度而进入弹塑性阶段所引起的损伤，在工程结构的正常使用阶段通常是不被接受的。因此，考虑材料弹塑性受力性能的结构设计理论和方法，实际上只是利用材料塑性阶段的变形能力和承载潜力作为工程结构在意外事件时的安全储备。

弹塑性结构分析理论和方法固然是先进的，但又是复杂和困难的，也难以被工程技术人员很好地理解和掌握。如前所说，弹塑性结构分析理论和方法及其相应的结构设计理论和方法形成的背景是，在迄今为止的工程材料技术发展历史阶段，不能为大量土木工程结构提供价廉物美的高强度工程材料。由于工程结构弹塑性受力性能的复杂性，特别是对于可能遭遇如罕遇地震作用等意外事件情况的复杂工程结构，使得迄今为止也不能建立令人满意的工程结构抗震设计理论和方法。

随着工程结构材料技术的进步，目前高强高性能工程材料已经可以以合理的价格提供土木工程结构应用。然而，高强高性能工程材料的力学性能与传统混凝土和钢材的弹塑性性能有很大差别，其中一个重要的受力性能就是高强高性能工程材料的塑性变形能力较小，甚至是弹脆性的（如 FRP 材料）。已经熟悉结构弹塑性理论和方法的研究者们和工程技术人员，甚至不能接受这种塑性变形能力小的高强高性能材料作为工程结构材料，也就是说现有的基于弹塑性理论的设计方法和结构安全度理论，束缚了工程技术人员积极推进和推广高强高性能工程材料，成为高强高性能混凝土和高强高性能钢材工程应用的一个理论上的障碍，这实在是一种理论和技术的惯性阻力。也可以说，现有的基于弹塑性结构分析理论和设计方法已不适应现代高性能工程材料结构的发展了。

如果时间可以倒流，在 100 多年前的材料技术水平就能够为土木工程结构提供所需要的高强高性能工程结构材料，且这些材料的塑性变形能力很小，那么不知现在的工程结构分析理论、方法和设计理论是否还会涉及弹塑性？笔者认为，弹性理论和基于可靠度理论的容许承载力设计方法，这种简单而方便的方法，可能会成为工程结构设计的主要方法，尤其是工程结构的抗震设计方法就会变得更加简单（弹性反应谱方法就足够了）。

既然是这样一个结论，那么随着高强高性能材料的不断发展和应用，弹性理论和基于可靠度理论的容许承载力设计方法，就会成为未来先进的工程结构设计理论和方法。这不是历史的倒退，而是高强高性能工程材料技术的进步所带来的结构设计理论和方法的解放。

尽管如此，因为低强材料良好的塑性性能能够减小结构的动力响应，低强高性能材料在未来的工程结构中也有其特殊的作用，人们在弹塑性理论中所获得的结构知识依然具有重要意义。

本文首先介绍高强高性能混凝土和高强高性能钢材的发展现状，及其推进和推广应用高性能混凝土和高性能钢材的意义，并从结构体系角度阐述高强高性能混凝土材料和钢材的合理应用原理和方法，重点介绍高强配筋混凝土结构的受力性能和抗震性能，最后介绍可考虑不同受力性能的统一结构安全储备理论，并对结构设计分析理论的发展进行了论述。

2. 高性能混凝土和高性能钢材的发展现状

2.1 高强高性能混凝土

2.1.1 高强和超高强混凝土

高强高性能混凝土具有强度高、弹性模量高、耐久性好、耐磨性强、抗渗性强、抗冻性好，并具有流动性好、可泵性好、低坍落度损失等良好的施工性能。推进和推广应用高强高性能混凝土，可减小结构构件尺寸，有效减轻构件和结构自重，对发展高耸结构、高层结构具有重要意义，并可显著提高混凝土结构的耐久性，具有长期的综合经济性[1]。

目前我国一般将 C50 以上的混凝土称为高强混凝土，C80 以上的混凝土称为超高强混凝土。

由于高强高性能混凝土所具有的优异的力学性能和良好综合性能，上世纪 90 年代，国外发达国家积极开展高强高性能混凝土的研究和应用。德国钢筋混凝土协会于 1995 年颁布的《高强混凝土指南》，最高强度达到了C115，是目前国际上强度等级最高的技术标准。挪威在高强混凝土方面也走在世界前列，于1995年颁布的《混凝土结构设计标准》（NS3473）中使用的混凝土最高达到了C105[2]。在工程应用方面，1998年德国Rockkensnssra的Potash矿山使用了强度达105MPa的超高强混凝土[2]。1998年挪威建成的世界上最深的钻井平台——挪威Troll平台使用立方体抗压强度超过100MPa的超高强混凝土[3]。为了提高公路的耐磨性，北欧国家的许多高速公路也采用高强混凝土，抗压强度达到135MPa[2]。在建筑工程中，加拿大1983年建造的特利亚La Lanretienne高层建筑，采用120MPa超高强混凝土；1989年美国西雅图太平洋第一中心使用了130MPa的混凝土。此外，超高强混凝土还运用于核电站冷却塔及安全壳、大跨桥梁、地下隧道等。国外使用超高强混凝土的典型工程如表1。

国外高强及高强混凝土工程应用实例 表1

工 程 名 称	达 到 强 度	建 造 年
加拿大特利亚 La Lanretienne 建筑	120MPa	1983
德国 Rockensnssra 的 Potash 矿山	105MPa	1988
美国西雅图双联广场	135MPa	1988
瑞典 Abeton 城电杆	100MPa	1990
日本 Takenaka 箱形桁架步行桥	100MPa	1993
日本巴西利特21北区高层住宅	100MPa	1997
马来西亚吉隆坡双塔石油大厦	100MPa	1998

我国经过20世纪80~90年代的系统研究，C50以上的高强混凝土已得到推广应用，目前《混凝土结构设计规范》GB 50010—2002中的混凝土强度等级标准已达到C80。此后，尽管对超高强混凝土也有所研究，但缺乏系统研究，特别是在结构性能方面的研究十分缺乏，在工程应用方面就更为少见（见表2），与国外发达国家存在一定差距。主要原因是，一方面可供实际工程应用的超高强混凝土商品化还有一定的技术困难，另一方面超高强混凝土的力学性能脆性显著，按传统设计方法，考虑脆性折减系数后，在承载力方面的优势不显著。但这些原因均忽略了超高强混凝土在耐久性方面的优势。

国内超高强混凝土工程应用实例 表2

工 程 名 称	混凝土强度等级	工程建造年
北京财税大楼首层柱	C110	1998
沈阳皇朝万鑫大厦	C100	建设中
国家大剧院部分柱子	C100	建设中

目前，国内外的研究者已在实验室里研制出C150以上的混凝土，但是距离实际工程应用还需要做很多研究，包括材料层面的研究和结构层面的研究。尽管超高强混凝土具有高强度、高弹模、高耐久性和高耐磨性等综合优势，但其脆性特性成为阻遏其工程应用的一个力学缺陷。现有的基于弹塑性理论的设计方法和结构安全度理论，成为高强混凝土和

高强钢材工程应用的一个理论上的障碍。不过值得指出的是，钢筋混凝土结构的延性不是取决于混凝土，而是主要取决于钢筋的配筋率和延性。而且，从结构方面也有很多方法可以克服和改善高强混凝土的脆性，如约束混凝土、钢管混凝土、钢纤维混凝土等。事实上从解决工程实际问题来说，超高强混凝土的应用不一定是要直接利用它的高强度，而更在于利用它的高弹模、高耐久性和高耐磨性。

2.1.2 纤维混凝土

其实，混凝土材料的最大缺陷是其抗拉强度与其抗压强度不对等，抗拉强度远小于抗压强度。在混凝土中参入各种纤维形成纤维混凝土，可显著提高混凝土的抗拉强度和受拉延性。纤维可以是金属纤维、无机非金属纤维、合成纤维或天然有机纤维等[4]。纤维增强复合材料可上溯到我国古代使用稻草来砌筑土墙。

与普通混凝土相比，纤维混凝土的抗拉强度、抗折强度、抗剪强度均有显著提高，但一般情况下纤维对抗压强度的提高有限。更重要的是纤维混凝土开裂后的变形性能明显改善，材料韧性明显提高，极限应变有所提高，受压破坏时基体裂而不碎，适合于抗冲击和抗爆工程。此外，纤维混凝土还具有抗疲劳性，在耐久性、耐磨性、耐腐蚀性、耐冲刷性、抗冻融和抗渗性方面都有不同程度的提高[4]。

1979年美国学者研制出流动性砂浆渗浇钢纤维混凝土（SIFCON, Slurry Infiltrated Fiber Concrete），抗压强度达到238MPa，抗拉强度达到38.5MPa，其受压韧性达到普通混凝土的60多倍，主要用于保险柜、现浇混凝土路面、防爆结构等。1986年，丹麦研制成功的一种中等含量钢纤维混凝土，其抗压强度达到220MPa，抗拉强度达到10MPa，锚固强度是普通混凝土的3.5倍左右[5]。

我国近年来钢纤维混凝土的研究和应用已趋成熟，同时在钢纤维高强和超高强混凝土研究方面，1991年东南大学采用基体抗压强度80MPa混凝土，掺入体积率为2%的钢纤维，得到了抗压强度超过100MPa的超高强混凝土[6]。同年，湖南大学研究出了抗压强度在200MPa以上的钢纤维超高强混凝土[5]。

2.1.3 活性粉末混凝土RPC

20世纪90年代初，法国的Bouygues公司研制出一种超高强度、超高韧性和高耐久性的超高性能混凝土——活性粉末混凝土（Reactive Powder Concrete，简称RPC）[7]。RPC由级配良好的石英砂作骨料，以及水泥、硅灰、高效减水剂和一定量的钢纤维（后来也有参其他纤维）等组成，因去除了大颗粒骨料，并增加了组分的细度和活性而得名。RPC分为RPC200和RPC800两个强度等级：RPC200的抗压强度达170~230MPa，需经90℃蒸汽养护；RPC800的抗压强度可达500~800MPa，需经高温高压养护。RPC的密度大、空隙率低、抗渗能力强，耐久性显著提高，同时流动性大，比普通混凝土和现有高性能混凝土的性能有了质的飞跃。我国目前研制的RPC，抗压强度可达到140MPa。

RPC除具有超高强度和优异的耐久性外，还具有较高的韧性和良好的变形性能。以RPC200为例，抗压强度达到170~230MPa，是高强混凝土的2~4倍。抗拉强度可以达到50MPa，是高强混凝土的5倍。抗折强度达到30~60MPa，是高强混凝土的6倍左右[8]，其断裂韧性是普通混凝土的250倍（见图1）。

研究表明，RPC梁的抗弯强度与自重之比已接近钢梁，与高强钢绞线结合，加上所具有的良好的耐火性和耐腐蚀性，综合结构性能已可超过钢结构。图2为美国、加拿大、瑞士、法国共同开发建成的连接加拿大魁北克和美国的 RPC 桁架步行桥，采用 RPC200 预制构件，跨度达 60m，结构非常轻盈。构件尺寸与钢桥几乎相同。

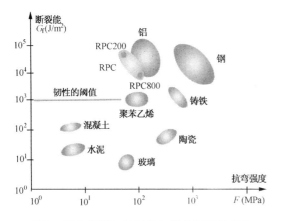

图 1　RPC 混凝土的韧性与其他材料的对比

图 2　用 RPC 建造的人行桥

2.1.4　工程化的纤维增强水泥基复合材料

由于粗骨料与水泥砂浆界面是混凝土中的最薄弱环节，因此近年来美国 Michigan 大学采用高性能纤维增强水泥砂浆，研制出一种工程化的纤维增强水泥基复合材料（Engineered Cementitious Composites，简称 ECC）。其生产工艺类似于纤维混凝土，但不使用粗骨料，纤维体积含量一般不超过 2%。由于 ECC 是基于细观层面的纤维增强机理，采用的是极细的高性能乙烯（PVA）纤维和聚乙烯（PE）纤维，基于材料细观力学设计理论和技术增强水泥砂浆，因此极大地改善了拉伸延性，甚至有类似金属材料的拉伸强化现象（见图 3），其极限拉伸应变可达 5%～6%，与钢材的塑性变形能力几乎相近，是具有像金属一样变形的混凝土材料（见图 4），这是由于高性能纤维使得裂缝分散及其细密，裂缝宽度仅 200μm，并不会因这些细小的裂缝而影响其承载能力，在很大的变形下，外观损伤很小（见图 5）[9]。由于缺少粗骨料，ECC 的抗压强度类似于混凝土，抗压弹性模量较低，但受压变形能力比普通混凝土大很多（见图 6）。此外，ECC 的耐火性和耐久性也被证明超过普通混凝土。

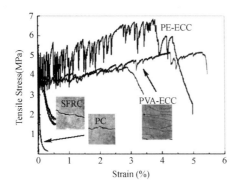

图 3　混凝土、钢纤维混凝土和 PVA-ECC 及 PE-ECC 的受拉应力—应变关系对比

图 4　ECC 弯曲性能试验，像金属一样变形

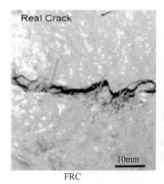 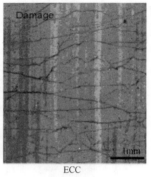 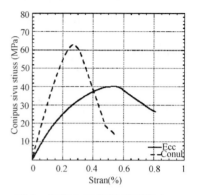

图5　普通纤维混凝土与ECC裂缝的对比　　　图6　混凝土与ECC受压应力—应变关系对比

可以认为ECC是一种具有高韧性的延性混凝土，具有很大的吸收能量的能力，因此ECC可以显著改善混凝土结构的抗震性能和抗剪性能，可用于抗震结构、抗冲击结构、结构裂缝控制和耐损伤工程结构[10]。ECC可用于结构中受力复杂的部位，如图7的S梁-混凝土柱节点，采用ECC后，其点抗震性能得到显著改善[11]。由于ECC具有与钢材基本一致的变形能力，ECC可用于混凝土结构中一些塑性变形较大的构件和部位，如在塑性铰区采用ECC，可在很大的塑性变形阶段保持塑性铰的完整性，使塑性铰具有更稳定塑性滞回耗能能力。图8是普通钢筋混凝土受弯构件和仅配纵筋ECC构件（无箍筋）在反复荷载下受力性能的对比，可见配筋ECC构件的塑性铰区在发生很大的塑性变形情况下仍保持较好的完整性，并具有稳定的滞回性能和滞回耗能能力，承载力退化小，且外观损伤很小[12]。这种"低损伤性"特别有助于抗震结构，使得结构能够在大地震的反复作用下也不遭受严重损伤，震后结构可免于修复或减少修复。日本首先将ECC用于高层建筑联肢剪力墙的连梁，利用其在大变形下和剪力作用下良好的耗能性能作为结构中耗能构件，减小大震下结构地震响应，提高结构的抗震性能，并可使得结构震后实现免修复（见图9）。

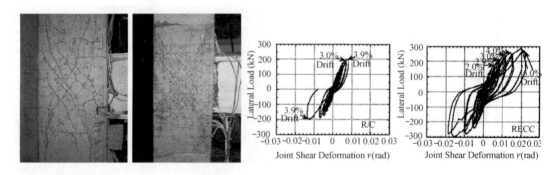

图7　S梁-RC柱节点与S梁-ECC柱节点抗震性能的对比

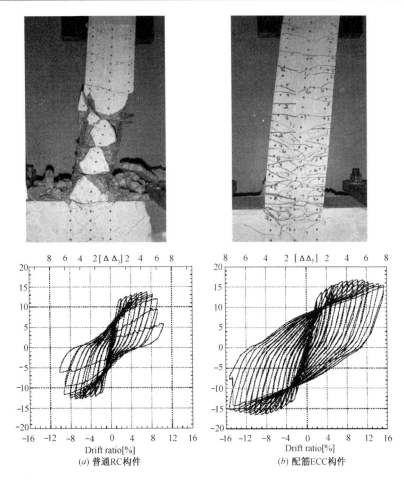

图 8 反复荷载作用下普通 RC 构件与配筋 ECC 构件受力性能的对比

2.2 高性能钢材和钢筋

2.2.1 高性能钢材

目前，高强高性能钢材的发展趋势主要有（图10）[13,14]：

(1) 高强度等级钢材和超厚板钢材，以满足建筑高层化和大跨距的发展需求；
(2) 低屈强比钢和极低强度钢，以提高结构的抗震性能；
(3) 高效焊接钢，提高钢材焊接性能，尤其是提高大厚度钢板的焊接性能；
(4) 耐火结构钢，可节省钢结构耐火被覆成本，提高钢结构抗火性能；
(5) 耐候钢，提高钢结构的防腐涂装和耐久性能。

随着建筑结构的高层化和大跨距的发展，高强度和大厚度钢板成为高性能钢材的首先发展目标。国外目前主要使用490MPa级和590MPa级钢材，780MPa级钢材也正在积极推广使用。与此同时，厚度超过40mm时还能保证钢材力学性能、并具有良好焊接性能的建筑用特厚钢板和特厚 H 型钢也已研制成功，钢板厚度达 90~100mm，H 型钢翼缘厚度达 70~90mm。国外高强高性能钢材的性能见表3。

我国现行的标准《低合金高强度结构钢》GB 1591—94 有 5 个强度等级，即 Q295、

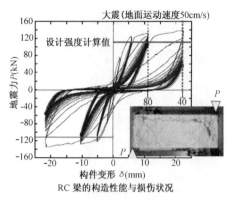
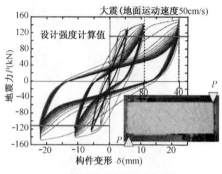

(a) 反复荷载下RC连梁与ECC连梁受力性能的对比

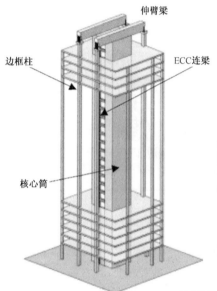

(b) ECC预制连梁及应用

图9 日本ECC连梁的研究、应用与施工

Q345、Q390、Q420、Q460，目前的主要建筑用钢为Q235和Q345（相当于490MPa级）。

除高强度和大厚度钢材外，高性能钢材还体现在：低屈强比、低强度高延性、高焊接性能、耐火性能、耐候性能，其中与结构受力性能有关的是低屈强比钢材和低强度高延性钢材。

屈强比是指钢材屈服强度与抗拉强度的比值。若近似认为地震作用下框架梁承受等梯度弯矩作用，则屈强比越小，梁端达到抗拉强度极限时，梁端进入屈服的塑性铰区长度就越大，结构的塑性耗能能力就越强。因此，低屈强比钢材能更好的发

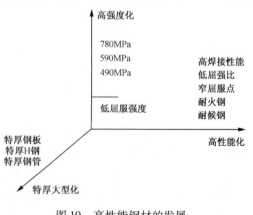

图10 高性能钢材的发展

挥钢材的塑性变形耗能能力,提高结构的抗震安全性。目前,日本已开发出屈强比小于 0.8 的 590MPa 和 780MPa 的高强厚钢板(80~100mm)(见表 3)[13]。为充分保证结构实现预期屈服机制,满足抗震结构在预期塑性铰区的耗能能力,在满足低屈强比条件下,屈服强度的变异性(窄屈服点)也是一个重要的方面,这需要对钢材原材料和生产工艺的严格控制[15,16]。

建筑用高强度钢的力学性能 表 3

强度级别	板厚(mm)	f_s(MPa)	f_b(MPa)	屈强比	δ(%)
590MPa	19~100	440~540	590~740	≤0.80	≥20
780MPa	25~100	≥620	780~930	≤0.85	≥16

在追求高强度钢材的同时,作为专门耗能用的极低屈服强度高延性钢材也在日本研制成功,其屈服强度约 100MPa,而延伸率可达到 50%~60%,其屈强比仅为 0.45~0.55(见表 4)[13]。这种钢材用于制作专门的滞迟型耗能阻尼器(见图 11)。

低屈服强度和极低屈服强度钢的力学性能 表 4

钢 号	规格(mm)	f_s(MPa)	f_b(MPa)	δ(%)
LY235	10~40	215~245	300~400	≥40
LY100	6~12	90~130	200~300	≥50

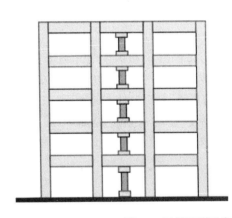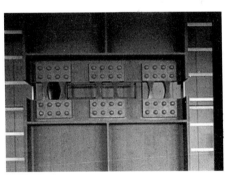

图 11 极低屈服强度高延性钢材耗能阻尼器

除上述在力学性能方面所具有的各种高性能钢材外,为提高钢结构的防火性能和耐腐蚀性能,20 世纪 80 年代日本通过在钢中添加微量的 Cr、Mo、Nb 等合金元素开发出了强度达 390~490MPa 的耐火耐候钢。这种钢材在 600℃高温下一定时间内(通常为 1~3h),其高温屈服强度为常温标准值的 2/3 以上;而在常温下,其各种性能与普通焊接结构钢相同,无论加工或焊割,还是在服役使用中表面擦撞或火灾后,其耐火耐候性不变。我国也研制出低屈强比高强度建筑用耐火钢,室温屈服强度达到 490MPa,屈强比 0.76，延伸率大于 17%；600℃温度时屈服强度达到 367.9MPa，大于 2/3 室温屈服强度[17]。

2.2.2 高性能钢筋

目前，国际上 200、300 级钢筋已基本淘汰，400 级钢筋成为普遍应用的品种，而发达国家钢筋的强度已达到 500 级或更高。同时，对钢筋的延性也提出分级要求，均匀伸长率（δ_{gt}）一般要求不小于 5%（强屈比不小于 1.08），抗震要求不小于 7%（强屈比不小于 1.15，也不大于 1.35），最低限度为 2.5%（强屈比不小于 1.05）。预应力配筋则全部采用中高强钢丝钢绞线，并向高强（1860MPa）、低松驰（25%）方向发展。

我国上世纪 50、60 年代使用低碳钢筋（HPB235）；70 年代通过低合金化（20MnSi）使强度提高 40%（HRB335）；80 年代进一步微合金化（20MnSiV）强度又提高 20%（HRB400）。目前，强度再提高 25% 的 HRB500 级钢筋已具备生产能力[18,19]。

低强度钢筋不仅导致配筋密集（例如在节点处），难以浇筑混凝土，而且导致安全度水平也难以进一步提高。实际上，随着强度的提高，钢筋的强度价格比逐渐提高，低强钢筋的经济效益反而最差。尽管在混凝土梁中应用高强钢筋，会因为裂缝和变形控制要求使得强度不能得到充分利用，但承载力储备却大为提高[20]。事实上，目前在工程应用中已经注意到，在同样的荷载水平下，其实混凝土梁的裂缝宽度并没有试验梁中裂缝宽度大，这是由于实际工程中混凝土梁两端还受到结构中其他构件的约束。从本文后面给出的关于结构体系中高性能材料的应用原理角度，框架梁中宜采用普通低强钢筋，使用高强钢筋主要是为了减少配筋密集程度（但还是需要配置一定的普通钢筋），且对于简支梁可提高其承载力安全储备程度。另一方面，在混凝土柱中应用高强钢筋，将可显著提高柱端塑性铰屈服转角，从而有利于避免形成柱铰屈服破坏机构，这一问题将在本文第 4 部分专门介绍。

20 世纪 90 年代，我国采用国际标准开始生产高强钢丝和钢绞线，强度达到 1570～1860MPa，2000MPa 以上的预应力钢筋也已试制成功。这些高效预应力筋（如三股钢绞线、螺旋肋钢丝等）不仅高强，而且有相当好的延性（均匀伸长率 δ_{gt} 大于 4%～5%）和锚固性能。其最明显的优势是高强度和高效率，强度价格比提高 40% 以上，而且不会发生脆性断裂破坏。表 5 和表 6 是我国目前钢筋和中高强钢丝、钢绞线的基本性能。

热轧钢筋的基本性能　　　　表 5

级 别	牌 号	规格 d（mm）	屈服强度 f_y（MPa）	抗拉强度 f_b（MPa）	伸长率 δ_5（%）
I	HPB235	8～20	235	370	25
II	HRB335	6～25, 28～50	335	490	16
III	HRB400	6～25, 28～50	400	570	14
IV	HRB500	6～25, 28～50	500	835	10

中高强钢丝钢绞线的基本性能　　　　　表6

类　型	牌　号	规格 d (mm)	屈服强度 $f_{0.2}$ (MPa)	抗拉强度 f_b (MPa)	伸长率 δ_{100} (%)
低合金钢丝	YD800 YD1000 YD1200	5 7 7		800 1000 1200	4 3.5 3.5
预应力钢丝	应力消除钢丝 刻痕钢丝 螺旋肋钢丝	4~9	620~1500	800~1770	3~4
预应力钢绞线	二股 三股 七股	5~12 6.2~12.9 9.5~15.2	1250~1580	1470~1860	3.5

2.3　高强高性能材料的应用与可持续发展

作为土木工程结构的主要材料，无论是钢材，还是混凝土，都是一种能源消耗性的材料，目前我国每年烧砖毁田 8000km^2、1t 钢铁消耗标准煤 1.66t；水 48.6m^3、1t 水泥消耗标准煤 178kg；同时放出约 1t 二氧化碳，消耗了大量自然资源，并在相当程度上污染了环境，对环境生态造成一定影响。此外，因达到耐久性极限或不能继续使用等原因，工程结构拆除也会造成大量的固体废弃物。在这样的背景下，从 20 世纪 90 年代开始，出于对有限资源的节约和对环境保护意识的不断增强，可持续发展成为政府的重要战略目标，并逐渐形成"可持续性工程"和"绿色工程"的概念。基于这一目标，有效降低资源和能源的消耗，提高我国工程结构的耐久性，节约工程结构材料的使用量，提高我国工程结构的安全水准，应积极推进和推广高强高性能材料的应用。

早期采用钢材和混凝土建造的许多工程项目，尤其是大型基础设施工程项目，由于各种原因出现了影响其耐久性和使用寿命的问题，大量的养护和加固费用支出使得人们认识到工程结构耐久性的重要性。同时，随着我国经济高速发展，大规模工程建设还将持续相当长的时间。由于所具有的独特性能和经济性，在未来相当长的时期内，混凝土和钢材将仍然是工程结构材料的主体。如果在这一时期不重视工程结构的耐久性和安全性，将在未来不长的时间内，因维修加固这些工程，造成大量的资源消耗。

高强高性能混凝土不仅强度高，并具有耐久性好、抗渗性强、抗冻性好，与高性能钢材的有效结合，一方面可适应现代工程结构向高层和大跨发展的需要，减小结构构件尺寸，有效减轻构件或结构自重，不仅可以节约材料用量，减少资源消耗和材料生产过程中的污染物排放，更重要的是可显著提高混凝土结构的耐久性，延长工程结构使用寿命，其所带来的长期经济效益和可持续发展是难以用具体指标来衡量的。

如果能将我国混凝土结构的主导受力钢筋强度提高到 400~500MPa（HRB400 级和 HRB500 级），则可节约钢筋用量 30%[19]。

我国在第六个五年计划时已完成 HRB400 级钢筋的研制和生产，但至 2002 年，上述钢筋的用量仍不足10%[19]。其原因,一方面有技术规范不配套的问题;另一方面有我国新技术推广应用体制方面的问题。还有一个问题就是,与高性能材料相适应的结构设计理论发展严重滞后,甚至一些观念还阻遏了高性能材料的推广应用。本文将针对这一问题进行讨论,并简要论述推广应用高强高性能材料的结构受力性能合理性和经济合理性。

3. 结构体系与材料性能的利用

3.1 概述

随着高强高性能工程结构材料的发展,高性能材料的多样化,为发展各种高效新型结构体系提供了新的途径。

从结构整体角度来说,结构应满足以下性能目标要求:
(1) 正常使用情况下的适用性要求;
(2) 长期使用条件下的耐久性要求;
(3) 意外情况下的安全性要求;
(4) 意外事件后的低损伤性要求。

一般来说,按照规范正常设计的工程结构通常是能够满足在正常使用情况下安全性要求。而采用高性能耐久性材料,如高强混凝土和耐候钢,以及利用高强钢筋和高强混凝土形成的高效预应力技术,可以极大地提高结构的耐久性。本文不重点讨论这方面的材料利用问题,而是主要讨论高强高性能材料利用与意外事件下结构的安全性要求和低损伤性要求。

人们对工程结构安全性的要求,通常是要求在遭遇到意外事件时,不产生与其原因不相称的垮塌,或造成不可接受的重大人员伤亡和财产损失。而对工程结构低损伤性要求,则是希望在意外发生后,工程结构不需修复或快速修复后,可以继续使用。

意外事件,如罕遇地震、爆炸、冲击等,往往属于极小概率事件,其量值难以预计,作用方式和形式也具有很大的随机性和不确定性,因此不可能要求像对待一般正常使用荷载作用那样要求工程结构在意外事件发生时仍然无任何损坏,通常是容许结构产生一定程度的损坏,但不能导致垮塌。

另一方面,意外事件作用通常具有显著的动力特性。动力作用对结构影响的重要特性是,其作用量值与结构自身的动力特性有很大关系。结构在动力作用下的性能与静力作用(一般荷载作用通常属于静力荷载)下的性能有很大差别。静力作用量值通常不会随结构的性能变化而变化,因此结构设计时应主要控制其不超过结构的承载力(但需要一定的可靠度)。而动力作用效应的量值会由于结构进入塑性而显著减小。这种动力效应减小来自于两方面,一是结构进入塑性后导致结构的自振周期增大,二是塑性滞回所形成的耗能作用。

因此,结构的塑性是结构能够经受意外事件动力作用所应具备的重要特性要求。但这并不是说整个结构都要采用塑性材料。因为,塑性变形和塑性滞回耗能意味着结构构件的损伤,如果整个结构都进入塑性,即使能够避免意外事件作用下的倒塌,这与结构的"低损伤性"要求也是不一致的。而不同的高性能工程结构材料在结构体系中的合理应用,则可以很好解决这个矛盾。

其实,由结构动力学原理可知,一个阻尼很大的弹性结构,其动力响应也很小。而弹性结构体只要不超过其材料强度极限，则不会产生损伤破坏。在动力作用下，材料进入塑性阶段所形成的滞回耗能等效于增大了结构的阻尼。因此在结构体系中，高强高性能材料和低强高性能材料的合理利用，就能够同时实现意外事件作用下的安全性和低损伤性要求。我们提出的结构体系中高强高性能材料的合理利用思想是：结构主体结构构件尽量采用高强高性能材料，使得在意外事件作用下不超过其屈服强度，从而不引起涉及结构主体的损伤；结构次要结构构件尽量使用低强高性能材料，使得在意外事件作用下，一方面利用其进入塑性来改变结构的自振周期特性，另一方面利用其塑性滞回耗能减小意外事件的动力响应。尽管次要构件进入塑性阶段会造成一定程度的损伤，但这种损伤不会影响主体结构的安全性，且便于修复。以下将结合一些具体应用实例说明结构体系中高性能材料的合理利用。

3.2 结构体系的层次性与高性能材料的利用

本文作者在文献[20]中指出，结构应具有层次性，这种层次性既可包含在结构体系中，也可包含在结构构件中。

结构体系的层次性，一是指多重结构体系，二是指结构中不同重要性构件的层次性。

多重结构体系具有两个以上的整体型子结构，当其中一个整体型子结构在意外事件作用下遭受一定程度破坏时，其他整体型子结构依然具备一定的承载能力，能够保持整个结构体系的整体稳定性。显然，不同子结构具有不同的重要性，并应具有不同的安全储备度。对于重要程度高的子结构应采用高强材料，而对于次要子结构则可采用普通材料，且要求具有足够的延性和滞回耗能能力。

如尼加拉瓜美洲银行大厦（图12），采用筒中筒结构体系，其中核心筒又由四个小筒通过连梁连接构成，形成了多重结构体系。由于该结构具有很好的层次性，既保证整个结构在正常使用情况下具有良好的工作性能（风荷载下的舒适度），又在遭遇罕遇地震时利用连梁屈服后的塑性和滞回耗能性能减小地震动力响应，并形成良好的抗震耗能机制，大大减小了地震对主体结构的影响。前述日本将ECC高性能材料用于核心筒或剪力墙的连梁，就充分利用ECC良好的塑性变形能力和耗能能力进一步提高结构的抗震性能，并可使得连梁具有"低损伤性"，震后可免于维修。

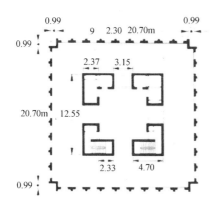

图12 尼加拉瓜美洲银行大厦结构平面

图13 北京电视中心

又如北京电视中心（图13），采用巨型框架结构体系，其抗震设防目标为：在多遇地震作用下，主框架和次框架都处于弹性状态；在中震作用下，构成巨型框架的桁架梁和巨型柱处于弹性状态，因此采用高性能钢材（Q345GJ），而附属次框架部分结构构件容许在中震下屈服；在罕遇地震作用下，次框架首先屈服消耗地震能量，因此采用普通钢材（Q345）[21]。

结构体系中，不同的构件，其重要性程度也是有层次性的。正确区分结构体系中的关键构件、一般构件和次要构件是保证结构在意外事件作用下具有足够安全性的前提。

所谓关键构件是指其破坏容易引起结构大范围的破坏或垮塌的构件。相对于关键构件，结构中的次要构件是指那些破坏后不会导致整个结构严重破坏的构件。次要构件的破坏甚至不会使得结构达到最大承载力或极限变形，或不会导致结构的承载力有很大降低，或者也不会使得结构形成几何可变体系。作为一种特殊的次要构件——赘余构件，将在后面专门讨论。除关键构件和次要构件以外，其他结构构件属于一般构件。一般构件的破坏对整体结构的承载力有一定影响，但不会导致整体结构的承载力产生急剧降低。通常，一定数量的一般构件破坏后才会导致整体结构的严重破坏。

明确结构体系中不同构件的重要性层次，合理采用相应的结构材料，可以使得结构整体具有更优异的结构性能。如，对于钢结构，可使用高强度钢（如490MPa、590MPa和780MPa钢，我国目前在建筑结构中尚无高强钢，但已推出Q235GJ、Q235GJZ和Q345GJ、Q345GJZ，比现有的Q235和Q345的设计强度要高）作为结构的关键构件（如前述图13的北京电视中心的巨型框架部分采用Q345GJ），使整个结构具有更高的承载力，在意外事件作用下具有较小的整体损伤，从而获得整体的合理经济性。对于一般构件，则可采用一般结构钢（如前述图13的北京电视中心的附属次框架部分采用Q345。笔者建议附属次框架梁可以采用Q235）。而作为结构中起耗能作用的次要构件（耗能构件）通常要求先于主体结构屈服（如连梁），可采用低屈服强度和高延性钢材，以保证结构在意外事件作用下塑性滞回耗能作用。

图14所示为本文作者参与设计的北京通用时代1号楼，四角采用钢支撑，是结构体系中的主子结构，其抗震设防目标要求在中震下基本处于弹性，但其中的支撑采用UBB耗能阻尼器，采用Q235钢材。

图14 北京通用时代1号楼

3.3 结构破坏模式与高性能材料的利用

结构在意外事件作用下的性能还在很大程度上依赖于结构的破坏模式。本文作者在文献[20]中指出，具有整体型破坏模式的结构体系对提高整体结构在意外事件作用下的抗损伤能力和抗倒塌能力才具有实际意义。具有整体型破坏模式的结构体系中，各种结构构件的层次性明确，即具有整体型关键构件、一般构件、次要构件和赘余构件，次要构件和赘余构件的破坏，乃至从结构去除，都不会对整体结构的安全性有重大影响。从结构在意外事件作用下的塑性滞回耗能（相当于结构的等效阻尼）角度来看，整体型破坏模式结构可以使得更多的次要构件或赘余构件进入塑性阶段，有利于更多地耗散动力输入能量，减小意外事件作用所产生的不利响应。

以抗震结构为例来说，对于普通框架结构，尽管采取了"强柱弱梁"等抗震设计概念和措施，但柱底塑性铰是难以避免的，同时由于地震作用对结构影响的随机性，其他楼层框架柱上端出现塑性铰的可能性也难以避免，因此即使是按"强柱弱梁"设计的框架结构，也难以避免会出现局部型破坏模式，至少形成这种局部型破坏模式仍达到一定的概率[22]。相比而言，剪力墙结构的破坏具有整体型破坏模式的特征，这是大量震害经验显示剪力墙结构抗震性能优于框架结构的重要原因之一，而两种结构关于地震作用大小的"刚柔"之争则相对是次要的。同样，筒体结构、束筒结构、巨型框架结构等也是具有整体型破坏模式特征的结构，且便于合理利用不同高性能材料，是值得发展的结构体系。

为了使得结构具有整体型破坏模式，可以将结构划分为不同的子结构。对于重要的子结构应优先采用高强高性能材料，以提高其承载力安全储备，并使得这些关键子结构对整体结构的破坏模式起到控制作用。如在北京通用时代1号楼工程设计中（图14），笔者在四个角部桁架部分就采用高等级钢材并提高安全等级的方法，以实现对结构破坏模式的整体控制。

应该指出的是，从结构体系整体角度而言，高强弹性材料对保证整体结构的承载力、低损伤性和可修复性具有重要意义，因为：

（1）高强弹性材料强度高，结构构件尺寸小，自重轻，便于实现超高度和大跨度工程结构，也有利于较小地震等意外事件作用的动力响应。

（2）高强弹性材料弹性范围大，在弹性范围无损伤，且具有良好的弹性回复能力，有利于结构在经受大变形后的复位。

（3）高强弹性材料弹性变形能力大，尽管高强弹性材料在达到其极限强度时往往具有脆性破坏特征，但其相应的变形能力与低强材料的延性是相适应的（如由图15中高强钢绞线的实测应力—应变关系可知，其极限延伸率已达到6%，而工程结构构件中普通钢筋的实际应变利用程度在3%~5%），结构体系中高强高性能材料构件与低强高延性材料构件的结合，有利于整个结构体系形成合理的损伤破坏机制，有利于减小和抵御意外事件的动力作用。

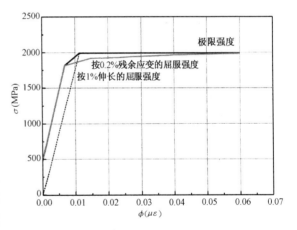

图15　高强钢绞线应力—应变关系

（4）在正常使用阶段，高强弹性材料的应力水平通常远低于其强度，高强材料结构构件的承载力安全储备高。从安全储备理论来分析，对于意外事件的作用，承载力储备要比塑性变形能力储备更有意义。塑性变形能力储备的最重要功能是改变结构自身的动力特性和耗散动力输入能量，减小结构在意外事件作用下的动力响应，并使结构尽快停止振动。

因此，对于整个结构体系，利用高强高性能材料作为主体结构和关键构件，以保证整个结构的整体性和承载力，及其低损伤性；利用低强高延性材料作为次要构件和赘余构件，利用其塑性变形和滞回耗能能力，减小意外事件作用引起的结构动力响应，是合理利用高性能材料，形成高性能结构体系的重要方法和发展方向。

3.4 赘余构件与高性能材料的利用

尽量形成超静定结构，特别是利用赘余构件增加结构的超静定次数，是提高结构安全储备，增强结构抵御意外事件作用能力的重要的结构上的措施。

赘余构件是一种特殊的次要构件，在正常使用情况下通常不起承载作用或只起很小的作用。但赘余构件可增加结构的刚度，以满足结构指出使用情况下的适用性要求，如抵抗风荷载下引起的结构振动。在意外事件作用下，赘余构件的破坏、甚至退出（从结构中去除）不会影响整个结构的完整性。赘余构件可以看作是结构在遭遇意外事件作用时的自动保险，即以赘余构件的损伤和破坏来达到保全和避免主体结构的严重震害和破坏。虽然赘余构件的采用可能违背工程经济与简洁的概念，但作为一种特殊的结构安全储备措施，对于结构抵御不可预测的意外作用具有重要作用[23]。许多消能减震结构，特别是采用位移型阻尼器的消能减震结构，位移型阻尼器实际上都是赘余构件。对于抵抗意外事件作用来说，合理设置赘余构件的概念，可能比计算设计更为重要。

根据上述赘余构件的功能，赘余构件应先于主体结构构件破坏，且赘余构件应具有足够的塑性变形能力，使得其破坏后仍可在一定程度上保持结构的整体性，并利用其塑性变形和滞回耗能能力来减小意外事件作用下的动力响应。由于要求赘余构件先于主体结构构件破坏，因此赘余构件的安全度不应提高，反而应该降低，这与前述不同重要程度构件层次性材料合理选择原则是类似的，即赘余构件应采用低强度高延性的高性能结构材料，但又可以采用一些更特殊的高性能材料，如：极低屈服强度钢材，就可以专门用于作为结构中的赘余构件；ECC材料也是作为混凝土结构中赘余构件的良好材料。如图16分别采用极低屈服强度钢材和ECC作为结构的位移型滞回耗能阻尼器，这些材料因与主体结构材料基本一致，不仅使得主体结构具有统一性，而且易于施工安装。日本也首先将极低屈服强度钢材用于无粘结耗能钢支撑（图17）。

为保证赘余构件能在预期的目标下屈服进入塑性阶段，其材料屈服强度的离散性必须得到控制，从而避免因不必要的材料强度偏大而导致赘余构件不能先于主体结构构件屈服。同时，赘余构件材料还必须具有很大的塑性变形能力和低周疲劳性能。这些都可以看作是对结构材料高性能的要求。

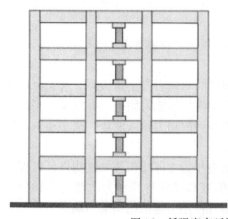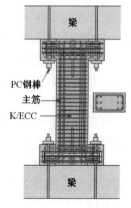

图16 低强度高延性采用作为结构中的赘余构件

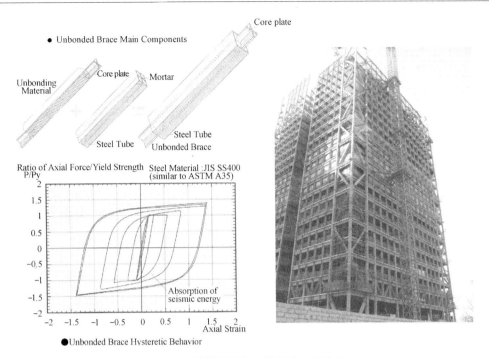

图 17　无粘结耗能钢支撑及其工程应用

3.5　结构构件性能与高性能材料的利用

除在整个结构体系中合理利用高性能材料外，也可根据结构构件的受力特点，合理利用高性能材料，使其具备和适应多种结构性能的要求。以下通过几个具体的例子给予说明。

3.5.1　高、低强钢筋组合配筋混凝土构件

普通钢筋混凝土构件，在钢筋屈服后表现为较好的塑性变形能力，但承载力基本维持不变或有所降低。将部分钢筋用高强钢筋替换，则在普通钢筋屈服后，因高强钢筋尚未屈服，构件的承载力可以继续增大。这可显著提高构件承载力安全储备，而构件的变形能力和塑性耗能能力不会降低，且这种替换也不会显著提高构件的造价。当采用极低强度钢筋与高强钢筋组合配筋，则可形成具有自身滞回耗能阻尼的钢筋混凝土构件。

图 18 所示为高、低强钢筋组合配筋混凝土柱受力性能数值分析结果的对比，高强钢筋为 1860 钢绞线，低强钢筋为普通 HRB335 级钢筋，按面积组合配筋比 $A_{ns}/(A_{ns}+A_{hs})$ 分别为 0（全部高强钢筋）、0.5（部分高强、部分普通钢筋）和 1.0（全部普通钢筋）三种情况进行了分析。柱最大侧移角为 1/25，由图 18（a）可见高强钢筋柱的侧移变形能力完全可以达到普通配筋柱的塑性变形能力；而由图 18（b）可见，随着高强钢筋配筋比例的增加，柱的滞回耗能能力在 $A_{ns}/(A_{ns}+A_{hs})=0.5$ 前呈线性增加。另一方面，随着高强钢筋配筋比例的增加，反复荷载作用后柱的残余变形随之减小，残余变形的减小主要来自于高强钢筋的弹性恢复。由此可见，采用组合配筋（其实全部采用高强钢筋配筋柱的滞回耗能能力最大，且残余变形最小，但配置一定的低强钢筋有利于结构动力特性的改变和良

好的综合耗能性能），可以使得钢筋混凝土柱具有更好的抗震性能。有关高强配筋混凝土结构的抗震性能分析将在以下第 4 部分专门介绍。

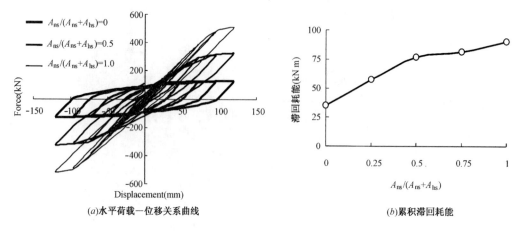

图 18　高、低强钢筋组合配筋混凝土柱在反复荷载作用下的受力性能

3.5.2　分布式滞回耗能阻尼结构构件

基于上述同样思想用于钢结构，将极低强度高延性钢材设置于钢结构梁柱构件端部，可在结构中形成分布式滞迟型阻尼（图 19）[24]，从而可大大提高整个结构的耗能能力，且不影响结构的使用空间。

3.5.3　外包 ECC 混凝土构件

混凝土的低抗拉强度和低断裂应变，通常是造成结构构件外观损伤的主要原因。利用 ECC 的高韧性和高抗拉延性，将其外包于混凝土构件外部，从而可减小构件的外观损伤，显著提高构件的耐久性，同时可相应减小混凝土构件的配箍。如果与前述高、低强钢筋组合配筋相配合，则可形成既具有滞回阻尼耗能，又具有"低损伤性"的混凝土构件。

3.5.4　钢管混凝土叠合柱

高强混凝土通常脆性显著，但采用横向约束可显著改善其脆性破坏特征，尤其是钢管高强混凝土是一种既具有高抗压承载力，又具有良好弹塑性变形能力的构件。但这种构件往往仅在轴压荷载下性能优异，而在弯曲荷载下与普通混凝土构件的承载性能相

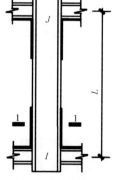

图 19　钢结构梁柱端设置极低屈服强度钢作为分布式阻尼器

差不大（当然变形能力可以大大提高）。钢管混凝土叠合柱就是利用这一特性，在施工阶段采用钢管高强混凝土，使得结构的大部分自重产生的压力由钢管高强混凝土承担（并因其高模量而可以分担更多后期增加的轴力），而外包钢筋混凝土则主要承担水平荷载引起内力。这一组合构件方法在沈阳某高层建筑中应用（图 20），既获得了良好的结构性能，又取得了良好的综合经济效益。

3.5.5　CFRP—铝合金组合构件

铝合金具有类似低碳钢的屈服特性，且延伸率可达到 15% 以上。CFRP 纤维方向的强度是普通铝合金的 10~15 倍，弹性模量为其 2~3 倍，但 CFRP 基本为线弹性材料，延伸

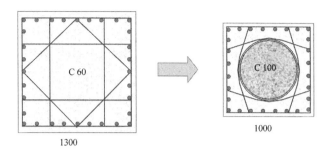

图 20　钢管混凝土叠合柱

率仅为 1.5% 左右。这两种结构材料都具有质量轻、抗腐蚀性好的优点，通过组合能够更好地发挥它们的共同的优势。同时，借助铝合金的机械连接方法可解决 CFRP 构件连接困难的问题，使 CFRP 的利用更为有效。

CFRP—铝合金组合构件与前述高低强配筋混凝土构件和高低强钢材组合构件具有类似的受力性能，在正常使用阶段，两种材料共同受力，随着受力增大，铝合金部分因先进入塑性变形阶段而使得结构的动力特性改变，并形成塑性滞回耗能能力，而 CFRP 部分仍保持弹性，可使得结构能够在意外超载时具有可持续增大的承载能力，并可使得意外超载后结构的的变形得到恢复。同时，利用铝合金的塑性变形能力还可避免纯 CFRP 构件的脆性破坏。图 21 所示为纯铝受弯构件和 CFRP—铝合金组合受弯构件的荷载—挠度试验曲线对比，可以看 CFRP—铝合金组合构件的承载力和刚度得到增大，且残余变形显著减小。

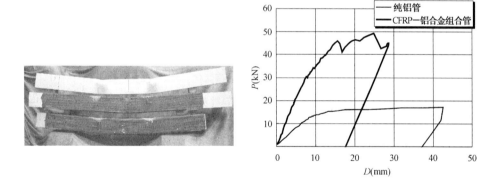

图 21　CFRP—铝合金组合受弯构件及其受力性能

4. 高强配筋混凝土框架结构抗震性能分析

4.1　概述

本节采用数值分析方法，对配置高强钢筋与普通钢筋的混凝土框架结构的抗震性能进行了系统全面的对比分析，深入阐明采用高强材料对结构抗震性能的改善，希望研究结果将有助于高强高性能材料在其他工程结构中的应用。

对于普通钢筋混凝土框架结构，尽管采用体现了构件层次性的"强柱弱梁"抗震设计原则，期望使塑性铰限制在梁端，但如 Paulay 和 Priestley 指出[25]，柱底塑性铰是不可避免的。另一方面，由于普通框架结构的梁柱均采用普通钢筋，框架柱的屈服变形与框架梁的屈服变形相近，加之地震作用的随机性和结构进入弹塑性阶段结构动力特性变化的不确定性，因此其他楼层框架柱上端出现塑性铰的可能性依然存在[22]。因而对于普通钢筋混凝土框架结构，即使采用了"强柱弱梁"抗震设计原则，仍有可能导致形成楼层破坏机构[26]。此外，Fischer 和 Li（2003）还指出[26]，因底层柱底部塑性变形过大，即使震后结构不倒塌，结构震后的残余位移也会很大，这将给结构修复工作带来更多困难。因此，减少震后结构残余变形也是结构抗震性能的一个重要目标。

针对普通钢筋混凝土框架结构抗震性能方面所存在的上述问题，本文作者建议通过在框架柱中引入了高强钢筋，来提高框架柱的抗弯承载力和相应的弯曲屈服变形能力，从而增大框架柱和框架梁之间屈服变形差，在变形层次上实现"强柱弱梁"。同时，利用高强钢筋的弹性变形恢复能力，减小震后结构的残余变形，以改善结构的震后可修复性。目前，1860 级高强钢绞线已可以以合理的价格提供使用，且仅在框架柱中引入高强钢筋，对整个结构的造价增加很少，但从下面的分析结果可知，其对整个结构的抗震性能将有很大的改善。

事实上，在预应力混凝土结构抗震性能的研究中，人们就已经发现其残余变形远小于混凝土结构，具有复位性能。但由于认识所限，一直认为这种复位性能使得预应力混凝土结构的塑性变形和滞回耗能能力小于钢筋混凝土结构，因而将这种现象认为是预应力混凝土结构抗震性能的一个不利因素。然而如前所述，在一个结构体系中，结构的塑性滞回耗能能力完全可利用结构其他低强高延性构件来实现。在这个思路下，高强高性能钢筋的应用已不再仅仅局限于提供预应力，而是利用其高强度和弹性恢复能力作为控制结构性能的一种措施。同样，将这种思想引入预应力混凝土结构，将会改变传统预应力混凝土结构抗震性能不如钢筋混凝土结构的观点。另外，需要进一步强调的是，高强钢筋的塑性变形小，并不代表其变形能力小，只要高强钢筋的极限变形能力能够与结构形成破坏机制时结构的极限变形能力需求相适应，就可以认为满足要求，至于结构对于意外事件动力作用所要求的塑性变形和滞回耗能能力，则可以通过结构体系中的其他低强高延性结构构件来提供。

Ikeda, S 和 Zater, W. A. 等在对混凝土桥柱中施加预应力后的抗震性能的研究中[27~29]，首先意识到采用高强预应力钢筋能够减小桥柱的残余变形。随后进行了专门研究，并主要集中到对用无粘结预应力筋方法来减小桥柱残余变形的问题进行分析和研究（见图 22）[30~32]，并且发现高强预应力钢筋混凝土桥柱在地震反复荷载作用下不会出现反应漂移现象，即正反方向水平加载时，位移反应幅值相差不会很大[32]。Kwan W. P. 和 Billington S. L 利用数值方法对配置在柱子边缘处的无粘结预应力筋不同用量下的单桥柱和顶部相连的双桥柱分别进行了模拟分析[33,34]，指出单纯配置无粘结预应力筋或普通钢筋的方式都是不好的，应采取混合配筋的方式，可使得在减小残余变形和保持适量耗能能力之间找到无粘结预应力筋的最佳用量。

本文作者认为，上述采用预应力的混凝土柱，其复位能力主要来自于高强预应力筋的弹性变形恢复能力，采用无粘结形式只是增强了对整个柱的弹性变形回复能力。因此，本节对高强钢筋配筋混凝土框架结构，只是简单将1860高强钢绞线按普通钢筋的方式配置于框架柱中，即可极大地改善抗震性能。

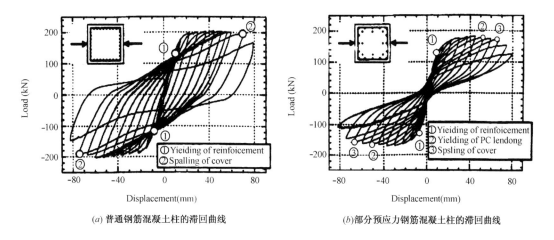

(a) 普通钢筋混凝土柱的滞回曲线　　　　(b) 部分预应力钢筋混凝土柱的滞回曲线

图22　W. Zatar 等的桥柱拟静力试验结果[30]

4.2　分析模型和数值方法简介

分别对图23所示6层和10层的两跨混凝土框架结构进行分析。框架梁、柱截面和总配筋率见表7。框架梁上均布荷载为30kN/m。框架柱中采用高强钢筋配筋的框架简称为高强配筋框架（标记为PF，Passive Frame）。高强配筋框架和普通框架（标记为OF，Ordinary Frame）的惟一不同点是在框架柱中使用的纵筋不同。普通框架的柱中配置屈服强度为400MPa普通钢筋，高强配筋框架的柱中配置屈服强度为1860MPa高强钢筋。材料特性见表8。

框架结构分析采用纤维模型建模，并借助MSC.MARC有限元分析软件进行分析[35]。框架梁、柱截面均划分成36条混凝土纤维和处于截面每个角上的

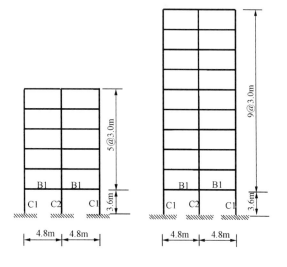

图23　框架结构尺寸

4条钢筋纤维（图24）。混凝土采用原点指向型弹塑性应力—应变关系（图25），因在极限状态情况下混凝土抗拉强度的贡献很小，忽略不计。钢筋采用弹塑性应力—应变关系（图26）。钢筋和混凝土的弹性模量分别为200GPa和30GPa。混凝土峰值强度、峰值应变、破坏强度、破坏应变等参数，以及钢筋的强度和屈服应变参数，均见表8。纵筋的最大极限应变使用FEMA 356中的定义，对抗压取0.02，对抗拉取0.05。此外，在分析中认为框架梁柱的抗剪承载力足够，忽略剪切变形和剪切破坏的影响。

框架几何尺寸和纵筋配筋率　　　　　　表7

框架	楼层	柱 截面（宽度×高度）(mm)	柱 ρ^a（%）	梁 截面（宽度×深度）(mm)	梁 ρ^b（%）
六层	1~2层	C1（400×450） C2（400×500）	1.0 1.2	B1（250×450）	1.1
六层	3~5层	C1（400×400） C2（400×450）	1.0 1.2	B1（250×450）	1.0
六层	6层	C1（400×400） C2（400×450）	1.0 1.2	B1（250×450）	0.9
十层	1层	C1（400×475） C2（400×550）	1.0 1.2	B1（300×450）	1.1
十层	2层	C1（400×475） C2（400×500）	1.0 1.2	B1（300×450）	1.1
十层	3~5层	C1（400×450） C2（400×500）	1.0 1.2	B1（250×450）	1.0
十层	6层	C1（400×450） C2（400×500）	1.0 1.2	B1（250×450）	0.9
十层	7~9层	C1（400×400） C2（400×450）	1.0 1.2	B1（250×450）	0.9
十层	10层	C1（400×400） C2（400×450）	1.0 1.2	B1（250×450）	0.8

注：ρ^a——钢筋总面积/截面毛面积；ρ^b——受拉钢筋面积/有效截面面积。

混凝土和钢筋材料参数　　　　　　表8

框架	混凝土 f_c（MPa） 梁	混凝土 f_c（MPa） 柱	混凝土 σ_u（MPa） 梁	混凝土 σ_u（MPa） 柱	ε_o	ε_u	钢筋 f_y（MPa） 梁	钢筋 f_y（MPa） 柱	ε_y
OF	25	30	15	20	0.002	0.004	400	400	0.002
PF	25	30	15	20	0.002	0.004	400	1860	0.0093

注：f_c——混凝土抗压强度；δ_u——混凝土极限强度；ε_o 和 ε_u——混凝土峰值应变和极限应变；
f_y 和 ε_y——纵筋屈服强度和屈服应变。

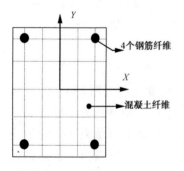

图24　截面纤维划分

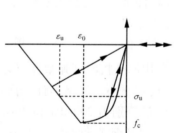

图25　混凝土应力—应变关系

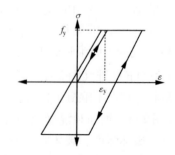

图26　钢筋应力—应变关系

分别采用静力弹塑性分析方法和弹塑性动力时程分析方法对普通框架和高强配筋框架的抗震性能进行对比分析。静力弹塑性分析方法按倒三角水平分布荷载施加侧推荷载。弹塑性动力时程分析的地震输入对6层框架取Northridge地震加速度记录［图27（a）］，对10层框架取Superstition地震加速度记录［图27（b）］。时程分析方法的时间步长为0.01s，阻尼比5%的瑞雷阻尼。6层和10层框架结构的振动基本周期分别为0.91s和1.48s。

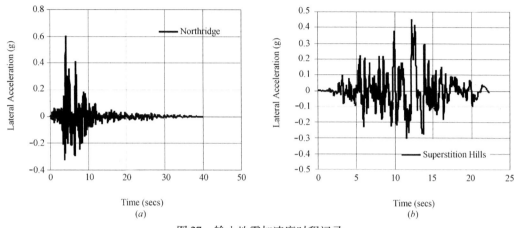

图27 输入地震加速度时程记录

4.3 破坏程度和破坏机构

为比较两种框架结构的损伤程度，采用框架梁、柱端塑性铰区的变形状态来标记。两种框架的框架梁相同，因此框架梁的损伤程度和损伤标记的定义相同，见表9。相对于框架梁，框架柱对框架结构整体的重要性程度高，因此同等损伤程度普通框架柱柱端的变形程度要小于框架梁，见表10。而对于高强配筋框架柱，考虑到混凝土框架柱的损伤主要是钢筋屈服造成的，同时由于高强钢筋的屈服强度和屈服应变远大于普通钢筋，钢筋达到屈服时柱端截面变形很大，因此按表11定义高强钢筋框架柱的损伤程度和损伤标记。

OF 和 PF 中梁的损伤标记和损伤程度 表9

损伤标记	截面材料应变		损伤程度	修复	结构安全	修后性能可靠度
	普通钢筋	混凝土				
1	$\varepsilon \ll \varepsilon_y$	$\varepsilon \ll \varepsilon_o$	轻微	不用修	安全	满意
2	$\varepsilon \leq \varepsilon_y$	$\varepsilon \leq \varepsilon_o$	轻度	可修		
3	$\varepsilon_y \leq \varepsilon \leq 0.015$	$\varepsilon \leq \varepsilon_o$	中度	可修		
4	$0.015 < \varepsilon \leq 0.03$	$\varepsilon_o < \varepsilon \leq \varepsilon_u$	比较严重	可修		
5	$0.03 < \varepsilon \leq 0.05$	$\varepsilon \geq \varepsilon_u$	严重	过度		

普通框架柱的损伤标记和损伤程度 表10

柱端损伤标记	截面材料应变		损伤程度	修复	结构安全	修后性能可靠度
	普通钢筋	混凝土				
1	$\varepsilon \ll \varepsilon_y$	$\varepsilon \ll \varepsilon_o$	轻微	不用修	安全	满意
6	$\varepsilon_y \leq \varepsilon \leq 0.005$	$\varepsilon \leq \varepsilon_o$	轻度	可修		
7	$0.005 < \varepsilon \leq 0.01$	$\varepsilon_o \leq \varepsilon \leq \varepsilon_u$	中度	可修		
8	$0.01 < \varepsilon \leq 0.015$	$\varepsilon_o \leq \varepsilon \leq \varepsilon_u$	比较严重	过度	不安全	不满意
9	$0.015 < \varepsilon \leq 0.02$	$\varepsilon \geq \varepsilon_u$	严重	不可修		

高强配筋框架柱的损伤标记和损伤程度　　　　　　　表 11

柱端损伤标记	截面材料应变		损伤程度	修复	结构安全	修后性能可靠度
	高强钢筋	混凝土				
1	$\varepsilon \ll \varepsilon_y$	$\varepsilon \ll \varepsilon_o$	微度	不用修	安全	满意
10	$\varepsilon < \varepsilon_y$	$\varepsilon_o \leq \varepsilon \ll \varepsilon_u$	轻度	可修		
11	$\varepsilon < \varepsilon_y$	$\varepsilon_o \leq \varepsilon \leq \varepsilon_u$	中度	可修		

4.4 静力弹塑性推覆分析结果

4.4.1 抗震承载力、变形能力和破坏机构

推覆分析得到的两种框架结构基底剪力-顶点位移曲线（侧推曲线）的对比如图 28 所示。在侧推曲线上，标记出了普通框架结构出现严重破坏或不可修复破坏时的侧移，以及出现破坏机构时的侧移，并标记出了高强配筋框架在相应侧移时的损伤程度。此外，高强配筋框架结构的框架柱钢筋达到屈服极限时的侧移也做了标记。图 29 给出了普通框架结构形成破坏机构时梁、柱端的损伤程度，并同时给出了相应侧移情况下高强配筋框架梁、柱端的损伤程度。以下给出具体分析结果对比。

对于 6 层框架，由图 28 可见，普通框架结构第 1 层柱底出现严重损伤时（图 28 中标记 9 的点），其侧向推力为 369kN，顶端位移达到 312mm；在此位移下，高强配筋框架结构的侧向推力为 406kN，仅梁端截面损伤较严重，损伤程度标记仅为 4。当普通框架结构侧向推力达到 375kN，顶端位移达到 642mm 时，因首层柱底截面和 4 层柱顶截面均形成塑性铰，且其间的框架梁端也均形成塑性铰，整个结构已形成破坏机构 [见图 29（a）左图，图 28 中标记为※]；但在此位移时，高强配筋框架结构的侧向推力为 475kN，首层柱底没有出现塑性铰，但梁端截面出现了严重损伤，整个结构尚未形成破坏机构 [见图 29（a）右图]，其破坏程度标记为 5。此后，普通框架结构已无法继续承受侧向变形，而高强配筋框架结构因柱底纵筋尚未屈服，侧向推力仍可继续增加。在侧向推力达到 513kN，侧向位移达到 896mm 时高强配筋框架结构的柱底才达到屈服，这一侧向推力几乎是普通框架结构出现破坏机构时侧向推力的 1.4 倍。

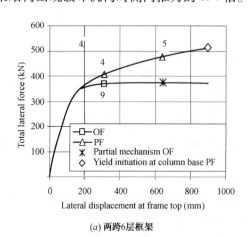

(a) 两跨6层框架

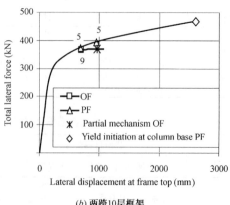

(b) 两跨10层框架

图 28　两种框架的侧推曲线对比

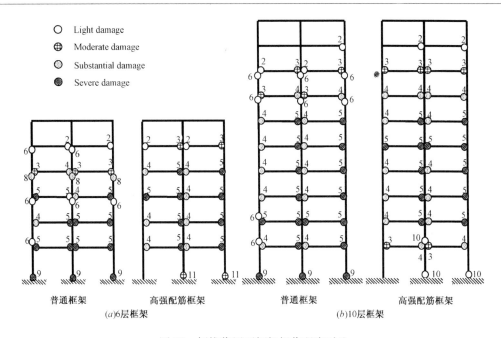

图 29 侧推作用下框架损伤程度对比

以上对比分析结果表明，当 6 层普通框架结构形成破坏机构时，高强配筋框架结构仍具有足够的抗侧承载力，且整体结构的抗震性能仍然是安全可靠的。由图 29（a）两种框架结构在普通框架结构形成破坏机构时梁柱端的损伤程度的对比可见，普通框架由于首层柱底和 4 层柱顶出现塑性铰并已达到屈服严重破坏而形成了部分楼层破坏机构；然而在相同的侧向变形的情况下，高强配筋框架结构的首层柱底没有屈服，只有中等程度的损伤。另一方面，两种框架结构梁端的损伤有些差异：普通框架结构中，第 1、第 2 层框架梁损伤的程度更大，在第 4、第 5 层情况却恰好相反，但由于框架梁的损伤对整个结构并不是致命的，因此从整体结构的损伤程度来说，高强配筋框架结构的损伤程度显著小于普通框架结构。同时，从图 29（a）可见，高强配筋框架结构的屈服机制是典型的所希望的梁铰屈服机制。

图 28（b）为两种 10 层框架结构侧推曲线的对比，可以很明显的看出，高强配筋框架结构比普通框架结构具有更大的承载能力和变形能力。普通框架结构侧推曲线上损伤标记 9 点所对应的承载力为 366kN，相应顶点侧移为 696mm，此时首层框架柱底已产生严重损伤。而在同样侧向位移情况，高强配筋框架结构仅框架梁端出现严重损伤（见图中标记 5），侧向承载力为 373kN。随着侧向位移的增加，普通框架结构的侧向承载力不再增加，但首层柱底的破坏程度却随变形增加而不断增加，在侧向荷载 369kN 和侧向位移 964mm 时，因出现足够多的梁端塑性铰而形成破坏机构［见图 29（b）左图］。此时，高强配筋框架结构的损伤仍然集中在框架梁端，侧向承载力达到 393kN，且可以继续承受侧向荷载和变形，直至达到普通框架结构侧向位移的 2.7 倍时，高强配筋框架结构首层柱底才出现屈服，整个结构刚开始形成破坏机构，此时的侧向承载力几乎是普通框架结构的 1.3 倍。此外，从图 29（b）还可以看到，高强配筋框架结构和普通框架结构中梁端损伤程度基本相近，且略偏小。

由以上分析可知，对于6层框架结构，高强配筋框架结构的抗侧承载力显著高于普通框架结构，且在相同侧移变形情况下，高强配筋框架结构的整体损伤程度显著小于普通框架结构；对于10层框架结构，普通框架结构形成破坏机构时，高强配筋框架结构仅框架梁为严重损伤，且因框架柱尚未屈服，结构远未达到其变形能力极限。

4.4.2 性能极限状态

为了进一步比较两种框架结构的抗震性能，根据 FEMA 356 建议的不同性能水平下梁柱塑性铰转角程度（表12），在两种框架的推覆曲线上将相应这些性能点标出，如图30所示。由图30可见，对于6层框架，普通配筋框架的生命安全和倒塌极限状态均发生于框架柱，而高强配筋框架的生命安全和倒塌极限状态均发生在框架梁，且相应的变形能力也大于普通框架；对于10层框架，普通配筋框架的生命安全和倒塌极限状态同时发生于框架梁、柱，而高强配筋框架的生命安全和倒塌极限状态仅发生在框架梁，且相应的变形能力也大于普通框架。不同性能水准下两种框架的顶点侧移和承载力比较见表13，可见在立即可用极限状态情况下，两种框架具有基本一致的承载力和变形能力，表面两种框架均可满足小震作用的抗震性能需求，但在生命安全和倒塌极限状态情况，高强配筋框架结构的承载力和变形能力均显著大于普通框架。事实上，由图28可知，即使在更大的变形情况下，高强配筋框架结构仍具有可持续增大的承载力，使得结构具有更高的抗震安全储备。

不同性能水准下的梁柱端塑性转角极限（rad）　　　　表12

构　件	性　能　水　准		
	立即可用	生命安全	倒　塌
柱　端	0.005	0.01	0.015
梁　端	0.01	0.02	0.025

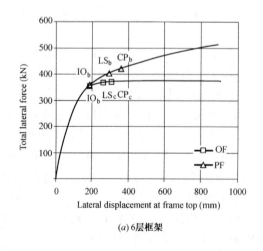
(a) 6层框架

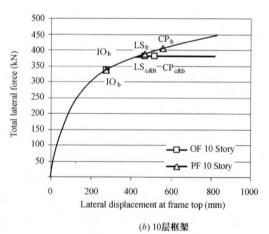
(b) 10层框架

图30　高强配筋框架与普通框架的不同性能点对比

不同性能极限状态下两种框架的侧移和承载力对比 表 13

框架		性能水准					
		立即可用		生命安全		倒塌	
		顶点侧移 (mm)	总水平力 (kN)	顶点侧移 (mm)	总水平力 (kN)	顶点侧移 (mm)	总水平力 (kN)
6层	普通框架	189	352	264	367	312	369
	高强框架	185	358	298	402	362	420
10层	普通框架	274	336	467	380	514	380
	高强框架	274	336	467	387	557	405

综上，高强配筋框架结构因首层框架柱底钢筋屈服明显推迟，相应出现塑性铰时的侧移变形能力比普通框架结构显著增大，形成破坏机构时的变形能力也大大增加，且因首层框架柱底的高强钢筋使得柱底塑性铰具有更大抗弯承载力，使得整个结构的侧向承载力比普通框架结构显著增加。需注意的是，由于框架柱抗弯承载力的增加，需相应增加框架柱的抗剪承载力，以保证框架柱为"强剪弱弯"，避免受剪脆性破坏，影响高强钢筋强度的发挥。

4.4.3 残余变形

由于高强钢筋具有更大的弹性恢复能力，高强配筋框架结构在经受较大的弹塑性变形后，其残余变形比普通框架结构要小。为此，进一步对两种框架施加相同侧推位移变形后卸载，卸载时间是结构中任一杆件的混凝土达到极限应变。

6层框架结构的加载卸载曲线如图31（a）所示，完全卸载的残余位移如图32（a）。从图32（a）可以清楚地看出，高强配筋框架结构的残余位移小于普通框架结构，其原因是：开始卸载时，两种框架的框架梁端都达到了混凝土极限应变，但普通框架首层柱底纵筋已经屈服，产生较为严重的损伤（损伤标记8），具有不可恢复的塑性变形；而高强配筋框架首层柱底纵筋尚未屈服，但混凝土已超过峰值应变，属于轻微损伤程度（损伤标记10），纵筋仍具有弹性恢复能力。

10层框架结构的加载卸载曲线如图31（b）所示，完全卸载的残余位移如图32（b）所示。可以清楚地看出，两种框架卸载后的残余变形基本相同，其原因是：卸载时刻都是两种框架的框架梁端混凝土达到极限应变，此时普通框架的底层柱底纵筋虽然产生一定程度的屈服，但屈服程度不严重，属于中等损伤，损伤标记为7；而高强配筋框架首层柱底纵筋未屈服，只有轻微损伤，损伤标记为1。可以认为，如果框架柱屈服程度不大，两种框架结构的残余变形几乎相同。但如果变形继续增加，则会出现前面6层框架结构的情况。

在下面的弹塑性动力分析结果中，可以进一步看到高强配筋框架结构在震后具有更小的残余变形。

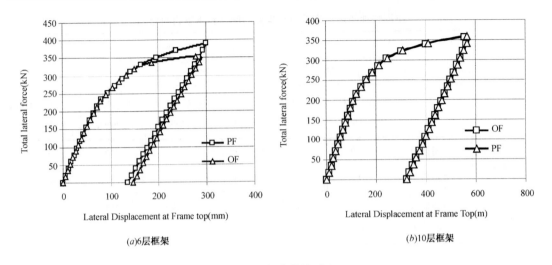

(a) 6层框架　　　　　　　　(b) 10层框架

图31　加卸载曲线对比

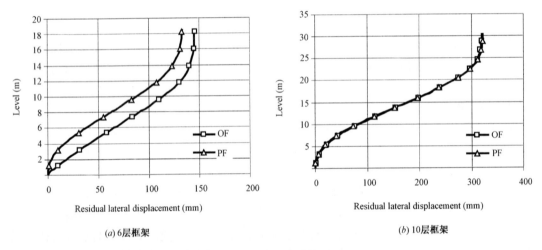

(a) 6层框架　　　　　　　　(b) 10层框架

图32　完全卸载时残余位移对比

4.5　地震动力响应分析结果

4.5.1　损伤情况对比

图33（a）所示为两种6层框架结构在同样地震作用下的损伤情况对比。对于普通框架，底层柱底塑性铰的损伤程度在中等以上，右柱柱底塑性铰已达到不可修复的严重损伤；同时4层柱顶也均出现中等损伤程度的塑性铰，整个结构已接近形成柱铰破坏机构。而对于高强配筋框架，由于框架柱采用高强钢筋，其柱端抗弯承载力比框架梁端的抗弯承载力大很多，因此整个框架的损伤基本均集中在框架梁端，所有的框架柱端钢筋都没有发生屈服，仅底层柱底混凝土超过受压强度，有中度损伤，属于可修复损伤。需要注意的是，高强配筋框架的第2层框架梁端损伤产生较严重的损伤，比普通框架第2层框架梁端的损伤程度严重一些，这是由于高强配筋框架中框架柱的屈服变形能力提高，对框架梁端

的弯曲变形要求也相应提高,但两种框架中所有框架梁的总体损伤程度基本相近,属于可修复损伤。

图33(b)所示为两种10层框架结构在同样地震作用下的损伤情况对比。由图可见,普通框架底层柱底严重损伤,已不可修复,上部楼层框架柱端也多处出现轻度损伤,整体结构接近形成部分柱铰破坏机构。而高强配筋框架结构的所有框架柱端都没有出现屈服,仅底层柱底混凝土超过受压强度,有轻度到中度的损伤,属于可修复损伤,整个结构未形成破坏机构。同样需注意的是,高强配筋框架结构在5层和8层框架梁端的损伤程度高于普通框架。

由以上强震作用下两种框架损伤情况的对比可知,对于普通框架结构,由于框架柱的严重损伤,整个结构已处于形成破坏机构状态,且几乎不可修复;而对于高强配筋框架结构,因框架柱都没有达到屈服,整个结构可避免破坏机构的形成。

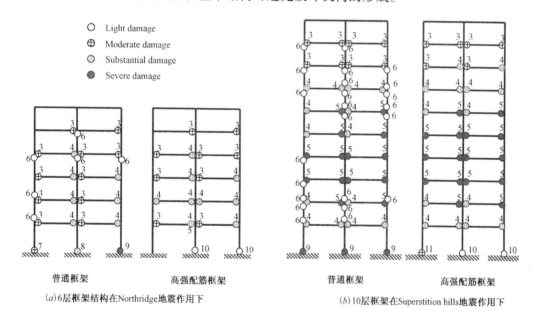

图33 地震作用下的损伤程度对比

4.5.2 侧移响应、残余侧移和可修复性

两种框架在地震作用下的顶点侧移时程比较如图34所示。由图可见,两种框架的最大侧移基本相同,但最大位移时刻以后,普通框架因柱中纵筋过渡屈服,塑性变形很大,整个结构严重偏向一侧;而高强配筋框架结构因柱中纵筋屈服程度较小,且有较大弹性恢复能力,结构偏移程度明显小于普通框架结构。

图35所示为两种框架结构地震结束后的残余侧移比较,可见高强配筋框架结构的残余侧移显著小于普通框架结构,这是因为高强钢筋的弹性使得结构具有一定的复位能力。

除结构在地震作用时需具备足够的抗震能力,避免倒塌破坏外,结构震后的可修复性也是结构抗震性能的一个重要方面。由以上损伤分析结果和残余侧移情况可知,普通框架损伤严重,残余侧移大,不利于结构的震后修复,尤其是当残余变形过大而导致无法修复,有时只能拆除,造成很大经济损失。

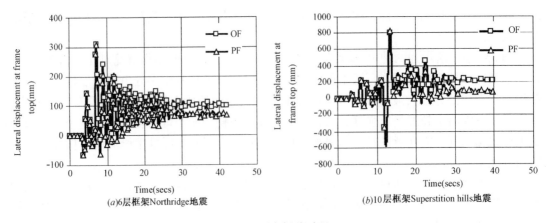

(a) 6层框架Northridge地震　　(b)10层框架Superstition hills地震

图 34　顶点侧移时程

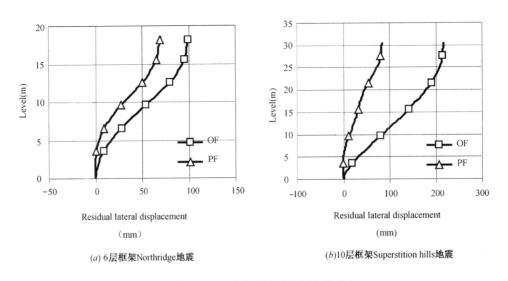

(a) 6层框架Northridge地震　　(b)10层框架Superstition hills地震

图 35　地震结束时的残余侧向位移

对于高强配筋框架结构，除因底层框架柱有轻度和中度损伤外，其他楼层的框架柱基本没有屈服和损伤，如果底层框架柱端能够提供更多约束，则就可以避免整个结构的大范围修复；而普通框架结构，即使通过约束可提高柱端塑性铰的塑性变形能力，但由于框架柱在强震下不能避免的钢筋屈服，残余变形大，其修复难度也会随之增大。

4.5.3　层间位移和楼层加速度

图 36 所示为两种框架结构的最大层间位移响应的比较。由图 36 可见，普通框架结构的底层层间位移显著大于高强配筋框架结构，这是因为普通框架结构首层框架柱底部损伤严重所致；两种框架结构的最大层间位移楼层基本都在结构高度中部，但高强配筋框架的层间位移略大于普通框架，这主要是由于中部框架梁端塑性变形较大引起的；而两种框架顶部楼层层间位移基本接近。需注意的是，上部楼层的层间位移还包含了楼层转动引起的非受力刚体变形。

图 37 所示为两种框架最大楼层加速度响应的比较。由图可见，两者相差不是很大，

第六章　高强高性能工程结构材料与现代工程结构及其设计理论的发展　　239

在结构高度中部，高强配筋框架的最大楼层加速度响应略大于普通框架，这是由于高强配筋框架塑性程度较小所致。

从以上最大层间位移和楼层加速度响应的分析结果可知，高强配筋框架与普通框架结构的地震响应基本一致，不会因为高强配筋框架柱中的高强钢筋没有屈服而导致楼层加速度响应增加，并导致楼层层间位移增大。

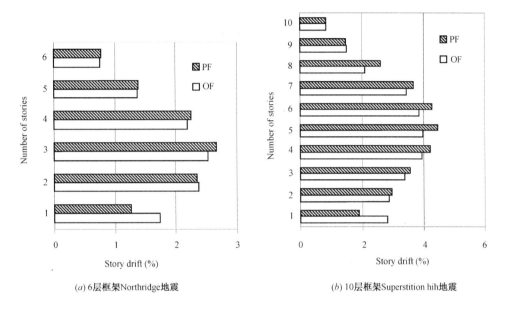

(a) 6层框架Northridge地震　　　　　(b) 10层框架Superstition hih地震

图 36　层间侧移响应

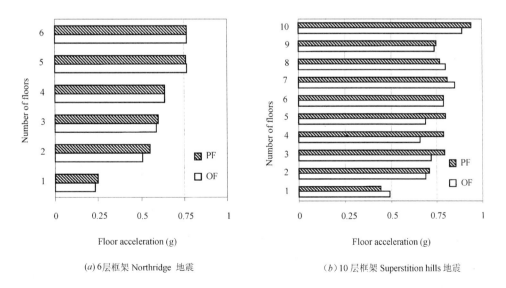

(a) 6层框架 Northridge 地震　　　　　(b) 10层框架 Superstition hills 地震

图 37　楼层加速度响应

5. 结构设计理论的发展

5.1 概述

结构设计理论包括结构分析理论和安全储备理论。这两方面的理论都与结构材料的力学性能有很大关系。目前的土木工程结构设计理论主要是基于传统具有弹塑性性质的普通混凝土和钢材建立起来的。尽管基于弹塑性理论的结构分析方法依然可适用于高强高性能材料结构，但一些结构设计概念和安全储备理论已不能适应高强高性能材料工程结构的发展。以下针对高性能工程材料结构的受力性能特点，探讨相应的结构安全储备理论，并对目前的基于弹塑性的结构设计理论及其有关设计概念中不适合高性能材料特性的一些问题进行讨论。

5.2 结构安全储备新指标

5.2.1 安全储备的概念

一般来说，按照规范正常设计的工程结构通常是能够满足在正常使用情况下安全性要求。但目前工程结构的安全储备理论是针对采用普通混凝土和钢材、具有理想弹塑性的受力性能结构建立起来的。对于高强高性能材料结构，由于不像传统材料结构具有弹塑性受力特征，现有的结构安全储备理论显然不适用了，而且还会因为错误使用现有的安全储备理论，可能导致一些错误的设计概念，或导致不合理的设计结果。

工程结构的安全储备是指结构和构件的极限破坏状态与设计目标状态的力学性能指标比值。目前，工程结构的安全储备指标大多采用承载力安全储备系数，即

$$S = \frac{S_u}{S_k} \quad (1)$$

其中，S_u 和 S_k 分别是结构（或构件）的最大承载力和设计目标状态时的受力（如图38所示）。

对于具有近似理想弹塑性受力性能的传统工程结构，基于承载力安全储备指标及其安全储备理论，已能够足够解决人们对工程结构安全性的问题，因为：其一，只要承载力安全储备系数 S 足够大，即可以保证工程结构在正常使用情况下不发生因承载力不足而导致的结构损伤；其二，在意外事件发生时，结构所具有足够的塑性变形能力可避免结构的垮塌。

由于塑性变形分析和塑性变形能力确定的困难性，事实上工程师们在结构设计中往往只进行承载力计算，而不进行塑性变形计算，只是通过规范规定的构造措施来保证结构具有一定的塑性变形能力。同时，长期以来（其实也只有100多年），面对所能够使用的传统工程材

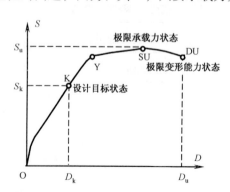

图38 力—变形关系曲线

料和长期的工程教育,工程师们(甚至一些研究者们)都认为没有延性的材料是不能作为工程结构使用的,也甚至忘记历史上,人们曾经使用铸铁、石头和木材等这些近似弹脆性材料所建造的大量工程结构,仍然具有足够的安全储备。显然,在利用这些没有什么塑性变形能力的弹脆性材料建造工程结构时,所采用的承载力安全储备系数 S 应比有塑性变形能力材料的结构取得更高一些,如经典材料力学中所指出的,通常取 $S=2.0$。根据我国现行的《混凝土结构设计规范》GB 50010—2002 和《钢结构设计规范》GB 50017—2003 所采用的结构可靠度体系估算,传统的具有近似理想弹塑性受力性能结构的承载力安全储备系数约为 1.25((根据文献[37]中承载能力极限状态分项系数的规定,恒载分项系数一般取 1.2,活载分项系数一般取 1.4,考虑活载与恒载的比例 1:3,不考虑材料分项系数,则承载力安全储备系数 $a=1.25$)。由此可以看到,经典的承载力安全储备系数 2.0 和目前的工程结构普遍实行的安全储备系数 1.25 之间的差别,主要体现在所使用的材料力学性能上。也就是说,具有塑性变形能力的工程结构,其承载力安全储备系数可以取得小一些,而在钢筋混凝土和钢出现以前的采用脆性材料建造的古代工程结构,其承载力安全储备系数应取得大一些。实际上,这种不同受力性能结构的承载力安全储备系数之间的差别,主要是为了保证工程结构在不可预见的意外事件发生时能够避免垮塌的一种安全储备。也就是说,当所使用的材料使得结构具有塑性变形能力,就可以利用其塑性变形能力的储备来代替承载力安全储备。因此,为正确评价结构的安全储备能力,应充分考虑结构的变形能力储备。

鉴于以上的认识,目前使用的具有塑性变形能力的普通工程材料结构,其塑性变形能力也是反映结构(或构件)在意外事件作用下安全储备的一个重要因素。这种考虑变形能力安全储备的指标可表示为,

$$D = \frac{D_u}{D_k} \tag{2}$$

其中,D_u 和 D_k 分别是结构(或构件)达到破坏时的最大变形能力和设计目标状态时的变形(图 37)。

以图 37 中普通强度材料结构的力—变形关系曲线为例,图中 K 点为设计目标状态点(即正常使用状态),相应的力和变形记为 S_k 和 D_k;SU 点为最大承载力点,相应最大承载力为 S_u;SD 点为破坏极限点,相应最大变形能力为 D_u。显然,对于一般的工程结构,最大承载力点 SU 和最大变形能力点 SD 并不是同一极限状态点。

由以上定义可知,承载能力安全储备系数 S 反映了结构相对于设计目标状态的承载能力安全储备程度;而变形能力安全储备系数 D 反映了结构相对于设计目标状态的变形能力安全储备程度。然而,由图 37 可知,结构的安全储备应同时包括承载能力安全储备和变形能力安全储备,即应同时考虑极限状态时承载能力和变形能力相对于设计目标状态的承载力和变形安全储备,只用其中一个指标是不能全面反映结构的安全储备程度。比如,一个具有承载能力安全储备系数 $S=2$ 和变形能力安全储备系数 $D=2$ 的弹性结构,与承载能力安全储备系数 $S=1.25$ 和变形能力安全储备系数 $D=4$ 的弹塑性结构相比,两者的安全储备程度是无法简单用这两个系数进行评价和比较,还需要进一步考虑结构的荷载特征。对于超载程度较大的静力荷载情况,采用承载能力安全储备系数 S 更为合适,因为此时变形能力储备对超载无任何意义。而当考虑结构遭受意外冲击荷载时,则采用变形能力安全

储备系数 D 较为合适,因为延性对于缓解冲击引起的动力作用更有意义。不过,应该注意的是,尽管延性可以缓解冲击动力作用效应,但同时具有足够的承载力安全储备依然十分重要,这将在以下分析中可以看到。

5.2.2 等效承载力安全储备系数

由于安全储备系数 S 和 D 不能同时全面反映结构在承载力和变形两个方面的安全储备程度,需要进一步考虑引入更合理的安全储备指标。文献[36]建议采用以下两个指标来评价结构(或构件)的安全储备:

变形能安全储备系数 $$Y = \frac{E_u}{E_k} \tag{3}$$

综合性安全储备系数 $$F = S^m D^n \tag{4}$$

其中,E_k 为设计目标状态时(K 点)结构的变形能,即图 38 中 OK 曲线下的面积;E_u 为结构(或构件)在达到破坏时的最大变形能,即图 38 中整个曲线下的面积;$m + n = 2$,m、n 为根据使用荷载情况设定的权重系数。

变形能安全储备系数 Y 反映了结构(或构件)吸收和耗散能量能力的安全储备;F 指标综合反映了承载力和变形能力的安全储备,通过调节指数 m 和 n 的大小来体现荷载特性结构对承载力安全储备和变形性安全储备需求的权重,如对于静力荷载情况,权重系数 m 大一些,对于动力荷载情况,权重系数 n 大一些。

以上四个安全储备指标比较全面地包含了安全储备的各个方面,针对不同的情况可使用不同的指标。对于理想线弹性材料来说,这四个安全储备指标系数是一致的。

尽管 Y 指标和 F 指标能够较好反映结构或构件在承载力和变形能力两方面的综合安全储备程度,但 Y 指标不便于直接反映安全储备的工程概念;而 F 指标只是将 S 指标和 D 指标简单相乘,没有明确的物理意义和解释。为此,本文提出以下等效承载力安全储备系数,

$$S_{eq} = \frac{Y}{D} \tag{5}$$

图 39 所示为不同具有弹塑性受力性能结构的力—变形关系曲线,图中的力—变形关系曲线是以设计目标状态 K 点的承载力和变形为基准标准化给出的(即取 $S_k = 1$,$D_k = 1$),因此极限状态点的承载力值就是 S 指标,变形值就是 D 指标。图 38 中 OYCU3 为线弹性结构的力—变形关系曲线,根据式(5)的定义,指标 S_{eq} 即等于承载力安全储备系数 S。

图 39 中 OYU1 为理想弹塑性结构力—变形关系曲线,这种受力性能是目前普通材料结构所经常近似采用的。图 39 中 OYU2 为强化型弹塑性结构的力—变形关系曲线,这种关系曲线是高强高性能材料结构所常见的,其中 Y 点为屈服点,U2 点为破坏极限状态点。屈服点 Y 相对于设计目标状态点 K 的承载力储备系数为 $S_y/S_k = S_y$,其中 S_y 为屈服承载力储备系数;AE 段斜率与 OA 段斜率之比为 k,则由式

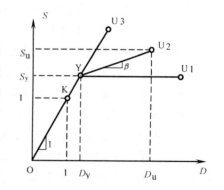

图 39 不同受力特性的力—变形关系曲线

(5) 定义可得 OBAE 弹塑性结构（或构件）的等效承载力安全储备系数 S_{eq} 为，

$$S_{eq} = \frac{2S_y D - S_y^2 + \beta (D - S_y)^2}{D} \quad (6)$$

对于理想弹塑性的情况，屈服后的强化刚度系数 $\beta = 0$，此时若取 $S_y = 1$，即设计目标状态与屈服状态一致，则由式（6）可得，

$$S_{eq} = 2 - \frac{1}{D} \quad (7)$$

由上式可知，当 $S_{eq} = 2$ 时，$D \to \infty$。这是一个十分有意思的结果。如前所述，对于具有弹脆性特征的结构，按经典材料力学中所指出的，安全系数通常取 2.0，即 $S = 2.0$、$D = 2.0$。如果取此时的设计目标点作为结构的屈服点，即屈服承载力储备系数 $S_y = 1.0$，将结构改为理想弹塑性结构，则由式（7）可得，变形能力安全储备系数 D 要求无限大，即意味着在设计目标状态点屈服的结构，其延性需求为无穷大才能满足传统经典线弹性结构的承载力安全储备系数等于 2 的需要。这从另一个方面说明了经典线弹性结构安全度的工程意义，即对于以静力荷载为主的结构（经典结构设计理论尚未发展到考虑动力荷载），塑性变形是没有意义的，只有依靠承载力储备。同时，对于符合经典线弹性特性的结构，由于结构在正常使用情况下处于弹性阶段，因此其结构分析理论只需要借助于弹性分析方法。

对于弹塑性结构，基于正常使用情况下的承载力安全储备的考虑，不可能在设计目标状态下使结构达到屈服。通常要求结构的屈服承载力应高于设计目标时的承载力，即屈服承载力储备系数 $S_y > 1$，本文称 S_y 为"基本承载力安全储备系数"。这一安全储备系数具有重要的工程意义。事实上如前所述，对于目前工程中常用的钢筋混凝土和钢这类弹塑性材料的结构，按我国工程结构设计规范，基本承载力安全储备系数 S_y 约为 1.25。

图 40 中 AB 曲线表示不同弹塑性受力性能结构具有相同基本承载力安全储备系数 $S_y = 1.25$ 和等效承载力安全储备系数 $S_{eq} = 1.9$ 的极限状态曲线［按式（6）计算得到结果］。由图 40 可见，利用等效承载力安全储备系数 S_{eq}，可较好地统一不同弹塑性受力性能结构在承载能力和变形能力两方面的安全储备程度，且 S_{eq} 指标与传统线弹性结构的承载力安全储备概念一致，便于工程理解。需指出的是，图 40 中 AB 曲线是针对特定超载荷载特性的极限状态曲线，如果超载荷载特性不同，AB 曲线也会不同，如对于撞击荷载，到达极限状态 AB 曲线的各种弹塑性力—变形关系曲线下的面积相等，即变形能相等；对于地震作用，则可根据地震过程中能量耗散能力相等的条件得到极限状态 AB 曲线。

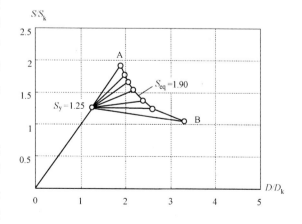

图 40 $S_y = 1.25$，$S_{eq} = 1.9$ 的极限状态点

5.3 结构的承载力和延性与结构的安全储备

在正常使用情况下,结构上的荷载作用往往以静力形式为主,通常超限幅度有限,材料小于预期设计强度的离散程度也很有限,且荷载作用超载程度和材料强度的偏低程度都是可以预期的,并通常采用概率方法给出荷载作用和材料强度的设计取值(包括标准值和分项系数)。因此,一般来说,按照规范正常设计的工程结构通常是能够满足在正常使用情况下的安全性要求。所以,事实上人们对工程结构安全性的要求,往往是针对可能遭遇到意外事件作用下的工程结构不产生与其原因不相称的垮塌,或造成不可接受的重大人员伤亡和财产损失的需求。对于意外事件,如罕遇地震、爆炸、冲击等,其作用的特性一方面是其量值难以预计,二是具有显著的动力特性。在意外事件发生时,通常是容许结构产生一定程度的损坏,但不能垮塌。

本文提出的将结构安全储备分解为基本承载力安全储备系数 S_y 和等效承载力安全储备系数 S_{eq},符合上述结构正常使用情况下的安全储备和意外事件情况下的安全储备这两种不同的安全储备需求,使得工程结构的安全储备具有更明确的工程意义。

基本承载力安全储备系数 S_y 主要解决结构在正常使用情况下的荷载作用超载和材料强度可能偏低的情况,可以通过合理的概率统计分析方法获得相应的失效概率,因为不超过结构的屈服强度,其计算分析理论仅需依据弹性方法。

对于意外事件,通常无法采用统计方法获得作用的量值及其相应的概率,往往需要工程师根据长期工程积累,凭借经验把握确定(前一段关于结构安全系数凭经验确定还是基于概率可靠度理论确定的争论在这里得到了统一[39])。同时,结构在意外事件作用下通常是容许结构产生一定程度的破坏,即结构将进入弹塑性阶段,需要同时考虑结构弹塑性阶段的强化所增加的承载力和弹塑性阶段的变形能力。结构在破坏阶段的塑性变形能力通常用延性大小反映。结构的延性对结构抵御意外事件作用的意义有以下几方面:

(1)实际意义上结构的破坏是以结构达到极限变形能力为依据的。延性是结构抗破坏能力的重要指标,足够的延性能力有利于避免结构的突然倒塌,因此应将延性大小纳入结构安全储备予以考虑,才能全面反映结构的安全储备程度。

(2)对于超静定的结构,次要构件和赘余构件的足够延性,有利于结构的充分内力重分布,也有利于提高整体结构的承载力,增加整体结构的安全储备。

(3)意外事件作用往往具有动力特性,结构的塑性变形能力和滞回耗能能力对缓解结构动力响应具有有利影响,这是延性对结构抵御意外事件的重要作用。

(4)从结构冗余度观点来看,脆性构件的破坏通常导致与该构件相关联的所有冗余度均丧失,而延性构件的破坏则不会导致与该构件相关联的所有冗余度同时丧失,即延性构件对于维持整体结构的冗余度具有重要作用。

需引起注意的是,在讨论结构延性问题时,不能仅仅局限于延性系数,而要将结构的延性与结构的破坏模式联系起来。从结构整体抵御意外事件的能力和保证整个结构的安全储备能力来说,整体结构的延性大小更为重要。整体结构的延性与结构中构件的延性既有联系,又有区别,它反映的是整体结构在某种荷载下的宏观变形能力。具有整体型破坏模式的结构,结构中大部分构件的延性得以充分发挥,结构的耗能能力大;而局部破坏模式,即使局部破坏部位构件的延性很大,结构的安全储备也不大,相应结构的耗能能力也小。比如说，延性系数达到 6 的框支结构或形成柱铰机制的框架结构，其抗震性能不可能好于延性系数只有 3 的剪力墙结构。

长期以来，人们将承载力安全储备和变形能力安全储备简单的割裂，而没有从两方面同时予以考虑。即通常在讨论安全储备时往往只考虑承载力储备（即基本承载力安全储备），而在讨论延性时又是指在承载力基本保持不变情况下（通常指屈服承载力）的变形能力。这种在仅满足基本承载力安全储备的情况下，仅根据延性大小来评价结构在意外事件作用下的安全性是存在较大的问题。如考虑两个具有同样等效承载力安全储备系数 S_{eq} 的结构：一个结构的屈服承载力小，但延性大；另一个结构的屈服承载力高，但延性小；显然在意外事件作用下，屈服承载力高、延性小的结构的损伤程度要比屈服承载力小、延性大的结构的损伤程度要小。因此，强调提高结构的承载力安全储备对结构安全性具有以下几方面的意义：

（1）对于主要构件，特别是整体型子结构，提高承载力安全储备比提高变形能力安全储备更重要，因为这些构件一旦达到其屈服承载力，即使其随后的变形能力再大，也难以避免结构的整体破坏，且破坏后果往往是较严重的，至少是难以修复的。而对于次要构件和赘余构件，增加延性则是十分重要的。

（2）现行的结构安全储备理论和结构设计理论，是在传统低强材料结构的基础上发展起来的。在如罕遇地震等意外事件作用下，仍要求低强材料结构处于弹性状态是不经济的。因此，现行的结构抗震设计理论容许结构在罕遇地震下产生一定程度的损坏，以利用损坏结构构件的塑性变形能和滞回耗能来耗散能量，减小意外事件作用引起的动力响应，避免结构的倒塌。随着材料技术的发展，高强高性能材料已经可以以合理的价格应用于工程结构，在这样的背景下，没有理由限制高强结构材料在结构中的应用。高强材料的应用可以使得结构（特别是结构中的关键构件）具有更高的承载力安全储备，有利于减小结构在意外事件作用下的损伤程度。如前述在框架柱中采用高强钢筋的框架结构在地震作用下，可以完全避免框架柱中出现塑性铰，形成真正意义上的梁铰破坏的整体型破坏模式，并且结构的整体损伤程度和残余变形显著减小，有利于震后的结构修复。

5.4 结构设计理论和设计概念的改进

目前基于传统结构材料的弹塑性结构设计理论及其有关设计概念中，存在着许多不适合高强高性能材料结构发展的一些问题，并可能阻遏工程结构设计理论的发展。

应该认识到，以往我们在讨论结构的承载力和延性的时候，往往将两者作为一个结构同时需要满足的两个指标，而没有从结构体系中不同结构构件的功能和意外事件作用引起结构的效应去分析。事实上，根据结构动力学原理，理想的结构应该是具有高阻尼比的弹性结构。这种结构在动力荷载作用下，动力效应小，且最终可以恢复到初始位置（复位）。这样的结构，弹性结构体和阻尼是结构体系的两个不同的部分，可以用不同方式获得。在结构体系中，那些低强大延性的结构构件屈服后，通常就具有增加结构体系阻尼的效果（称为塑性滞回耗能阻尼），这些构件的屈服，甚至破坏，只要不影响弹性结构体的完整性，结构就可以有效地抵御意外事件的作用，而且弹性结构体的承载力越大就越好。显然，要使得弹性结构体具有更大的承载力，就需要采用高强高性能材料。

另一方面，对于一个具有高阻尼比的弹性结构，其结构分析理论和设计方法，尤其结构动力分析理论和结构抗震设计方法将会变得十分简单,因为只需要弹性理论和基于反映谱的振型分解方法即可。因此,基于高强高性能材料发展的现代工程结构,结构设计概念和设计理论将会变得更为明确和简单。

基于高性能材料的现代工程结构的设计概念是:在整个结构体系中,根据不同结构构件在结构体系中的作用和功能,采用高强高性能材料或低强高性能材料。高强高性能材料一般用于结构的主要子体系或主要构件,使主体结构尽可能保持弹性;低强高性能材料则主要用于结构体系中赘余构件和次要构件,以增大结构在意外事件情况下的结构阻尼,减小意外事件发生时的结构动力响应。

需注意的是,高强高性能材料的变形能力并不比低强高性能材料小很多。况且,可以利用合理的结构形式,将低强高性能材料用于结构体系中相对变形较大的部位,从而更充分地发挥低强高性能材料的塑性耗能作用。

结构的主要子结构体系或主要构件采用高强高性能材料,不仅可减小主体结构部分的结构构件尺寸,而且利用其高强弹性性能可使结构整体受力性态基本表现为弹性特征,这一方面可以使得结构的整体损伤程度较小(与普通结构材料相比,结构的自身损伤和残余变形都显著减小),另一方面也使得结构分析可近似采用弹性方法,从而简化结构设计计算。

基于高性能材料的现代工程结构的设计理论是:由于结构的主要子体系或主要构件采用高强高性能材料,当在意外事件作用下,次要构件或赘余构件屈服进入塑性阶段后,整体结构的受力性能更像前面所述的高阻尼弹性结构,因此借助于弹性理论进行简化分析即可。而正常使用荷载作用下,结构次要构件或赘余构件不屈服,依然是按照弹性理论进行结构分析。需注意的是,结构在正常使用情况和在意外事件作用下结构的受力特性是不同的,在正常使用阶段结构的刚度更大一些。

此外,由于结构的整体性态受到高强高性能材料为主体结构的弹性性能的控制,也更易于实现结构性能目标的控制,符合目前基于性能结构设计发展的需要。

6. 结语

随着近年来材料技术的发展,高强高性能工程结构材料已可以以合理的价格提供给工程结构应用,各种高强高性能材料的研究和应用也已经有了长足的发展。随之而来的是问题,如何正确并合理应用各种不同性能的高性能材料,同时应注意到其受力性能与传统低强工程材料工程结构的差别,以及由此对基于传统低强工程材料发展成熟并已为广大结构工程师们和研究者们所熟悉的结构设计理论和方法不适用于高强高性能工程结构体系的挑战。本文结合各种高性能工程材料的特性和工程结构功能的需求,论述了发展合理工程结构的途径,以及工程结构设计理论的发展,取得以下主要成果:

(1)应根据结构体系的层次性,将高强高性能材料用于结构主体子结构,而将低强高性能材料用于结构次要子结构,而且这种设计概念可以引申到结构构件层次。由此,可以使得结构具备同时具有抵御意外事件作用所需要的能量耗散能力和整体结构的低损伤性要求,并通过在框架柱中采用高强钢筋的高强配筋混凝土框架结构的详细算例阐明了这一特性。

(2) 针对高强高性能工程结构的受力特性不同于传统工程结构的问题，本文发展了结构安全储备理论，提出了基本承载力安全储备和等效承载力安全储备的两层结构安全储备体系，以分别解决正常使用情况下的结构安全储备问题和意外事件情况下的结构安全储备问题，指出了正常安全储备系数确定的科学性和意外事件下安全储备系数确定的经验性。所建议的安全储备理论体系，不仅合理考虑了结构塑性变形能力对抵御意外事件的作用，也具有与传统承载力安全储备相一致的概念。

　　(3) 基于以上两个成果，高强高性能工程结构可简化为高阻尼弹性结构，从而使得结构的分析理论和设计理论也得到简化。这种简化是基于高强高性能材料的发展和合理应用的结果。虽然从某种程度上可以说，这一发展结果将结束人们长期以来无高强材料可用、而只能使用低强材料所不得不引入弹塑性结构分析理论和方法的历史。不过，弹塑性结构分析理论和方法所揭示的结构塑性在抵御意外事件动力作用方面的贡献，也被融合到这种所发展了的适用于高强高性能材料结构的弹性分析和设计理论中。可以说，弹性结构分析理论在高强高性能结构中的应用，印证了科学理论的螺旋型发展的模式。

参考文献

[1] 陈肇元. 高强混凝土及其应用. 北京：清华大学出版社，1992
[2] 蒲心诚. 超高强高性能混凝土 [M]. 重庆：重庆大学出版社，2004
[3] 杜婷，郭太平，林怀立，刘中心，周志强. 混凝土材料的研究现状和发展应用 [J]. 混凝土，2006 (5)：7-9
[4] 黄承逵. 纤维混凝土结构 [M]. 北京：机械工业出版社，2004
[5] 林小松，杨果林. 钢纤维高强与超高强混凝土 [M]. 北京：科学出版社，2002
[6] 孙伟，严云. 钢纤维硅灰高强混凝土的力学行为及界面特性 [J]. 中国科学 A 辑，1992，(2)
[7] 刘小平. 活性粉末混凝土的特性及其发展前景 [J]. 混凝土与水泥制品，2006，(3)：17-18
[8] 覃维祖，曹峰. 一种超高性能混凝土——活性粉末混凝土 [J]. 工业建筑，1999，V29 (4)：16-18
[9] Fischer, G. and V. C. Li. Influence of Matrix Ductility on Tension-Stiffening Behavior of Steel Reinforced Engineered Cementitious Composites (ECC). ACI Structural Journal, Vol. 99, No. 1, Jan.-Feb., 2002, pp 104-111
[10] Victor C. Li. Engineering Cementitious Composites for Structural Application. Journal of Materials in Civil Engineering-May 1998.
[11] Gustavo Parra-Montesions and James K. Wight. Seismic Response of Exterior RC Column to Steel Beam Connections. Journal of Structural Engineering-October 2002.
[12] Gregor Fischer, Hiroshi Fukuyama and Victor Li 2003. Effect of Matrix Ductility on the Performance of the Reinforced ECC Column Members under Reversed Cyclic Loading Conditions. ACI Structural Journal, Vol. 99, No. 6, 2002, pp 781-790
[13] 孙邦明，杨才富，张永权. 高层建筑用钢的发展. 宽厚板，2001，Vol. 7，No. 3，1-6
[14] 王锡钦. 高功能结构用钢板的发展 [J]. 建筑钢结构进展，2002，(1)：16-22
[15] 侯宝隆. 建筑钢材的发展趋势 [J]. 钢结构，2001，(1)：61
[16] 柴昶. 我国建筑钢结构用钢材的现状与展望 [J]. 钢结构，2001，(1)：1-6
[17] 贾大朋等. 低屈强比高强度建筑用耐火钢控轧控冷试验. 轧钢，2003，Vol. 20，No. 3，1-4
[18] 徐有邻. 我国混凝土结构用钢筋的现状及发展. 土木工程学报，1999，Vol. 32 No. 5，3-9

[19] 徐有邻. 混凝土结构用钢筋的合理选择 [J]. 建筑结构, 2000, 30 (7): P52-54

[20] 叶列平, 陆新征, 冯鹏等. 简论结构抗震的鲁棒性. 第十届高层建筑抗震技术交流会论文集, 2005年12月, 广州

[21] 束伟农, 柯长华等. 北京电视中心主楼巨型结构设计. 建筑结构, 2006, Vol. 36, No. 6, 1-5

[22] 蔡健, 周靖, 方小舟. 基于结构性能系数的抗震设计方法. 城市与工程安全减灾研究与进展（周锡元主编），北京：中国科学技术出版社, 2006, 335-339

[23] M. J. N. 普瑞斯特雷, F. 塞勃勒, G. M. 卡尔维. 桥梁抗震设计与加固. 北京：人民交通出版社, 袁万城等译, 1997

[24] 彭凌云, 周锡元, 闫维明. 建筑结构的分布式阻尼减震方法. 东南大学学报, 2005, Vol. 35, 增刊(I), 45-48

[25] Paulay, T. and Priestley M. J. N. Seismic Design of Reinforced Concrete and Masonry Buildings. John Wiley and Sons, Inc., 1992, pp. 98-106

[26] Gregor Fischer and Victor C. Li. Intrinsic Response Control of Moment Resisting Frames Utilizing Advanced Composite Materials and Structural Elements. ACI Structural Journal, V. 100, No. 2, March-April 2003

[27] Ikeda, S.. Seismic behaviour of reinforced concrete columns and improvement by vertical prestressing, Challenges for Concrete in the Next Millennium. *Proceedings*, *XIIIth FIP Congress*, Vol. 1, Rotterdam, pp 879-884, 1998

[28] W. A. Zatar and H. Mutsuyoshi. Dynamic response behavior of partially prestressed concrete piers under servere earthquake. *Annual Meeting of JSCE*, 1998

[29] W. A. Zatar and H. Mutsuyoshi. Inelastic response behavior of partially prestressed concrete piers. *Proceedings of the Seventh East Asia-Pacific Conference on Structural Engineering &Construction*, August 27-29, Vol. 2, pp 1217-1222, 1999

[30] W. A. Zatar and H. Mutsuyoshi. Control of residual displacements of RC piers by prestressing. *Seminar on Post-Peak Behavior of RC Structures Subjected to Seismic Loads*, October 25-29, Vol. 2, pp 305-319, 1999

[31] W. A. Zatar and H. Mutsuyoshi. Reduced residual displacements of partially prestressed concrete bridge piers. *WCEE* 12, paper No. 1111, 2000

[32] W. A. Zatar, H. Mutsuyoshi. Residual Displacements of Concrete Bridge Piers Subjected to Near Field Earthquakes. *ACI Structural Journal*, November/December, 2002, pp 740-749

[33] Kwan W. P. and Billington S. L.. Unbonded post-tensioned bridge piers. I: monotonic and cyclic analyses. *J. of Bridge Engineering*, ASCE, March/April, pp 92-101, 2003

[34] Kwan W. P. and Billington S. L.. Unbonded post-tensioned bridge piers. II: seismic analyses. *J. of Bridge Engineering*, ASCE, March/April, pp 102-111, 2003

[35] 陆新征, 缪志伟, 江见鲸, 叶列平. 静力和动力荷载作用下混凝土高层结构的倒塌模拟, 山西地震, 126 (2), 2006: 7-11

[36] 冯鹏, 叶列平, 黄羽立. 受弯构件的承载力、延性及变形性指标的研究. 《工程力学》, 2005, Vol. 22, No. 6, 28-36

[37] 中华人民共和国国家标准,《建筑结构可靠度设计统一标准》GB 50068—2001. 北京：中国建筑工业出版社, 2001

[38] 叶列平, 林旭川, 冯鹏. 高强钢筋混凝土梁受弯承载力的安全储备及其经济性分析.《建筑结构》增刊, Vol. 36 (S1), 2006, 5-72~5-75

[39] 陈肇元, 杜拱辰. 结构设计规范的可靠度设计方法质疑.《建筑结构》, 2002, 32 (4):, 64-69

第七章 | Chapter 7

现代钢管混凝土结构研究的若干关键问题

韩林海[1]　陶　忠[2]

1. 清华大学土木工程系，北京，100084　E-mail: lhhan@tsinghua.edu.cn
2. 福州大学土木工程学院，福州，350002

摘　要：钢管混凝土能适应现代工程技术发展的需要，正被越来越广泛地应用于各类工程结构中。本文简要回顾了钢管混凝土的应用和研究现状，探讨了钢管混凝土结构应用中存在的若干关键问题，如一次加载和长期荷载作用下的性能、滞回性能，抗火设计和火灾后的力学性能，钢管初应力的影响以及混凝土浇筑质量的影响等方面的基本原理和方法等。本文还简要介绍了一些新型组合构件，如薄壁钢管混凝土、钢管高性能混凝土、中空夹层钢管混凝土和FRP约束钢管混凝土等的基本原理。本文最后简要论述了钢管混凝土结构节点、采用钢管混凝土的框架结构和混合结构力学性能研究方面的一些新进展。

关键词：钢管混凝土，力学性能，设计理论，构件，节点，框架，混合结构

SEVERAL KEY ISSUES ON MODERN CONCRETE-FILLED STEEL TUBULAR STRUCTURES

L. H. Han[1]　Z. Tao[2]

1. Department of Civil Engineering, Tsinghua University, Beijing, 100084, P. R. C
2. College of Civil Engineering, Fuzhou University, Fuzhou, 350002, P. R. C

Abstract: Concrete-filled steel tubular (CFST) structures have several structural and constructional benefits, which are widely recognized and have led to an increased use in modern engineerings. A brief introduction to application and research on concrete-filled steel tubes are introduced in this paper, several key issues on designing CFST structures are presented, including members subjected to static loading and long-term loading; seismic behaviours; fire resistance design and damage evaluation after exposure to fire, pre-stress effect in steel tubes and concrete compaction effect. Some new types CFST members, such as CFSTs with thin-walled steel tubes or high-performance concrete used, concrete-filled double skin steel tubular (CFDST) as well as FRP-confined CFST members and etc. are introduced. Finally, some recent reseach developments on CFST column-to-beam connections, frames and hybrid structures using CFST members are generally described in this paper.

Keywords: concrete-filled steel tubes (CFST), behaviour, design theroy, member, connection, frame, hybrid structures.

1. 前言

钢管混凝土（CFST）是指在钢管中填充混凝土而形成，且钢管及其核心混凝土能共同承受外荷载作用的结构构件，按截面形式不同，可分为圆钢管混凝土，方、矩形钢管混凝土（如图 1 所示）和多边形钢管混凝土等。

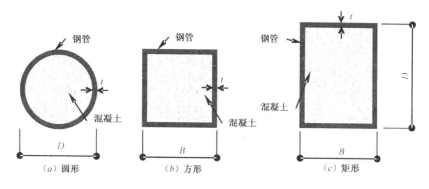

图 1　钢管混凝土截面示意图

钢管混凝土诞生已有一百多年的历史。在 20 世纪 60 年代左右，人们对钢管混凝土压弯构件的力学性能，以及钢管和混凝土之间粘结问题等的研究已取得了较大进展，但当时的工作多以试验研究为主。七八十年代，研究者们开始较多地研究该类结构的抗震性能和耐火极限，以及长期荷载作用的影响等问题。90 年代以来，对钢管混凝土结构抗震性能的研究进一步深入，对采用高性能材料的钢管混凝土构件，及薄壁钢管混凝土工作性能和设计方法的研究也有不少报道。在这一阶段，研究者们还较多地开展了压弯剪和压弯扭构件性能的研究，对钢管混凝土工作机理的理论研究得到较快发展，使人们对这类组合构件力学实质的认识逐渐深入。

Schneider（1998）对钢管混凝土轴压力学性能的研究现状进行了较为透彻的综述和分析；Han（2002）在 Schneider（1998）综述的基础上，进一步给出了其他钢管混凝土轴压性能方面的研究结果；Gourley 等（2001），Shams 和 Saadeghvaziri（1997），Shanmugam 和 Lakshmi（2001）较全面总结了各国研究者在钢管混凝土结构力学性能研究方面取得的一些进展。韩林海和杨有福（2004）则较系统地归纳和总结了钢管混凝土结构节点方面的研究成果。

对于钢管混凝土构件的研究存在各种不同的方法，对其钢管和核心混凝土之间组合作用的认识程度会有所不同，因而所获计算方法和计算结果也就会有所差异。但无论采用哪种办法，都有其特点，其目的都是为了寻找钢管混凝土结构合理科学的设计理论，都是值得借鉴的。

钢管混凝土具有一系列力学性能和施工等方面的优点，因此已被较广泛地应用于各个工程建设领域，如冶金、造船、电力等行业的单层或多层工业厂房、设备构架柱、各种支架、栈桥柱、送变电杆塔、桁架压杆、桩、大跨和空间结构、商业广场、多层办公楼及住宅、高层和超高层建筑以及桥梁结构等。

我国是研究和应用钢管混凝土较多的国家之一（蔡绍怀，2003；韩林海，2000，2004，2006；钟善桐，1994，2003）。自20世纪60年代以来，有关钢管混凝土结构的科学研究、设计和施工等方面均取得较大进展。七八十年代以前，以研究和应用圆形截面的钢管混凝土居多。此后，对方、矩形截面钢管混凝土的研究取得了较大进展，近十几年来，工程应用也逐渐增多（韩林海，2006）。

国内外已有一些有关钢管混凝土结构的设计规程（ASCCS，1997）。日本目前在钢管混凝土房屋建筑方面的设计规程有 AIJ（1997）等；美国的设计规程 ACI（2002）和 AISC（2005）都给出了钢管混凝土结构设计方面的规定；英国的设计规程有 BS5400（1979）。此外，还有欧洲规范 EC4（2004）等。

我国近十几年来先后颁布了几本有关钢管混凝土结构设计方面的规程，例如国家建筑材料工业局标准 JCJ01—89（1989），中国工程建设标准化协会标准 CECS28∶90（1992）（该规程目前正在修编中），中华人民共和国电力行业标准 DL/T 5085—1999（1999）等都给出了圆钢管混凝土结构设计方面的规定。中华人民共和国国家军用标准 GJB 4142—2000（2001）给出了方钢管混凝土结构设计方面的条文。2003年颁布实施的福建省工程建设标准《钢管混凝土结构技术规程》DBJ13—51—2003（2003）可适合于圆形和方、矩形钢管混凝土结构的设计计算。2003年颁布的天津市工程建设标准《天津市钢结构住宅设计规程》DB29—57—2003（2003）中也给出了钢管混凝土结构设计计算方面的规定。中国工程建设标准化协会标准 CECS159∶2004 给出了方、矩形钢管混凝土结构设计方面的规定，福建省工程建设标准《钢—混凝土混合结构技术规程》DBJ13—61—2004（2004）给出了不同截面情况下钢管混凝土核心混凝土收缩的计算方法以及构件恢复力模型等的确定方法等。由于具备了上述设计依据，可更好地促进钢管混凝土结构在我国的推广应用。

以往钢管混凝土理论研究成果和工程实践经验的取得为该类结构的进一步发展创造了条件。在国内外学者和有关技术人员的共同努力下，钢管混凝土结构理论与设计成套技术正逐步趋于成熟，钢管混凝土正逐步形成为一个系统而完整的新学科。

现代结构技术和钢、混凝土材料的不断发展，对钢管混凝土结构理论的发展提出了新的要求，这就需要科技人员用发展的眼光和科学的态度对原有的研究成果不断进行完善和改进，且要对一些新型钢管混凝土结构进行专门深入的研究，此外，还需要对工程技术中出现的新问题和难点问题不断探索。只有这样，才可能适应这一现代技术实际发展的需要，才可能促进钢管混凝土结构学科向更高层次的发展。

近些年来，随着现代科学技术的进步，试验科学、计算机技术及分析计算手段的发展，都为更细致和深入地研究钢管混凝土结构的工作机理及其设计理论创造了条件。随着我国国民经济的健康发展和社会的快速进步，使土木工程技术得到了前所未有的发展机遇。在这一过程中，诞生了不少采用钢管混凝土结构的典型工程。不断的工程实践不仅提出了不少需要解决的新问题，而且也促进了对原有研究结果的完善和提高，同时也使现代钢管混凝土结构技术快速趋于成熟。作者及其合作伙伴们适逢盛世，有幸在这一发展过程中进行了一些力所能及的研究工作。

在国家各类科学基金、科研项目及工业界等的支持下，作者及其合作伙伴们得以有机会先后进行了一千多个钢管混凝土构件在各种荷载作用下的典型试验研究，其中的不少试验补充了实际工程中常见的，但尚没有经过试验验证的工况。例如大体积核心混凝土的水化热和收缩试验，大轴压比下构件的滞回性能、耐火性能及火灾后性能的试验等。通过这些典型的试验研究，不仅积累了宝贵的第一手资料，而且也进一步增强了人们对这类构件在不同工况下工作特性的感性认识。这些实测结果与国内外同行们进行的大量试验结果一起共同奠定了钢管混凝土结构学科发展的必要基础，也为更为全面和扎实地开展这类结构的理论分析工作创造了条件。

作者及其课题组还先后采用纤维模型法和有限元法等数值方法计算分析了钢管混凝土构件受力全过程。尤其在近些年，课题组采用大型非线性有限元软件建模，深入地研究了不同截面形式、不同加载路径情况下钢管混凝土构件在压（拉）、弯、扭、剪及其复合受力状态下的工作机理，较为细致地分析了钢管及其核心混凝土之间的相互作用，较全面和透彻地揭示了该类组合构件的力学实质。此外，课题组还初步解决了钢管混凝土受局压荷载时及考虑火灾全过程影响情况下构件力学性能的理论分析等问题，从而使力学建模能更真实地反映构件在实际结构中的工作情况。

根据钢管混凝土结构技术发展的需要，作者及其课题组成员们近年来还对钢管混凝土结构节点、平面框架结构和由钢管混凝土框架—钢筋混凝土剪力墙组成的混合结构体系等的力学性能进行了理论分析和试验研究。

本文拟简要归纳和总结作者及其课题组合作伙伴们在钢管和混凝土结构研究领域取得的一些新的研究结果，具体内容包括：一次加载和长期荷载作用下的性能、滞回性能、抗火设计和火灾后的力学性能、钢管初应力的影响以及混凝土浇筑质量的影响等方面的基本原理和方法。本文还拟介绍一些新型组合结构，如薄壁钢管混凝土、钢管高性能混凝土、中空夹层钢管混凝土和 FRP 约束钢管混凝土等的基本原理；本文最后拟介绍钢管混凝土结构节点、采用钢管混凝土的框架结构和混合结构的力学性能及一些研究新进展等。

2. 钢管混凝土结构构件设计中的系列关键问题

钢管混凝土是一种组合构件，在受力过程中，钢管及其核心混凝土协同互补、共同工作，二者之间存在组合作用，因此，如何合理地认识和深入了解这种相互作用的"效应"是钢管混凝土理论研究的热点课题。从应用部门的角度，不仅希望这一问题在理论上取得较透彻的解决，而且更希望能进一步提供便于应用的实用方法。从研究者的角度来说，在工程技术领域从事科学研究，其最终目的也应该是更好地为实际应用服务，要把理论成果放到工程实践中去接受检验。

为了实现把理论成果推进到实用化程度的目标，需要有定量的结果，还需要考虑到钢管混凝土结构几何参数、物理参数、荷载参数等一系列实际影响因素，这会大大增加研究工作的深度和难度，但这样做的结果反过来也增强了研究成果的理论价值。当然，从事这样的理论研究工作，往往需要花费更大的精力和更长的时间。近年来，课题组对钢管混凝土结构构件的系列关键问题，如静力性能、滞回性能、抗火性能和工程应用中的一些关键问题等进行了研究，每个问题的研究大致都经历了三个阶段，即首先确定组成钢管混凝土的钢材及混凝土的应力—应变关系模型，在此基础上，采用数值计算方法，分析钢管混凝土构件在静力、动力和火灾作用等情况下的荷载—变形全过程关系曲线，并和已有的国内外的试验结果进行了对照，结果总体上令人满意（韩林海，2004，2006）。为了进一步更加全面验证以上理论分析结果的准确性，在第二阶段针对国内外以往进行过的试验研究状况，有计划地进行一系列钢管混凝土构件在一次加载、长期荷载、往复荷载、火灾作用下和火灾作用后等情况下的试验研究。

充分考虑组成钢管混凝土的钢管和混凝土之间相互组合作用的分析方法自然是比较系统和完善的，且得到大量试验结果的验证，计算结果也较精确。但从实际应用的角度考虑，这种分析方法显得还是比较复杂。如何从上述理论成果出发，搭起必要的桥梁，过渡到便于广大设计人员应用的实用方法，是一项十分有意义的工作，这也是在第三阶段研究中拟解决的主要任务。

为了实现这一目标，在充分考虑到工程实际应用的情况下，对影响钢管混凝土性能的基本参数（包括物理参数、几何参数和荷载参数等）进行了系统的分析，并考虑各种可能的影响因素，然后对所得大量计算结果进行统计分析和归纳，考察钢管混凝土力学性能的变化规律；最后进行高度的概括，提出钢管混凝土构件在各种荷载作用下的设计方法（韩林海，2004，2006；Han 和 Tao，2006）。

下面简要归纳作者及其课题组近年来在钢管混凝土结构构件研究方面取得的一些研究结果。

2.1 一次加载下的性能

在确定组成钢管混凝土的钢管及其核心混凝土本构关系模型的基础上，采用数值方法对圆形、方形和矩形钢管混凝土构件的荷载—变形关系进行了全过程分析，分析结果与试验结果总体吻合较好。最后在系统参数分析结果的基础上，提出了钢管混凝土基本构件的承载力简化计算公式。在收集国内外现有试验数据的基础上，对不同设计规程提供的方法进行了分析比较，结果表明，推导出的计算公式的计算结果与试验结果吻合较好，且总体上稍偏于安全（Han，2000a，2002，2004；Han 等，2001，2004c）。韩林海（2004）还介绍了钢管混凝土构件可靠度分析方面的结果。

受力全过程分析是深入认识钢管混凝土力学性能的重要前提。采用纤维模型法和有限元法都可方便地计算出钢管混凝土构件的荷载—变形全过程关系曲线（韩林海，2004，2006）。

韩林海（2004）介绍的钢管混凝土纤维模型法是建立在一些基本假定基础上，其中，如何合理地确定核心混凝土的应力—应变关系最为关键，且如何考虑钢管和核心混凝土之间的相互作用是核心问题。当分析长期荷载作用的影响时，需考虑混凝土收缩和徐变的影响；研究滞回性能时需确定加、卸载准则；研究火灾作用下结构的反应时，则需考虑升、降温的影响等。纤维模型法的特点是计算简便、概念直观。大量计算结果表明，纤维模型法可满足一次加载情况下最常见的压弯构件分析的要求，并被成功地推广应用到长期荷载、往复荷载、火灾下和火灾后、考虑钢管初应力影响等问题的研究中。但该方法的缺点是：对于更为复杂的问题，例如钢管混凝土构件在受压、受弯的同时承受扭、剪或进行复合受力分析时，其适用性差，此外，该方法不利于细致地分析钢管和混凝土之间的相互作用。

利用有限元法分析钢管混凝土受力特性的优点在于：（1）方法通用性强，可进行不同荷载参数、几何参数和物理参数情况下构件的计算分析；(2)适用性强，可应用于不同边界条件、不同初始缺陷等条件下构件的分析；(3)可较为细致地分析钢管和核心混凝土之间的相互作用，有利于深入揭示钢管混凝土构件的工作机理。但有限元法的缺点是：计算方法较复杂，计算时间长且不便于应用，尤其在需要进行大规模参数分析时的计算工作量较大。

在进行钢管混凝土构件力学性能的研究时，可根据实际问题的需要，因地制宜地选用合适的方法。但无论纤维模型法或有限元法，或者其他研究钢管混凝土力学性能的理论方法，都有其合理的适用范围，都应采用尽可能多的典型试验结果进行验证才能更为充分地说明其有效性。

除了轴压和压弯的受力情况外，实际工程结构中的钢管混凝土还可能处于压扭、压弯扭、压弯剪，甚至压弯扭剪等复杂受力状态。例如海上采油平台立柱以及在地震荷载作用下的轴压柱就可能产生压扭的受力状态；钢管混凝土用作建筑物的框架角柱、高速公路的曲线形桥和斜交桥的桥墩、海上采油平台的立柱，停机场的定向塔及螺旋楼梯的中心柱等，除承受轴向压力和弯矩外，在风荷载和地震等作用下，尚有扭矩的共同作用，当排架角柱采用平腹杆双肢柱时，还可能产生压弯扭的受力状态等。

国内外学者对不同截面形状钢管混凝土压弯构件及圆钢管混凝土压弯扭构件力学性能的研究已有一些研究结果，但对方钢管混凝土在压弯扭以及圆、方形钢管混凝土在压弯扭剪复合受力状态下工作机理和设计理论的研究尚有待于深入进行。

尧国皇（2006）采用 ABAQUS 软件（Hibbitt, Karlsson & Sorensen Inc, 2003），建立了钢管混凝土构件在压、弯、扭、剪及其复合受力状态下力学性能的分析模型。对纯扭、横向受剪、压扭、弯扭、压弯扭、压弯剪和压弯扭剪构件的荷载—变形关系曲线进行了全过程分析，研究了各阶段钢管和核心混凝土截面的应力状态及其相互作用，探讨了不同加载路径情况对钢管混凝土构件力学特性的影响规律。最终在参数分析结果的基础上推导了钢管混凝土构件在不同受力状态下的承载力实用计算方法。

研究结果表明，加载路径对钢管混凝土构件承载力极限状态影响不大。下面以弯扭构件为例简要进行说明。

为了便于说明问题，分析两种加载路径的影响，路径Ⅰ：加载方式是先作用弯矩 M，然后保持 M 的大小和方向不变；不断增加扭矩 T。路径Ⅱ：采用的加载方式是弯矩（M）和扭矩（T）按一定比例施加，即弯扭比 $m_o = M/T$ 保持不变。

图2和图3分别给出了圆、方钢管混凝土弯扭构件在达到极限承载力时钢管和核心混凝土截面剪应力 τ_{xz}（图中为S13）分布的比较，其中，算例的计算条件为：钢管直径或边长 $D(B) = 400 \text{mm}$；含钢率 $\alpha = 0.1$（$\alpha = A_s/A_c$，A_s、A_c 分别为钢材和混凝土的截面积）；构件长度 $L = 1200 \text{mm}$；C60混凝土；Q345钢。可见，对于不同的加载路径，当构件达到极限承载力时，钢管和核心混凝土截面剪应力的分布和数值基本相同。

图4给出加载路径Ⅰ（实线）和加载路径Ⅱ（虚线）情况下，钢管混凝土弯扭构件在受力过程中钢管与核心混凝土相互作用力（p）-转角（θ）关系曲线的比较。由于弯扭构件中钢管与核心混凝土相互作用沿构件截面和长度方向分布都不均匀，为方便起见，取构件中截面同一位置处的相互作用力进行比较。可见，由于加载路径的不同，p-θ 关系曲线的变化规律也不同，但构件达到极限承载力时的相互作用力基本相同。可见，加载路径

对钢管混凝土弯扭构件的承载力极限影响不显著。

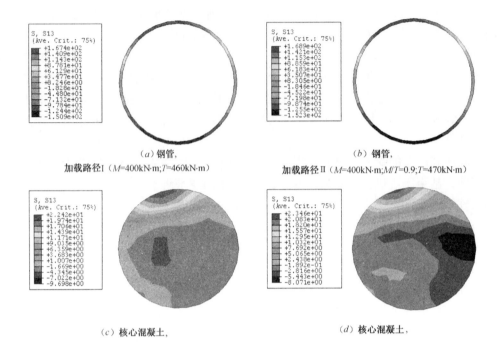

图 2 不同加载路径下的圆钢管混凝土弯扭构件截面剪应力分布

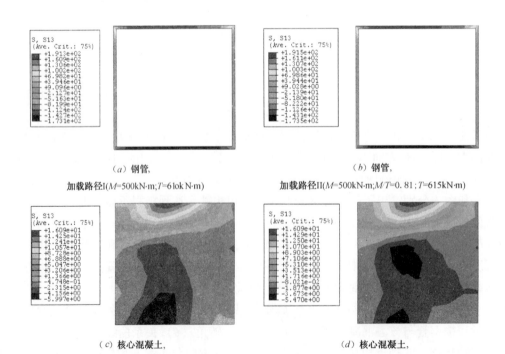

图 3 不同加载路径下的方钢管混凝土弯扭构件截面剪应力分布

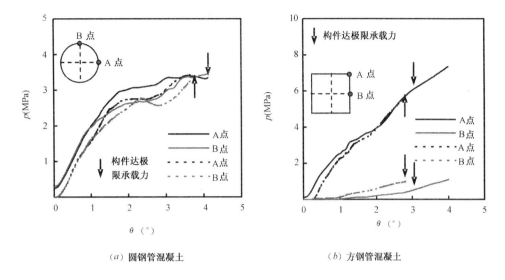

(a) 圆钢管混凝土　　　　　　　　　(b) 方钢管混凝土

图 4　不同加载路径下弯扭构件的 $p\text{-}\theta$ 关系曲线

基于研究结果，建议钢管混凝土压弯扭剪构件承载力相关方程可采用如下表达式：

当 $N/N_u \geqslant 2\varphi^3 \eta_0 \sqrt{1-\left(\dfrac{T}{T_u}\right)^2 - \left(\dfrac{V}{V_u}\right)^2}$ 时

$$\left(\dfrac{1}{\varphi}\cdot\dfrac{N}{N_u} + \dfrac{a}{d}\cdot\dfrac{M}{M_u}\right)^{2.4} + \left(\dfrac{V}{V_u}\right)^2 + \left(\dfrac{T}{T_u}\right)^2 = 1 \quad (1a)$$

当 $N/N_u < 2\varphi^3 \eta_0 \sqrt{1-\left(\dfrac{T}{T_u}\right)^2 - \left(\dfrac{V}{V_u}\right)^2}$ 时

$$\left[(-b\cdot\left(\dfrac{N}{N_u}\right)^2 - c\cdot\left(\dfrac{N}{N_u}\right) + \dfrac{1}{d}\cdot\dfrac{M}{M_u}\right]^{2.4} + \left(\dfrac{V}{V_u}\right)^2 + \left(\dfrac{T}{T_u}\right)^2 = 1 \quad (1b)$$

式中，N_u 和 M_u 分别为钢管混凝土轴压强度承载力和抗弯强度；T_u 为钢管混凝土抗扭强度；V_u 为钢管混凝土抗剪强度。a、b、c、d 等为计算系数（韩林海，2004；2006）。

式（1）的特点是，当左边某一项荷载为 0 时，即为另外三种荷载共同作用下的承载力极限状态；当某两项荷载为 0 时，即为另外两种荷载共同作用下的承载力极限状态；当某三项荷载为 0 时，即为单一荷载作用下的极限承载力计算公式。

刘威和韩林海（2006）研究了钢管混凝土受轴向局压荷载时的工作机理研究和承载力计算方法。通过采用 ABAQUS 建模进行系统的参数分析，发现局压的受力方式仅影响钢管混凝土部分区域的力学性能，局压影响区内的核心混凝土受到较大的横向压应力，局压承载强度与塑性性能有所提高，且局压面积比（局压计算底面积与局压面积之比）越大，局压影响区越小。通过分析不同参数的影响规律，发现随着含钢率和钢材强度的增加，核心混凝土中的横向压应力增大，钢管混凝土的局压承载力与塑性性能有所提高；随着混凝土强度的提高，钢管混凝土局压承载力提高，塑性性能降低；随着端板刚度的增加，核心混凝土端面的实际受力范围有所增加，受力更为均匀，钢管承担的纵向应力增大，局压受力趋近于全截面受力的情况。基于参数分析结果，刘威和韩林海（2006）提出了钢管混凝土局压承载力实用计算方法，算例计算结果与试验结果吻合较好。

2.2 长期荷载作用下的性能

和普通钢筋混凝土构件中的混凝土相比，长期荷载作用下钢管混凝土核心混凝土的工作具有如下特点：(1) 处于密闭状态，和周围环境基本没有湿度交换。(2) 沿构件轴向的收缩将受到其外包钢管的限制。因此，钢管混凝土其核心混凝土的收缩变形就有可能不如同条件下的普通混凝土显著。(3) 受力过程中，核心混凝土和其外包钢管之间存在着相互作用问题。在进行长期荷载作用下钢管混凝土构件变形性能的研究时应适当考虑上述特点。

基于 ACI（1992）提供的混凝土徐变和收缩模型，提出一种适合于长期荷载作用下钢管混凝土构件变形的理论分析模型，计算结果和试验结果基本吻合。在此基础上，系统分析了加载龄期、持荷时间、轴压比、构件截面含钢率、钢材牌号、混凝土强度等级和荷载偏心率等因素对长期荷载作用下钢管混凝土构件变形性能和承载力的影响规律，最终提出考虑长期荷载作用影响时钢管混凝土构件承载力的实用验算方法（Han 等，2004b；Han 和 Yang，2003）。

2.3 滞回性能

在确定了组成钢管混凝土的钢材和核心混凝土应力—应变滞回关系模型的基础上，对钢管混凝土构件的弯矩（M）—曲率（ϕ）和侧向荷载（P）—位移（Δ）滞回关系曲线进行了计算和分析，计算结果得到国内外大量试验结果的验证（韩林海，2004；Han 等，2003b；Han 和 Yang，2005）。

分析结果表明，钢管混凝土压弯构件弯矩（M）—曲率（ϕ）骨架曲线的特点是无显著的下降段，转角延性好，其形状与不发生局部失稳钢构件的性能类似，这是因为钢管混凝土构件中的混凝土受到了钢管的有效约束，在受力过程中不会发生因混凝土被压碎而导致构件破坏的现象；另外，由于混凝土的存在可以避免或延缓钢管发生局部屈曲，从而可以保证其材料性能的充分发挥，这样，由于组成钢管混凝土的钢管和其核心混凝土之间相互贡献、协同互补、共同工作的优势，使其弯矩—曲率滞回曲线表现出良好的稳定性，基本上没有刚度退化和强度退化，曲线图形饱满，呈纺锤形，没有明显的捏缩现象，耗能性能良好。

分析结果还表明，影响钢管混凝土弯矩（M）—曲率（ϕ）滞回曲线骨架线的因素主要有：含钢率、钢材屈服强度、混凝土强度和轴压比等。影响钢管混凝土侧向荷载（P）—位移（Δ）滞回曲线骨架线的主要因素是：含钢率、钢材屈服强度、混凝土强度、轴压比和构件长细比等。在理论分析和试验研究结果的基础上，提出了钢管混凝土压弯构件弯矩（M）—曲率（ϕ）和侧向荷载（P）—位移（Δ）滞回模型，以及位移延性系数的简化确定方法（韩林海，2004）。

2.4 耐火性能和抗火设计方法

在确定了高温下组成钢管混凝土的钢材和核心混凝土的应力—应变关系模型的基础上，根据高温下截面温度场的分析结果，提出了计算钢管混凝土柱耐火极限及火灾下强度与变形的理论模型，并对已有的试验数据进行验证，理论结果与试验结果吻合良好。在此基础上，利用该理论计算模型，进一步分析了荷载大小，材料强度，截面含钢率，截面尺寸，构件长细比，荷载偏心率，防火保护层厚度等参数对钢管混凝土柱耐火极限的影响，深入了解了钢管混凝土柱在火灾下的工作机理。最后，在系统参数分析结果的基础上，提出钢管混凝土柱耐火极限和防火保护层厚度的实用计算方法（Han，2001；韩林海等，2002；Han 等，2003a，2003c，2003d）。

2.5 火灾后的力学性能

随着钢管混凝土在实际工程中的应用日益增多，深入研究其火灾后的力学性能和承载力损伤规律，对于合理确定火灾后该类结构的修复加固措施非常必要，以往国内外尚缺乏该方面研究的报道（韩林海，2004）。

课题组先后进行了钢管混凝土构件在恒高温以及火灾（例如 ISO—834 标准火灾曲线）作用后的理论分析和试验研究。在确定了钢材和混凝土在高温作用后应力—应变关系模型的基础上，利用数值分析方法计算了 ISO—834 规定的标准升温曲线作用后圆钢管混凝土和方、矩形钢管混凝土压弯构件荷载—变形关系曲线，理论计算结果和试验结果基本吻合。在系统分析了火灾持续时间、构件截面含钢率、钢材屈服强度、混凝土强度等级、荷载偏心率和构件截面尺寸等因素对火灾作用后钢管混凝土构件承载力影响规律的基础上，提出火灾作用后钢管混凝土构件承载力的简化计算方法。分析结果表明，火灾作用对钢管混凝土构件的火灾后承载力有较大影响，且影响因素主要是构件截面尺寸、长细比、受火时间和防火保护层厚度（Han 和 Huo，2003；Han 等，2002a，2002b，2005a）。最近，还开展了火灾作用后钢管混凝土柱抗震性能的研究工作（Han 和 Lin，2004）以及利用纤维增强复合材料（FRP）或钢套管修复加固火灾后钢管混凝土构件的力学性能研究（Han 等，2006a；Tao 等 2006a，2006b）。

2.6 钢管初应力的影响

多、高层建筑中采用钢管混凝土柱时，往往是先安装空钢管，然后安装梁，再进行楼板的施工。为了加快施工进度，提高工作效率，通常是先安装若干层空钢管柱，然后再在空钢管中浇灌混凝土。这样，在混凝土硬化、与其外包钢管共同组成钢管混凝土构件之前，由于施工荷载和湿混凝土自重等因素，可能会在钢管内产生纵向初压应力（以下简称钢管初应力）。此外，钢管混凝土拱桥施工时，往往也是先安装空钢管拱肋，然后再浇筑钢管内的混凝土，同样也会在钢管中产生初应力。上述初应力对钢管混凝土构件力学性能的影响问题一直是工程界关注的热点问题之一。合理确定钢管初应力对钢管混凝土构件力学性能的影响将对更安全合理地应用钢管混凝土结构和进行施工组织设计具有重要意义。以往对该方面的研究工作尚不够深入。Han 和 Yao（2003b）考虑钢管初应力的影响，进行了钢管混凝土压弯构件力学性能的试验研究，然后对构件的荷载—变形关系曲线进行计算分析。在此基础上，分析初应力系数、构件长细比、截面含钢率，荷载偏心率、钢材和混凝土强度等因素的影响规律。理论分析和试验结果均表明，钢管初应力对钢管混凝土构件的承载力有影响，其存在可使钢管混凝土构件的极限承载力最大降低 20% 左右。因此应合理考虑钢管初应力对钢管混凝土构件承载力的影响。在系统参数分析结果的基础上，提出了工程常用参数范围内考虑钢管初应力影响时承载力实用验算公式。

2.7 混凝土浇筑质量的影响

钢管混凝土由外包钢管及其核心混凝土共同组成，在进行核心混凝土的施工时，混凝土的质量应符合混凝土有关施工验收规范的要求，但是，由于核心混凝土为外围钢管所包覆，从而导致对混凝土浇筑质量控制问题的特殊性和难度。组成钢管混凝土的钢管及其核心混凝土间的协同互补作用是钢管混凝土具有一系列突出优点的根本原因，因此，研究者们自然会很关心钢管和核心混凝土二者的共同工作问题。因此，有必要研究混凝土浇筑质量对钢管混凝土构件的承载和变形能力的影响问题。

通过对不同混凝土浇筑质量情况下钢管混凝土力学性能的研究表明，混凝土密实度对钢管混凝土构件的力学性能影响很显著，且混凝土浇筑方式对钢管混凝土中核心混凝土的密实度有较大影响。在进行钢管混凝土中混凝土浇筑质量问题的研究时，要充分考虑控制混凝土的强度和密实度，前者可以保证混凝土达到设计强度，后者则可以保证钢管和核心混凝土二者相互协同作用的充分发挥。工程实践是丰富多彩的，进行钢管混凝土中混凝土的施工方法也会有所不同，但无论采用哪种方法，不仅要保证混凝土的强度，还要保证其有良好的密实度（Han，2000b；Han 和 Yang，2001；Han 和 Yao，2003a）。

上述研究结果使钢管混凝土结构学科进一步趋于完善，并受到工程界的重视。有关静力和滞回性能方面的部分结果被国家电力行业标准《钢—混凝土组合结构设计规程》（DL/T 5085—1999，1999）、国家军用标准《战时军港抢修早强型组合结构技术规程》（GJB4142—2000，2001）、福建省工程建设标准《钢管混凝土结构技术规程》（DBJ13—51—2003，2003）和《钢—混凝土混合结构技术规程》（DBJ13—61—2004，2004）等采用；有关耐火性能方面的部分结果为福建省工程建设标准《钢管混凝土结构技术规程》（DBJ13—51—2003，2003）、天津市工程建设标准《天津市钢结构住宅设计规程》（DB29—57—2003，2003）、中国工程建设标准化协会标准《矩形钢管混凝土结构技术规程》（CECS159:2004，2004）、中国工程建设标准化协会标准《建筑钢结构防火设计规范》（CECS200:2006）和浙江省工程建设标准《建筑钢结构防火技术规范》（送审稿，2003）等采用。

3. 新型钢管混凝土结构构件

如前所述，钢管混凝土结构学科是不断发展的。出于工程建设的需要，近年来一些新型钢管混凝土结构构件引起了工程界的重视，如薄壁钢管混凝土、钢管高性能混凝土、中空夹层钢管混凝土、FRP约束钢管混凝土柱、劲性钢管混凝土以及 FRP—混凝土—钢管组合柱等（陶忠和于清，2006）。下面对这些新型钢管混凝土结构构件的工作特点及其力学性能作一简单介绍。

3.1 薄壁钢管混凝土

薄壁钢管混凝土是相对通常壁较厚的普通钢管混凝土而言的。薄壁钢管混凝土由于采用了薄壁钢管，通常需要考虑局部屈曲对构件力学性能的影响。本文所论述的薄壁钢管是指截面直径与厚度的比值（圆钢管）或者宽度或高度与厚度的比值（方、矩形钢管）超过钢结构对其局部屈曲控制的限值或者钢管壁厚小于 3mm 的钢管。工程实践表明，在钢管混凝土工程中采用薄壁钢管，可以减少钢材用量，减轻焊接工作量，达到降低工程造价的目的。

薄壁钢管混凝土构件的承载力会受到局部屈曲的影响。这种影响主要体现在两个方面：一方面是使得屈曲部位的钢管部分截面提前退出工作，另一方面是降低了钢管对混凝土的约束作用。Han 等（2005c）针对薄壁钢管混凝土轴压构件开展的相关研究工作表明，对于薄壁圆钢管混凝土，极限荷载对应的峰值应变 ε_u 值随钢管径厚比（D/t）的增大而减小，当 D/t 达 125 时，其 ε_u 值接近 $3300\mu\varepsilon$（即普通混凝土的极限压应变），此时钢管对核心混凝土的约束效果较弱；对于薄壁方钢管混凝土，ε_u 随截面宽厚比 B/t 的变化规律与圆钢管混凝土类似，当截面宽厚比超过 100 时，ε_u 值小于钢材屈服时的屈服应变，原因在于钢管的钢材还没有进入屈服阶段，钢管就发生了局部屈曲，从而降低了试件的极限承载力。当截面宽厚比更大时，在极限状态时钢材甚至还处于弹性阶段，说明材料没有充分发挥作用。

由此可见，在进行薄壁钢管混凝土结构的设计时，根据其自身工作机理，应合理确定薄壁钢管 $D(B)/t$ 限值以及考虑钢管局部屈曲对钢管与核心混凝土组合作用的影响。韩林海和杨有福（2004）在总结了不同规程中对钢管混凝土截面的 D/t 或 B/t 限值规定的基础上，根据统计分析，建议了薄壁钢管混凝土的合理径（宽）厚比：当采用 DBJ13—51—2003（2003）中的方法计算薄壁钢管混凝土的承载力时，钢管混凝土中钢管的径厚比 D/t（对于圆钢管混凝土）或宽厚比 B/t（对于方钢管混凝土）限值可按照对应受压构件中空钢管局部稳定限值的 1.5 倍确定。在上述条件下，可暂不考虑钢管局部屈曲对钢管与核心混凝土组合作用的影响，按普通钢管混凝土构件的计算方法计算薄壁钢管混凝土的承载力。

采用薄壁钢管混凝土虽然可以达到降低工程造价的目的，但当管壁进一步趋于减薄，此时钢材的材料强度将不能充分发挥，且延性下降。为此，近年来研究者先后提出了一些抵消这种影响的构造措施，主要方法包括采用约束拉杆（陈宗弼等，2000）、角部隅撑（Huang，2002）和设置纵向加劲肋（Ge 和 Usami，1992；Tao 等，2005）等，分别如图 5(a) ~ (c) 所示。

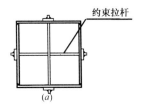
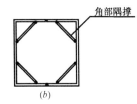
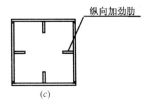

图 5 限制管壁局部屈曲的构造措施

上述三种方法各有优缺点，但采用纵向加劲肋的方法具有制作板件少的优点，且加劲肋和钢管焊成一整体，如设计合理，完全可以使其发挥抗压、抗弯和抗剪的作用。

为考察在纵向设置加劲肋对薄壁钢管混凝土力学性能的影响，课题组先后进行了轴压和偏压两批构件的试验。

所进行的轴压试验结果表明，对于加劲的钢管混凝土，其局部屈曲的发生一般要晚于非加劲的钢管混凝土，且加劲肋的刚度越小，其局部屈曲发展越迅速。此外还发现对于加劲充分的构件，其钢管表面的局部屈曲明显小于非加劲构件，且分布较为均匀，体现了增大加劲肋刚度的有效性。外部加劲的钢管混凝土比内部加劲的钢管混凝土更易产生局部屈曲。典型试件在轴压试验结束后的破坏情况如图6所示（Tao等，2005）。

（a）空钢管　　　　　（b）钢管混凝土　　　　（c）带肋钢管混凝土

图6　典型试件的破坏形态

对于所有加劲的钢管混凝土柱，其实测的极限承载力均高于钢管、混凝土和加劲肋三项承载力叠加值，同时也高于未加劲的钢管混凝土柱承载力（扣除加劲肋直接承担的荷载）。由此可见，加劲肋不仅延缓了钢管的局部屈曲，同时还提高了钢管对混凝土的约束作用。钢管的高厚比越小，这种约束作用的提高越明显。

课题组所进行的薄壁方钢管混凝土偏压试件的试验包括6个带加劲肋的试件［截面形式如图5（c）所示］和12个不带加劲肋的试件。对于不带加劲肋的试件，其中各有6个分别浇筑了普通混凝土和钢纤维混凝土。

通过试验观测发现是否设置加劲肋或在混凝土中添加钢纤维与否对试件整体破坏形态影响不大，但当在试件内部设置加劲肋时，由于加劲肋对管壁的支撑作用，其局部屈曲在试件接近达到极限承载力时才开始出现，可见加劲肋很好地起到了延缓钢管局部屈曲的作用。在混凝土中添加钢纤维与否对试件的承载力影响不大，但在混凝土中添加钢纤维可一定程度上提高试件的延性。

针对薄壁钢管混凝土一般延性稍差的特性，目前课题组还在进行有关提高薄壁钢管混凝土延性的措施研究工作，相关试验工作正在进行当中。

3.2　钢管高性能混凝土

自密实混凝土在自重或少振捣的情况下就能自密实成型。在钢管中灌自密实高性能混凝土可形成钢管高性能混凝土。这样不仅可更好地保证混凝土的密实度，且可简化混凝土振捣工序，降低混凝土施工工作强度和费用，还可减少噪声污染。一些高层建筑和地铁工程中的钢管混凝土柱中采用了自密实混凝土技术，取得了较好的效果。

Han等（2005c，2006b），Han和Yao（2004）进行了100多个钢管高性能混凝土轴压、纯弯和压弯构件力学性能的试验研究，结果表明，钢管高性能混凝土构件的力学性能和钢管普通混凝土基本类似，对普通钢管混凝土构件的计算方法适用于钢管高性能混凝土。

Han 等（2005d）进行了采用高强高性能混凝土的圆形和方形钢管混凝土压弯构件滞回性能的试验研究，试验参数包括轴压比（$n = N_0/N_{u,cr}$，N_0 为施加在试件上的恒定轴压力；$N_{u,cr}$ 为试件轴心受压时的极限承载力）：0~0.6；混凝土强度：90.4~121.6MPa；钢材屈服强度：282~404MPa。结果表明，所进行的钢管高强高性能混凝土构件的滞回性能与钢管普通混凝土的滞回性能基本类似。

3.3 中空夹层钢管混凝土

中空夹层钢管混凝土是在两个同心放置的钢管之间浇筑混凝土而形成的构件。它是在实心钢管混凝土的基础上发展起来的一种新型的钢管混凝土构件形式。如变换内外钢管的截面形式组合，可形成多种不同截面形式的中空夹层钢管混凝土，如图 7 所示。

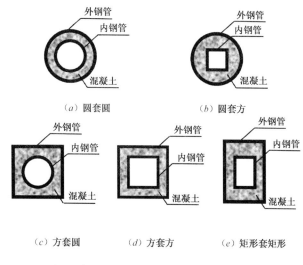

图 7 常见中空夹层钢管混凝土柱截面示意图

中空夹层钢管混凝土柱除了具备实心钢管混凝土柱的基本优点外，尚有自重轻和刚度大的特点，且由于其内钢管受到混凝土的保护，使得该类柱可具有更好的耐火性能。由于中空夹层钢管混凝土具有上述特点，在某些工程领域有其潜在的应用优势，如用作桥墩、海洋平台结构的支架柱、建筑物中的大直径柱以及其他有关高耸构筑物或其柱构件。此外，中空夹层钢管混凝土还可用作大尺寸的灌注桩等。

近十年来，美国、日本、澳大利亚和我国台湾省的学者先后开展了一些有关中空夹层钢管混凝土柱力学性能的研究工作。但目前国内外研究的中空夹层钢管混凝土构件多为如图 7（a）所示的圆套圆截面形式，且以针对轴压构件的试验研究居多，而有关纯弯和偏压构件的力学性能以及压弯构件的滞回性能研究较为少见（陶忠和于清，2006）。为此，作者及其合作伙伴们选择了几种较有应用前景的构件截面形式，较系统地研究了其轴压构件、纯弯构件和偏压构件的静力性能以及压弯构件的滞回性能，并采用数值方法探讨了其耐火性能和抗火设计方法（Han 等，2004a；黄宏，2006；黄宏等，2006；Tao 和 Han，2006；陶忠和于清，2006；Tao 等，2004；杨有福，2005；Zhao 和 Han，2006）。以下简要归纳和总结相关研究工作。

3.3.1 静力性能

针对图7（a）、（c）和（e）所示的中空夹层钢管混凝土（本文分别称之为圆中空夹层钢管混凝土、方中空夹层钢管混凝土和矩形中空夹层钢管混凝土），作者近年来先后进行了系列的试验研究，共包括37个轴压、13个纯弯以及42个偏压构件的试验（Han等，2004a；黄宏，2006；陶忠和于清，2006）。结果表明中空夹层钢管混凝土的工作性能和实心钢管混凝土基本类似，二者的差别取决于内管能否不过早地发生局部屈曲。在内管径厚比（或高厚比）较小的情况下，此时内管可对混凝土提供足够的支撑作用，使得构件的整体工作行为和实心钢管混凝土类似；否则构件的延性就会低于相应实心钢管混凝土的延性。

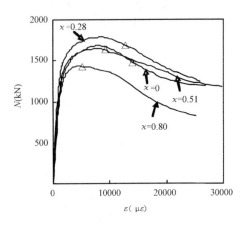

图8 空心率χ对轴压试件N-ε关系曲线的影响

图8所示为空心率对典型的如图7（a）所示的圆中空夹层钢管混凝土轴压柱的轴压力（N）—平均纵向应变（ε）关系的影响。对于圆中空夹层钢管混凝土，空心率的定义为$\chi = D_i / (D_o - 2t_o)$，其中$D_i$和$D_o$分别为内管和外管的外径；$t_o$为外管的管壁厚度。从中可见，除空心率为0.80的试件其延性稍差外，其余试件的N-ε关系变化规律和$\chi=0$的实心钢管混凝土基本一致。试验结果还表明，所有中空夹层钢管混凝土试件由于混凝土延缓了钢管的局部屈曲，因而总体力学性能均明显优于相应空钢管对比试件。

基于上述试验结果，假设由于内钢管对混凝土的支撑作用，中空夹层钢管混凝土与具有相同外钢管的实心钢管混凝土其二者的核心混凝土所受到的约束作用相同，这样在利用纤维模型法进行中空夹层钢管混凝土构件的荷载—变形关系分析时，可借用适用于实心钢管混凝土的核心混凝土的应力—应变关系。由此对作者及他人试验结果进行计算分析，结果表明计算的荷载—变形关系曲线和试验曲线基本吻合。

此外，采用ABAQUS软件（Hibbitt, Karlsson & Sorensen Inc, 2003）建立了有限元分析模型，对中空夹层钢管混凝土轴压N-ε关系进行了有限元计算（黄宏，2006；陶忠和于清，2006）。总体而言，有限元计算的构件荷载—变形关系曲线与纤维模型法的计算结果基本一致，但对径厚比（或宽厚比）较大或空心率较大的试件而言，有限元法计算的N-ε关系曲线下降段与试验结果要更为吻合。

在荷载—变形关系分析的基础上，根据参数分析，对于中空夹层钢管混凝土轴压、纯弯和压弯构件，其承载力简化计算公式可基于实心钢管混凝土承载力计算公式进行适当修正。

3.3.2 滞回性能

针对中空夹层钢管混凝土的滞回性能，黄宏（2006）共进行了12个圆中空夹层钢管混凝土和16个方中空夹层钢管混凝土构件滞回性能试验研究，此外，还进行了5个空钢管的对比试验。构件设计时所考虑的变化参数为轴压比（$n=0\sim0.65$）和空心率（$\chi=0\sim0.77$）。

试验结果表明，圆形和方形中空夹层钢管混凝土试件的外观破坏形态和相应实心钢管混凝土试件基本一致。随着试件横向位移的逐级增大，在刚性夹具与试件连接处受压区开始出现局部的微凸曲。随后随着位移的不断增大，局部凸曲的范围逐渐增大且渐渐沿环向发展。在往复荷载作用下，截面在上下部位都有鼓曲现象发生。试件接近破坏时，这种鼓曲急剧发展，最后，在刚性夹具两侧与试件连接处形成两条沿圆周方向的灯笼状鼓曲。

试验实测的弯矩（M）—曲率（ϕ）和荷载（P）—位移（Δ）滞回关系结果表明，无论是圆形截面，还是方形截面，所有中空夹层钢管混凝土试件的弯矩—曲率和荷载—位移滞回关系均较为饱满，没有明显捏缩现象，试件的耗能能力良好。通过对比分析，发现空心率不同的中空夹层钢管混凝土试件，与具有相同轴压比的实心钢管混凝土试件相比，其 P-Δ 滞回曲线的形状和变化规律基本一致。而空钢管试件由于受局部屈曲的影响，和相同轴压比下的中空夹层钢管混凝土及实心钢管混凝土试件有所不同，其 P-Δ 滞回曲线的骨架线具有较陡的下降段，曲线后期趋于不稳定，循环次数较少。

为了考察截面空心率对试件 P-Δ 滞回骨架线的影响，对不同试件的 P-Δ 滞回骨架线的荷载纵坐标除以各自的峰值荷载 P_{ue} 进行归一化处理。图 9 所示为归一化处理后的典型圆截面试件的 P/P_{ue}-Δ 关系曲线，图中同时绘出了空钢管及空心率为 0 的实心钢管混凝土试件的相应曲线。从中可见，中空夹层钢管混凝土试件的滞回骨架线和相应的实心钢管混凝土试件基本一致，表明在本次试验参数范围内，空心率对骨架线形状的影响较小。而相应空钢管试件由于没有填充混凝土，因而钢管较易发生局部屈曲，其延性明显比钢管混凝土试件的延性差。

同样采用纤维模型法可对中空夹层钢管混凝土的 M-ϕ 和 P-Δ 滞回关系进行分析，计算结果表明其和试验结果二者基本吻合。根据参数分析，结果发现和实心钢管混凝土类似，中空夹层钢管混凝土的 M-ϕ 和 P-Δ 滞回模型骨架线也可采用三线型模型来表示。

3.3.3 耐火性能和抗火设计

以往国内外的研究者针对钢管混凝土的耐火性能和抗火设计方法已开展了大量研究，并取得不少研究结果（韩林海，2004）。随着新

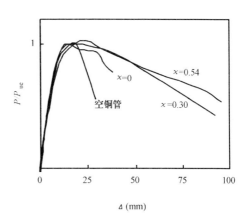

图 9 空心率对 P-Δ 滞回骨架线的影响
（轴压比 n = 0.21、0.22）

型中空夹层钢管混凝土逐渐被人们所接受，有关其耐火性能和抗火设计的研究也已开始引起人们的重视，因而有必要开展相关研究，以供有关工程实践和规程制定提供参考。

根据钢材和混凝土的热工性能及热力学性能，杨有福（2005）采用非线性有限元方法计算了中空夹层钢管混凝土截面温度场的分布规律，在此基础上分析了典型参数对中空夹层钢管混凝土的内、外钢管温度—时间关系曲线的影响规律。结果表明，保护层厚度和外截面尺寸对外钢管温度—时间关系曲线的影响规律与实心钢管混凝土类似，随着保护层厚度的减小、截面尺寸的减小及空心率的增大，内、外钢管的温度均升高，内钢管管壁厚度对内、外钢管的温度—时间关系曲线影响均较小。

在确定钢材和混凝土高温下应力—应变关系模型的基础上，杨有福（2005）采用数值

方法对中空夹层钢管混凝土柱荷载—变形关系曲线及承载力进行了计算分析。在此基础上，分析了各参数对"火灾有效荷载"作用下中空夹层钢管混凝土柱耐火极限的影响规律，结果表明，各参数对圆中空夹层钢管混凝土柱和方中空夹层钢管混凝土柱耐火极限的影响规律类似，且与实心钢管混凝土的规律类似，其中，外截面尺寸、构件长细比、空心率和保护层厚度是影响中空夹层钢管混凝土柱耐火极限的主要因素，其他各参数的影响则不明显。根据系统参数分析结果，杨有福（2005）给出了不同耐火极限情况下防火保护层厚度的实用计算方法。利用该方法，在进行新型钢管混凝土柱的抗火设计时，只需进行适当的防火保护，即可达到《高层民用建筑设计防火规范《GB 50045—95》（2001）对柱构件所要求的耐火极限。

3.4 FRP 约束钢管混凝土

FRP 约束钢管混凝土柱是在钢管混凝土柱外包 FRP 材料，从而使钢管内的核心混凝土处于 FRP 和钢管的双重约束之下。FRP 约束钢管混凝土是 FRP 约束混凝土和钢管混凝土二者的有机结合，利用 FRP 约束钢管混凝土，不仅可提高钢管混凝土的承载力，还可利用钢管混凝土具有延性较好的特点，弥补 FRP 约束混凝土这方面的不足。利用 FRP 约束可对既有钢管混凝土结构进行修复加固。图 10 所示为典型的 FRP 约束钢管混凝土截面形式。

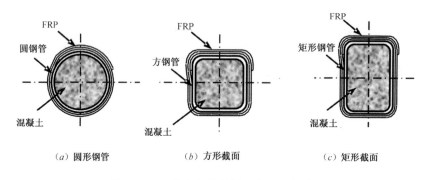

图 10 FRP 约束钢管混凝土截面示意图

随着钢管混凝土的广泛应用，其发生火灾的危险性也不断增加；此外，由于腐蚀、设计或施工考虑不周以及建筑用途发生改变等都对钢管混凝土结构提出潜在的修复加固要求。目前，国内外利用 FRP 对混凝土或钢结构进行修复加固已成为当前国内外土木工程界研究和应用的热点（Teng 等，2002）。相比较而言，采用 FRP 材料对钢管混凝土构件进行修复加固的研究还极为少见（陶忠和于清，2006）。事实上，利用 FRP 约束钢管混凝土，不仅可提高钢管混凝土的承载力，还可利用钢管混凝土具有延性较好的特点，弥补 FRP 约束混凝土这方面的不足，由此形成的 FRP 约束钢管混凝土组合柱是 FRP 约束混凝土和钢管混凝土二者的有机结合。

3.4.1 FRP 约束钢管混凝土的轴压性能

Tao 等（2006b）共进行了 9 个轴心受压钢管混凝土构件的试验研究，其中有 6 个试件为圆钢管混凝土，3 个试件为矩形钢管混凝土。试验参数为钢管截面形状（圆形和方形）、截面尺寸（100～250mm）和包裹 FRP 的层数（1 层和 2 层）等，其中共有 6 个试件包裹了 FRP。

通过试验发现，和 FRP 约束混凝土类似，截面形状对 FRP 约束效果的发挥影响较大，FRP 对圆形钢管混凝土的约束效果要明显优于对矩形钢管混凝土的约束效果。随着包裹层数的增加，构件达到峰值荷载所对应的峰值应变有所提高。对圆钢管混凝土而言，包裹 FRP 的层数越多，构件的承载力提高越大；但本次试验包裹 2 层 FRP 的矩形钢管混凝土较包裹 1 层 FRP 的矩形钢管混凝土其承载力未见有提高。在 FRP 破坏阶段，达到同样纵向应变时，包裹 FRP 试件其承载力一般均高于未包裹 FRP 试件的承载力。图 11 所示为典型圆试件在包裹 FRP 及未包裹 FRP 时的荷载—应变关系的对比情况，其中试件编号的最后一位数字为包裹 FRP 的层数。

图 12 所示为受到不同约束情况下的 150mm 直径的混凝土圆柱体的体积应变 ε_v 随纵向应变 ε_c 的变化情况。其中编号为 CC 的试件代表素混凝土圆柱体，其后的数字为包裹 FRP 的层数。体积应变的定义为 $\varepsilon_v = \varepsilon_c + 2\varepsilon_h$，其中 ε_h 为环向应变。应变以受压为正，受拉为负，因而当体积应变为正时代表体积压缩，为负时代表体积膨胀。从图 12 可见，所有试件都经历了体积压缩到体积膨胀的变化过程。值得注意的是，所有 FRP 约束钢管混凝土柱的曲线在初期和无 FRP 约束的钢管混凝土柱的曲线基本重合。但在膨胀阶段，由于 FRP 约束的存在，随着 FRP 层数的增加，体积膨胀的速率降低。FRP 约束钢管混凝土的体积膨胀速率也低于 FRP 约束素混凝土。

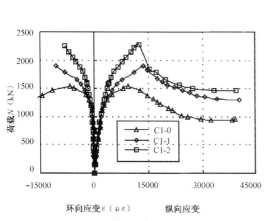 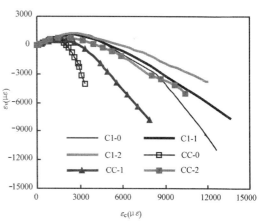

图 11　FRP 约束钢管混凝土典型试件荷载—应变关系　　图 12　体积应变随纵向应变的变化情况

3.4.2　FRP 加固火灾后钢管混凝土静力性能

为了进一步考察用 FRP 加固火灾后钢管混凝土柱的可行性，课题组先后进行了相应的 8 个轴压、8 个纯弯和 28 个偏压构件的试验，且每类构件试验中都包含了相应常温、火灾后未加固及火灾后加固构件的试验（Tao 等，2006a；陶忠和于清，2006）。对于需受火的试件，升温按 ISO—834（1975）规定的标准升温曲线进行控制。升温时试件为四面受火，升温时间设定为 180min，升温时试件不承受荷载作用。试件加固的方法为采用单向 CFRP 沿试件的环向进行包裹。

对于轴压试件，经受火灾作用后试件的承载力大大降低，同时刚度下降明显，但试件下降段要平缓得多（Tao 等，2006a）。由于火灾的影响，试件承载力损失均超过一半以上。随着加固层数的提高，构件的峰值荷载逐渐增加。同时，构件刚度也有提高的趋势，但提高幅度不大。本次试验加固构件的承载力提高幅度为 12% ~71%，表明即使在加固 2 层 FRP 的情况下也未能使构件承载力恢复到未受火以前的状态。其原因有三：一是由于采用了高性能混凝土，未受火的试件其混凝土强度从 28 天时的 48.8MPa 提高到承载力试验时的 75MPa，而受火试件其混凝土强度在受火下降后基本不发生变化；二是由于本次试件受火时间较长，试件承载力损失严重；三是加固 FRP 的层数尚不足以使承载力大幅提高。

和轴压试件类似，火灾后钢管混凝土纯弯试件的承载力有显著降低（Tao 等，2006a）。对于圆形和方形截面的钢管混凝土试件，经受火灾作用后，其极限弯矩 M_{ue}（取钢管受拉区最外边缘应变达 $10000\mu\varepsilon$ 时所对应的弯矩值）分别下降了 48.6% 和 46.4%，均接近一半左右。此外，火灾也造成试件抗弯刚度的下降。经过加固后，受火试件的承载力和刚度都略有提高，但提高作用总体有限，如包裹两层 FRP 的试件其抗弯承载力最多提高为 29%。因而对于纯弯构件建议采用双向纤维布或配合其他有效措施进行构件的修复与加固。

对于偏压试件，与未经受火灾试件相比，经受火灾的试件其强度和刚度都有明显降低，但下降段趋于平缓（陶忠和于清，2006）。由于火灾影响，试件的峰值荷载 N_{ue} 值均降低了 60% ~70% 左右。图 13 所示分别为典型的经受火灾与否对圆形截面钢管混凝土试件轴力 N—跨中挠度 u_m 关系曲线的影响，图中编号为 CUC 的试件代表常温试件，编号为 CFC-0 的试件代表火灾后未加固试件。

图 14 所示为加固与否对火灾后典型圆形截面试件 N-u_m 关系曲线的影响。图中编号 CFC 后的数字表示加固 FRP 的层数，从中可见，加固后的试件与未加固试件相比，试件的承载力和刚度都有所提高。但上述提高作用随着长细比的增大和偏心距的增大，加固效果有降低的趋势。总体而言，加固偏压试件效果比加固轴压短试件的效果要差，除个别试件外，其余加固偏压试件的承载力较未加固对比试件承载力的提高最大仅为 34.5%。

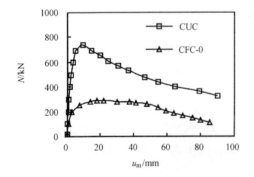 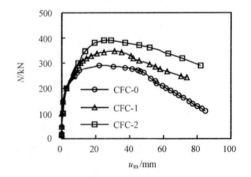

图 13　经受火灾与否对典型圆试件 　　图 14　加固与否对火灾后圆试件
　　　N-u_m 关系曲线的影响　　　　　　　　N-u_m 关系曲线的影响

3.4.3 FRP加固火灾后钢管混凝土滞回性能

前述有关FRP约束钢管混凝土静力性能的研究结果表明，经受火灾后的钢管混凝土柱通过合理设计，采用FRP进行加固可有效提高其承载力，可见当柱受火灾影响较小时采用FRP进行加固可达到预期的目的，尤其是在长细比和偏心率较小的情况下。但要在实际工程中应用FRP对钢管混凝土进行修复加固，还有必要针对修复加固后的柱开展抗震性能研究。

为此，课题组先后进行了20个钢管混凝土压弯试件在低周往复荷载作用下的滞回性能试验，试件按截面形状分为圆形和方形各10个试件，每组试件包括3个常温、3个火灾后未加固及4个火灾后加固的试件，其他试验参数还包括轴压比（0~0.78）和包裹FRP布的层数（1层和2层）（陶忠和于清，2006）。

在试验加载过程中，未加固的圆试件在夹具两侧钢管均产生局部屈曲，剖开钢管后发现其混凝土存在轻微压碎现象，而加固后的圆试件无明显鼓曲，剖开钢管后发现混凝土也无明显压碎现象。但对于方试件，加固及未加固试件的钢管均在夹具两侧产生局部屈曲，屈曲处的混凝土均存在较为明显的压碎现象。

试验结果表明，无论是火灾后加固或未加固的试件，其 $M\text{-}\phi$ 和 $P\text{-}\Delta$ 滞回关系曲线均较为饱满，无明显捏缩现象产生。相对于常温试件而言，火灾后未加固及加固试件的滞回曲线饱满性要更好。受火后试件的承载力下降较多，降幅达40%~60%。但经过加固以后，随着加固的FRP层数增加，试件的刚度、极限承载力和延性均有所提高。但和前述静力构件类似，由于本次试验构件的受火时间较长，加固后试件的承载力也未恢复到常温下试件的承载力水平。

3.5 其他新型组合结构柱

除上述薄壁钢管混凝土、钢管高性能混凝土以及中空夹层钢管混凝土外，近年来还有不少其他类型的新型组合结构柱被提出，如劲性钢管混凝土、FRP—混凝土—钢管组合柱以及钢管再生混凝土等，这些新型组合柱或者可以提高构件承载力及抗火性能，或者有利于实现废弃混凝土的资源化战略等。以下简要介绍作者及其合作者们针对这些新型组合柱所进行的研究工作。

3.5.1 劲性钢管混凝土组合柱

劲性钢管混凝土组合柱是以钢管混凝土为核心，在其周围配置钢筋并浇筑混凝土而形成的构件（见图15）。该类柱目前还被称为钢管混凝土核心柱、配有圆钢管的钢骨混凝土柱、钢管混凝土叠合柱等（洪哲和陶忠，2005）。

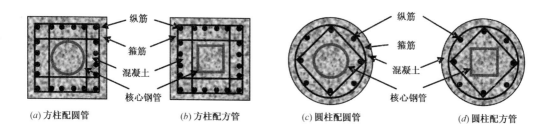

图15 劲性钢管混凝土组合柱截面形式示意图

采用劲性钢管混凝土柱可有效提高构件承载力、减小构件截面尺寸,并提高其抗震性能。此外,由于钢管埋置在混凝土内,也有利于提高构件的耐久性及抗火性能等。由于存在上述优点,劲性钢管混凝土组合柱近年来越来越受到国内有关研究者和工程技术人员的青睐(陈周熠,2002;李惠等,1999;林拥军等,2001)。

课题组共进行了9个劲性钢管混凝土柱的滞回性能试验,分别包括3个方柱配圆管、3个方柱配方管以及3个圆柱配圆管的构件,截面分别如图15(a)~(c)所示。试验结果表明,所有构件其整体破坏形态均表现为压弯破坏,而非脆性剪切破坏,在试件表面也未观察到斜裂缝出现。在轴压比较小的情况下,受拉区钢筋和核心钢管先后屈服,而后受压区混凝土才达极限抗压强度,试件延性总体较好;在轴压比较大的情况下,受压区混凝土先达到极限抗压强度,试件破坏较突然,表现出小偏压柱的破坏特征。图16所示为试验结束后的试件整体破坏形态。

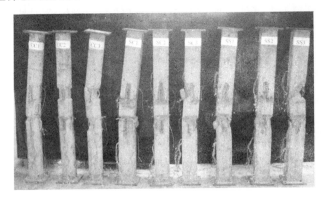

图16 试验后的劲性钢管混凝土柱

图17为截面如图15(a)所示的典型试件在不同轴压比情况下实测的侧向荷载(P)—位移(Δ)关系曲线。可见,随轴压比(n)的增大,试件的延性降低较为明显。但相对普通钢筋混凝土柱而言,劲性钢管混凝土柱仍具有较高的承载力,较好的延性及抗震能力。在较高的轴压比下,劲性钢管混凝土柱由于有核心钢管混凝土的存在,提高了劲性钢管混凝土柱受压塑性铰的转动能力,因而能有效地延缓脆性破坏的发生。

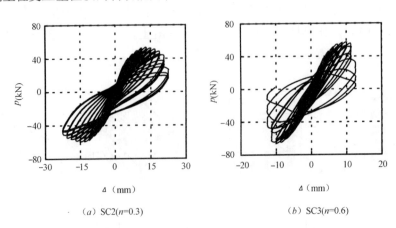

(a) SC2(n=0.3)　　　　(b) SC3(n=0.6)

图17 典型劲性钢管混凝土柱的侧向荷载(P)—位移(Δ)关系曲线

3.5.2 "FRP—混凝土—钢管"组合柱

FRP—混凝土—钢管组合柱是Teng等（2004）给出的一种新型组合柱构件。它是由内层空钢管、外层FRP管（或布）以及位于两者之间的夹层混凝土所构成。图18所示为该类柱的两种典型截面形式。该类柱的提出旨在利用FRP、钢材和混凝土三种材料的优点，形成的构件可具有如下特点（Teng等，2004）：（1）自重轻；（2）由于FRP不会受压而发生局部屈曲，所以它能给混凝土提供很好的约束，同时混凝土可以阻止内钢管的向外局部屈曲；（3）由于FRP和混凝土的存在，无须对钢管进行防腐蚀保护。

目前对该类组合柱结构滞回性能的研究尚少见有报道。为此，课题组最近进行了一批共8个该类柱的滞回性能试验，其中图18（a）和（b）所示截面构件各4个（徐毅和陶忠，2005）。试件外包的FRP均采用双向碳纤维布制作。主要试验参数为截面形式、轴压比和包裹FRP层数。

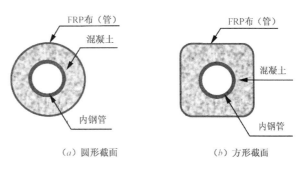

图18 典型FRP—混凝土—钢管组合柱截面形式

在试验过程中，对于所有试件，由于包裹了双向FRP布，在FRP纤维断裂以前未观察到混凝土开裂等现象。随着加载位移的增大，在刚性夹具与试件连接处的两侧，方试件受拉侧角部区域的局部纵向碳纤维开始被拉断；对于圆试件，纵向纤维的拉断则出现在拉区最边缘纤维处。试件承载力在出现纵向碳纤维拉断后开始迅速下降，此后截面边缘混凝土被压碎。

图19所示为典型的圆形和方形FRP—混凝土—钢管组合柱试件其实测的P-Δ滞回关系曲线。

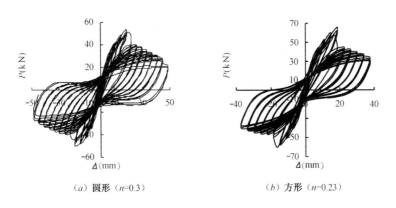

图19 典型FRP—混凝土—钢管组合柱的P-Δ滞回关系曲线

可见，由于混凝土不断开裂和裂缝闭合的影响，试件的P-Δ滞回曲线中部有轻微的捏拢现象，但总体而言，试件的滞回曲线均较为饱满。此外，还需要指出的是，试验构件在达到最大承载力后 FRP 纵向纤维断裂,此时构件承载力下降较为显著,从而影响构件的延性。但由于试件内部配有钢管,承载力在下降到一定程度后在后期趋于稳定。因而对于构件延性有较高要求时,在设计 FRP—混凝土—钢管组合柱时应合理确定 FRP、混凝土和钢管三部分的荷载分担比例,以避免由于 FRP 断裂而使构件承载力产生较大的突然降低。

3.5.3 钢管再生混凝土柱

目前,世界各国都在着手研究如何合理有效地实现废弃混凝土的再利用,其中应用途径之一就是将其应用到新建工程结构中,这对于有效地节约天然骨料和保护环境、实现废弃混凝土的资源化具有重要意义。

再生混凝土由于较普通混凝土力学性能有所降低,将其灌入钢管形成钢管再生混凝土后,由于受钢管的有效约束和保护,从而有利于提高再生混凝土的力学性能和工作性能。

已有研究成果表明,由于再生混凝土骨料表面粗糙、孔隙率大和弹性模量低,使再生混凝土的表观密度、强度和弹性模量、收缩和徐变、流动性、导热系数和脆性与普通混凝土均有所差别,从而使得钢管再生混凝土和钢管普通混凝土在力学性能上可能存在一定的差异。为全面考察钢管再生混凝土构件在一次加载下的静力性能,Yang 和 Han(2006a,2006b)共进行了 24 个轴压短柱、8 个纯弯构件和 24 个压弯构件的试验研究,同时进行了与钢管普通混凝土的对比试验研究。

研究结果表明,在进行的研究参数范围内,钢管再生混凝土构件与钢管普通混凝土构件的力学性能基本类似,但由于再生混凝土的强度和弹性模量均低于相同配合比的普通混凝土,致使钢管再生混凝土构件的承载力、轴压短柱弹性模量和纯弯构件刚度均低于相应的钢管普通混凝土构件,但差异并不十分显著。利用国内外现有主要的设计规程提供的公式进行承载力计算比较,发现福建省工程建设标准 DBJ13-51-2003(2003)较适合于钢管再生混凝土构件承载力的计算。

除上述进行的钢管混凝土力学性能研究外,Han 等(2005b)还研究了钢管约束混凝土压弯构件的静力和滞回性能。和钢管混凝土不同,钢管约束混凝土是指外加荷载仅作用在核心混凝土上,钢管不直接承受纵向荷载,只对核心混凝土起约束作用。通过研究表明,在轴压情况下,钢管约束混凝土试件的破坏形态与钢管混凝土不同,破坏时钢管表面没有出现明显的局部屈曲,且其承载力一般比钢管混凝土稍大;对于偏压构件,其破坏形态以及表现出的力学性能与钢管混凝土基本类似;在往复荷载作用下,钢管约束混凝土压弯构件也表现出良好的延性和耗能能力。

4. 钢管混凝土结构节点

在实际结构中,柱端和梁端一般均受到约束。如框架结构在外荷载作用下,框架梁和框架柱均将产生轴向变形,它将对框架柱本身和邻近杆件产生作用,从而引起结构内力的变化。可见,仅仅研究单个构件的力学性能还很不够。要深入全面地了解钢管混凝土结构在静力、动力和火灾等荷载作用下的力学性能,并基于此提供实用设计方法,深入研究钢管混凝土结构节点的工作机理非常必要(韩林海和陶忠,2005)。

以往国内外已在钢管混凝土结构节点的研究和应用方面取得不少成果，结合这些成果及一些工程实践所取得的宝贵经验，韩林海和杨有福（2004）、刘大海和杨翠如（2003），钟善桐（1999）等较系统地总结了有关研究进展，本文在此不再赘述。

目前有关钢管混凝土结构节点的研究尚存在下述一些急需解决的关键问题（韩林海和陶忠，2005）：

（1）合理的理论分析模型。建立合理的理论分析模型，是深入研究钢管混凝土结构节点机理的重要基础。其关键问题包括：①受力全过程分析；②影响参数分析；③组成钢管混凝土结构节点各材料受力状态的分析；④钢管和核心混凝土、钢管混凝土柱和钢梁之间等相互作用的认识；⑤节点刚度和延性的研究等。

（2）钢管混凝土结构节点的计算方法。实际进行钢管混凝土结构节点设计时，大致可分为节点设计原则的确定、节点形式的选取、节点计算和节点构造措施的确定等几个过程。其中，节点的计算一般包括以下内容，即：①节点连接强度的计算；②节点本身的抗弯和抗剪强度验算；③节点板件（例如加强环板等）的计算；④节点区钢管和其核心混凝土之间粘结强度的验算等。

根据钢管混凝土结构节点目前的应用和研究现状，作者及课题组成员们对钢管混凝土柱与钢梁及钢筋混凝土梁连接节点的设计理论进行了研究，下面简要介绍一些有关结果。

4.1 钢管混凝土柱—钢梁节点的力学性能

王文达等（2006b）共进行了10个钢管混凝土柱—钢梁外加强环板式中柱节点在低周往复荷载作用下的试验，其中有2个试件采用了圆钢管混凝土柱，其余采用方钢管混凝土柱。试验时以环板尺寸和轴压比为主要变化参数，分析其对极限承载力、滞回性能、延性、耗能及破坏形态等的影响。

以环板尺寸作为变化参数，主要是根据工程实践经验及一些研究结果，发现现有的加强环板式节点的环板尺寸设计方法偏于保守，容易造成环板尺寸偏大。经过计算和分析后发现，选取DBJ13—51—2003（2003）规程计算尺寸2/3的外环板节点可保证承载力满足规范要求。因此试验时分别选用按规程计算宽度、2/3以及1/3计算宽度的环板节点进行试验研究，并分别称之为Ⅰ型、Ⅱ型和Ⅲ型节点。

试验时在柱顶先施加恒定轴力，然后用MTS伺服加载系统在柱顶施加低周往复水平荷载。节点边界条件为：框架柱下端和左右框架梁均为铰接；框架柱顶端为随水平荷载的变化可在平面内自由转动的铰接，具体试验装置如图20所示。

通过试验观察发现，不同环板尺寸节点的破坏形式有所不同，以下对采用方钢管混凝土柱、柱轴压比均为0.6的三种类型典型节点的破坏形态进行分析，其具体破坏形态如图21（a）~（c）所示。试件总体的破坏形态特点是：

Ⅰ型节点SJ-13：破坏发生在环板与钢梁交接的过渡截面处，该处梁截面上、下翼缘首先出现鼓曲变形，然后腹板发生鼓曲，最后翼缘因变形过大而被撕裂。

Ⅱ型节点SJ-23：此类节点破坏时环板与钢梁过渡区域的较大范围内环板均发生显著的屈曲，此范围内的腹板也有屈曲现象发生，最后导致环板与钢梁交接截面以外的钢梁也

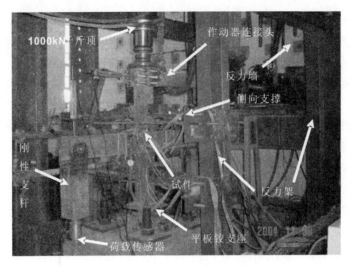

图 20 节点往复加载试验装置

（a）SJ-13(I型节点)

（b）SJ-23(II型节点)

（c）SJ-33(III型节点)

图 21 典型钢管混凝土柱—钢梁节点试件破坏形式

发生较严重的鼓曲变形，最终表现为钢梁破坏。

III型节点 SJ-33：节点破坏时，从环板和柱交接处截面到环板与钢梁交接处截面范围内的环板均发生明显的鼓曲变形，此时腹板并无显著变形，最后的破坏是典型的环板破坏，即在环板和柱交接处环板出现断裂裂缝。

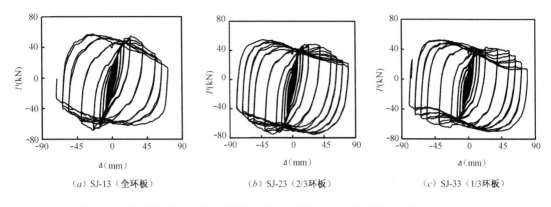

(a) SJ-13（全环板） (b) SJ-23（2/3环板） (c) SJ-33（1/3环板）

图22　不同环板尺寸时节点荷载（P）—位移（Δ）滞回关系曲线（$n=0.6$）

图22和图23所示分别为上述三个典型节点试件滞回曲线及其骨架线的对比情况。可见，不同环板宽度的钢管混凝土柱—钢梁外加强环板节点的荷载—位移滞回曲线较为饱满，没有明显的捏缩现象，刚度和强度退化不明显。随着环板宽度的减小，节点的水平承载力总体呈下降趋势，但由于Ⅱ型节点仍为钢梁破坏，试验表明其仍可满足结构抗震设计的承载力及刚度等要求，可考虑应用于工程实践当中，王文达等（2006b）建议了有关计算公式。而Ⅲ型节点则因环板发生破坏，不建议采用。

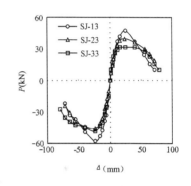

图23　不同环板尺寸时节点荷载（P）—位移（Δ）骨架曲线（$n=0.6$）

4.2 钢管混凝土柱—钢筋混凝土梁节点的力学性能

目前我国在多、高层钢管混凝土结构建筑中配合采用钢筋混凝土（RC）梁有一些应用，如泉州邮电中心和广州新中国大厦等（韩林海，2004，2006）。

CFST柱和RC梁的节点连接可有不同的形式，较典型的有：加强环式节点、钢筋混凝土环梁式节点、钢筋贯通式节点、钢筋环绕变宽度梁节点、劲性环梁节点等。这些节点形式各有其合适的应用范围（韩林海和杨有福，2004）。

Q_u等（2006）选取DBJ13—51—2003（2003）推荐并已在工程中广泛采用的钢筋环绕式CFST柱—RC梁节点进行往复荷载作用下的试验研究，考察柱截面形状及柱轴压比对节点抗震性能的影响。共制作了8个中柱节点试件，其中采用圆形和方形钢管混凝土柱的节点各4个，如图24所示。所采用的试

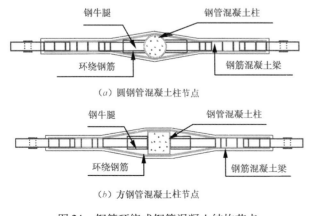

(a) 圆钢管混凝土柱节点

(b) 方钢管混凝土柱节点

图24　钢筋环绕式钢管混凝土结构节点

验装置和图 20 类似。

通过试验观测发现，无论是圆形还是方形钢管混凝土柱节点，在不同轴压比作用下，均表现为钢筋混凝土梁上下边缘裂缝沿牛腿逐步扩展形成通裂后，最后和节点核心区的主斜裂缝贯通而形成剪切破坏。通过试验观察可见，试验节点的破坏过程与钢筋混凝土结构节点总体类似，其破坏过程可分为以下四个阶段：

（1）初裂阶段：加载至屈服荷载的 0.5 倍左右时核心区混凝土开裂。开裂出现在钢牛腿外端梁底，为弯曲直裂缝。开裂荷载随着轴压比的增大而增大。在初裂阶段，节点总体尚处于弹性阶段。

（2）通裂阶段：随着反复荷载的逐级增大，节点核心区不断产生双向交错的新裂缝，裂缝长度沿对角线方向不断扩展，最终形成宽度大约为 0.28~0.5 mm 左右贯通节点核心区的主斜裂缝。在此过程中，钢管混凝土柱与钢筋混凝土梁沿柱在出平面外方向产生剥离，两侧梁顶混凝土沿翼缘板两侧开裂，形成"Π"形裂缝。通裂荷载对于圆钢管混凝土柱节点约为极限荷载的 70% 左右，对于方钢管混凝土柱节点约为极限荷载的 80% 左右。

（3）极限阶段：核心区混凝土出现"通裂"后，节点所承受的荷载仍能继续增加，直至达到峰值荷载。在这一阶段，节点核心区混凝土裂缝呈明显的交叉贯通状，裂缝宽度加大，最大裂缝宽度最终均超过 1mm，并伴随有轻微的混凝土劈裂声，核心区剪切变形明显增大。

（4）破坏阶段：节点承载力开始下降，破坏加剧，两侧梁顶和梁底节点处翼缘板混凝土完全开裂；圆钢管混凝土柱节点核心区沿主斜裂缝形成"灯笼"形区域破坏［图 25（a）］；方钢管混凝土柱节点由两侧梁的"Π"形裂缝与核心区主斜裂缝贯通形成主裂缝破坏［图 25（b）］。此后，核心区附近混凝土开始大块剥落，节点荷载逐渐降至极限荷载的 85% 以下，试验结束。

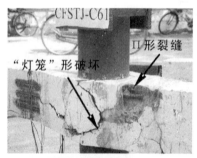
（a）圆钢管混凝土柱节点

（b）方钢管混凝土柱节点

图 25　钢筋环绕式钢管混凝土结构节点的破坏形态

在整个试验过程中，未见钢管混凝土柱发生明显变形。将核心区钢管剖开，发现其内部混凝土没有发生明显破坏。从整个试验过程看，节点的破坏模态受柱轴压比的影响不大。

图 26 所示为实测典型钢筋环绕式钢管混凝土结构节点的侧向荷载（P）—位移（Δ）滞回关系曲线。从中可见，无论是方形还是圆形钢管混凝土柱节点的滞回曲线均表现出为

钢筋和混凝土之间存在粘结滑移的滞回特征，但滞回曲线总体较为饱满，兼有梭形与滑移型的性质，抗震性能良好，且随着轴压比的增大，这种趋势趋于明显。同时还发现圆钢管混凝土柱节点由于外围环绕的纵向钢筋对混凝土的开裂和剥落具有较强约束作用，因而受力性能良好，节点的整体性优于方钢管混凝土节点，而方形钢管混凝土柱节点滞回曲线在加载后期出现较明显的捏缩和滑移现象。

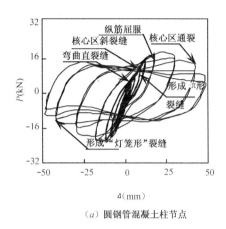 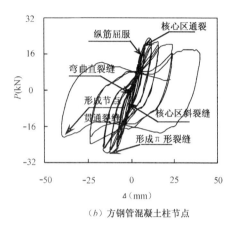

(a) 圆钢管混凝土柱节点　　　　　　　　(b) 方钢管混凝土柱节点

图26　典型钢筋环绕式钢管混凝土结构节点的 P-Δ 滞回关系曲线

通过计算发现，本次试验的8个节点的位移延性系数 μ 的变化范围为3.706~6.369，明显高于钢筋混凝土结构节点的平均值2（中国建筑科学研究院，1994），接近或高于型钢混凝土结构节点的平均值4（周起敬等，1991）。我国《建筑抗震设计规范》GB 50011—2001，2002规定：对多、高层钢筋混凝土结构弹性层间位移角 $[\theta_e]$ =1/550=0.0018，弹塑性层间位移角 $[\theta_p]$ =1/50=0.02；而本次试验的8个节点构件的弹性极限层间位移角 θ_y =（3~4.22）$[\theta_e]$，弹塑性极限层间位移角 θ_u =（1.06~2.14）$[\theta_p]$，因此进行的8个节点的位移延性和转角延性指标均能满足抗震规范要求。目前，对钢管混凝土结构节点的非线性分析工作正在进行中。

4.3　火灾作用对钢管混凝土结构节点的影响

随着钢管混凝土建筑高度的不断增加，火灾的危险性以及由火灾带来的生命财产损失和其他直接或间接经济损失，以及对环境的破坏等问题日趋严重，因而有必要研究火灾下钢管混凝土结构的抗火性能。对于钢管混凝土结构构件耐火性能的研究，目前已取得较大进展，但有待于向结构体系的性能研究上发展，因此首先需要解决的问题就是其结构节点在火灾下和火灾后的性能，这也是研究钢管混凝土框架乃至整体结构体系在火灾下力学性能的重要基础，是解决钢管混凝土结构抗火的性能化设计问题的重要前提（陶忠等，2005）。

如前所述，目前国内实际工程中应用钢管混凝土时多采用刚接节点。由于火灾下钢材的强度和刚度会随着火灾温度的升高而不断降低，使得常温下表现为刚接的节点其承载能力以及抗变形能力在火灾下会大大降低，因而火灾下钢管混凝土节点总体表现出一定的半刚性，从而使得结构的变形性能以及内力重分布的机制发生改变。由此可见，要研究火灾对钢管混凝土结构体系性能的影响，就有必要进行钢管混凝土梁柱节点火灾下力学性能的研究，目前课题组正在从事这方面相关的理论分析和试验研究工作。

研究钢管混凝土结构节点耐火性能和抗火设计的一些关键问题主要有：（1）试验装置和试验方法。（2）温度场的测试和模拟：在进行节点弯矩—转角关系试验时，要做好温度场的测试工作，并采用数值方法对其进行模拟，这是准确模拟节点弯矩—转角关系的前提条件。（3）弯矩—转角—温度关系研究：弯矩—转角关系是深入研究体系性能的基础。由于温度场的瞬态性，所以弯矩—转角关系要在不同的恒定温度下进行测试。（4）节点对子框架、框架及整体结构火灾下性能影响：由于火灾下结构的整体性对其抗火性能有着至关重要的作用，单个构件的试验并不能准确反映其在整体结构中的力学性能和反应。有必要开展子框架、框架和整体结构的火灾下性能研究，考虑结构整体性影响下的节点力学性能。

对于节点火灾作用后的性能，霍静思（2005）进行了一系列火灾作用后钢管混凝土柱—钢梁外加强环板式节点在恒定轴压力和低周反复荷载作用下的滞回性能试验研究，研究了常温下和火灾后、不同梁柱线刚度比和轴压比情况下，以及常温下与经过修复后节点力学性能的变化规律。修复方法为保留受火后的柱，更换新的钢梁和节点板。

试验结果表明几乎所有常温和火灾后试件都按照设计的强柱弱梁节点形式发生了类似图21（a）所示梁端屈服形成塑性铰的破坏形态。而对于火灾后修复节点，由于只更换了钢梁和环板，而柱仍处于受损状态，因而形成了强梁弱柱节点，结果该类试件有的发生了柱端屈服形成塑性铰的破坏形态［图27（a）］，有的则发生了节点核心区钢管屈曲、焊缝开裂的破坏形态［图27（b）］。由于节点的理想破坏形式为梁端屈服形成塑性铰，而柱端和节点核心区不发生破坏，因此对于节点的修复加固，应注意采取适当的措施，使得整体结构体系仍符合"强柱弱梁"和"强节点弱构件"的原则。

(a) 节点核心区破坏　　　　　　　　　　(b) 柱端屈服破坏

图27　火灾后修复节点的典型破坏形态

图28所示为火灾对钢管混凝土结构节点 $P\text{-}\Delta$ 滞回曲线的影响。通过对比分析发现，火灾作用对实测的 $P\text{-}\Delta$ 滞回曲线影响很大，即节点火灾后的承载力降低，弹性刚度也有所降低，骨架线趋于扁平，但滞回曲线更为饱满。

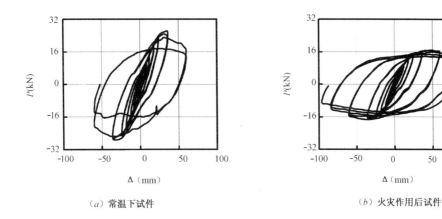

(a) 常温下试件　　　　　　　　(b) 火灾作用后试件

图 28　火灾作用对钢管混凝土结构节点 P-Δ 滞回曲线的影响

在确定了高温后钢材和混凝土材料本构关系模型，以及构件截面温度场的基础上，霍静思（2005）建立了节点非线性有限元模型，并编制了相关全过程分析的非线性有限元程序。利用该程序对常温下和火灾后钢管混凝土节点试验结果进行了计算比较，结果表明该模型具有较强的实用性和较好的计算精度。利用该计算模型，霍静思（2005）对钢管混凝土节点在恒定轴压力和水平往复荷载作用下的荷载—位移非线性特性进行了分析，并对有关影响参数进行了计算分析，在此基础上探讨了火灾作用后节点的承载力和 P-Δ 恢复力模型的简化计算方法。

5. 钢管混凝土框架结构的力学性能

要深入研究钢管混凝土结构体系的力学性能，就必须首先研究由钢管混凝土构件及节点组成的钢管混凝土框架结构的力学性能。在钢管混凝土框架结构中，钢管混凝土一般是作为框架结构中以承受竖向荷载为主的柱构件，因而必将受到其他周围构件及其连接性能的影响，其力学性能和工作机理与单个构件有一定的区别。

王文达等（2006a）共进行了 12 榀圆形及方形钢管混凝土柱—钢梁外加强环板式平面框架在恒定轴力和水平往复荷载下的试验研究，在此基础上，王文达（2006）利用 ABAQUS 有限元分析软件建立了分析模型，对钢管混凝土框架进行了二阶弹塑性分析。以下对有关研究结果作一简单介绍。

5.1　试验研究

在以往研究工作的基础上，为了系统地研究更大参数范围内钢管混凝土框架结构的力学性能，王文达等（2006a）进行了 4 组共计 12 个单层单跨圆形和方形截面钢管混凝土柱—工字钢梁平面框架的低周往复荷载试验研究，试件具体信息参见王文达等（2006a）。试验时考察的主要参数有：柱截面形状、柱截面含钢率、柱轴压比、梁柱线刚度比等。图29 所示为试验框架的加载装置和支撑布置情况。

通过试验观察发现，所有钢管混凝土框架试件最终均发生了强柱弱梁型破坏模式，即

图 29　框架试件试验加载情景及侧向支撑布置

首先分别在梁端出现塑性铰，最后分别在柱脚形成塑性铰而破坏。基于框架试件的设计和构造特点，所有试件的梁端塑性铰位置在加强环板之外约 50mm 范围内的钢梁截面上，而柱上形成的塑性铰位置在柱脚加劲肋板高度之上约 30mm 左右的位置。图 30 所示为方钢管混凝土和圆钢管混凝土框架试件的梁端及柱端典型破坏形态。

（a）圆钢管混凝土框架柱脚典型破坏形式

（b）圆钢管混凝土框架梁端典型破坏形式

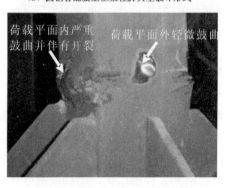

（c）方钢管混凝土框架柱脚典型破坏形式

（d）方钢管混凝土框架梁端典型破坏形式

图 30　钢管混凝土框架试件的梁端及柱端典型破坏形式

图 31 所示为典型钢管混凝土框架试件的荷载（P）—位移（Δ）滞回关系曲线。通过分析试验框架的滞回曲线、延性和耗能性能以及不同参数的影响规律，主要有以下一些发现：(1) 钢管混凝土柱—钢梁框架的滞回曲线总体较为饱满，具有明显的钢框架特征，没有明显捏缩现象，刚度和强度退化不明显，说明钢管混凝土框架具有良好的抗震耗能能力。(2) 总体上圆钢管混凝土框架的抗震耗能性能优于方钢管混凝土框架。(3) 轴压比和含钢率是影响钢管混凝土框架抗震性能的主要因素。随着轴压比的增大，框架的水平极限承载力下降，屈服状态提前，耗能能力和位移延性均有所降低。而含钢率的影响规律正好与之相反。(4) 框架梁柱线刚度比对其承载力和抗震性能也有一定的影响，随着梁柱线刚度比的减小，框架水平极限承载力下降，延性和耗能能力降低。

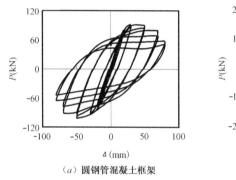 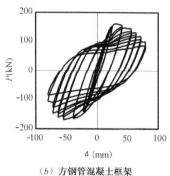

(a) 圆钢管混凝土框架　　　　　　(b) 方钢管混凝土框架

图 31　典型钢管混凝土框架试件的 P-Δ 滞回曲线

5.2　理论分析

作为"精确"解法，钢管混凝土框架的二阶弹塑性分析要考虑分布塑性。王文达（2006）中应用 ABAQUS 软件，基于混凝土塑性损伤模型，建立了钢管混凝土柱—钢梁框架的有限元分析模型，理论计算结果和试验结果吻合良好。下面以柱轴压比为 0.3 时的圆柱框架试件 CF-22 为例，简要介绍其分析结果。

由于框架柱首先承受了轴向荷载，此时的钢管和混凝土都处于全截面受压状态，当框架梁端作用有水平荷载时，框架柱为压弯构件，随着水平荷载的增加，框架柱核心混凝土截面由全截面受压开始出现了不均匀受压，进而出现了受拉区。当框架达到水平极限承载力时，框架柱脚塑性铰位置处截面核心混凝土及外钢管的纵向应力分布如图 32 (a) ~ (c) 所示，截面分为受拉区和受压区，其中图 32 (b) 为截面上混凝土纵向应力等值线，f'_c 为核心混凝土圆柱体抗压强度，可见，由于外钢管对核心混凝土的约束作用，混凝土的塑性性能和强度都得到了提高，在框架达到水平极限承载力时核心混凝土纵向应力最大值已达到 $1.56 f'_c$。同样，外钢管也出现受拉区和受压区，并且由于钢管中已有轴力引起的压应力叠加，使得钢管的最大压应力数值要高于最大拉应力。

图 32 (d) 和 (e) 分别为 CF-22 试件框架梁端塑性铰截面处在达到水平极限承载力时的纵向应力分布情况，由于框架梁存在反弯点，左侧截面为上翼缘受拉、下翼缘受压，而右侧截面为上翼缘受拉、下翼缘受压。塑性铰位置在节点加强环板之外的梁端范围内，试验测试和 ABAQUS 理论分析均反映了这个规律。当框架达到极限承载力时，梁端截面翼

缘纵向应力已超过其屈服强度，随着水平荷载（或位移）的继续增大，发生屈服的钢材逐渐向腹板内部扩展。相比而言，左端截面受压区钢梁的变形及应力小于右端截面，其主要原因是钢梁在传递水平荷载作用时由于钢梁具有有限刚度，在受力过程中要发生变形，只能传递部分水平荷载到右柱，从而使得右柱的水平位移总小于左柱。

图 32 (f) 给出了 CF-22 试件左右框架柱核心混凝土沿高度的纵向应力分布，选取了荷载作用平面位置，可见左右框架柱沿高度的纵向应力分布是不同的，右框架柱的反弯点位置要高于左框架柱。其主要原因是水平荷载作用在左框架柱侧的梁端，右框架柱所受到的水平荷载通过框架梁传递，而框架梁具有有限的刚度，且在梁端开始出现塑性铰后钢梁会发生局部屈曲，从而传递到右框架柱的荷载总是小于左柱所受到的荷载，左右框架柱的水平位移不同步，且受到框架梁的约束，因此右柱的反弯点位置要上移。

需要指出的是，实际工程中的钢管混凝土框架结构一般为空间受力，除了框架梁、柱、节点等构件外，还有楼板或者其他抗侧力构件与之共同工作，因而形成的钢管混凝土空间框架结构受力更趋于复杂。因此，在框架研究中如何合理考虑空间作用的影响，建立相应分析模型等问题都是今后需要进一步深入进行研究的课题。

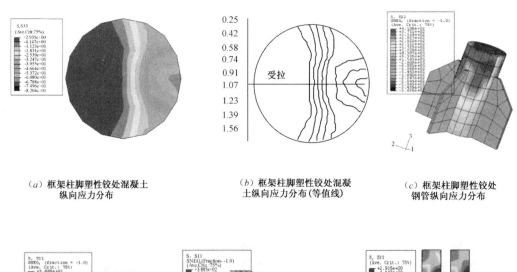

（a）框架柱脚塑性铰处混凝土纵向应力分布　　（b）框架柱脚塑性铰处混凝土纵向应力分布（等值线）　　（c）框架柱脚塑性铰处钢管纵向应力分布

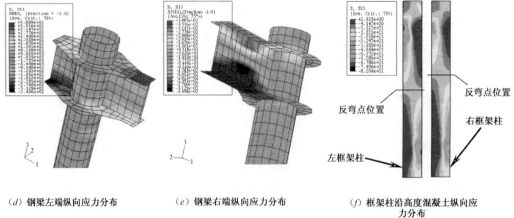

（d）钢梁左端纵向应力分布　　（e）钢梁右端纵向应力分布　　（f）框架柱沿高度混凝土纵向应力分布

图 32　钢管混凝土框架 CF-22 在达到水平极限承载力时不同截面位置应力分布

6. 钢管混凝土混合结构剪力墙

在不少钢管混凝土高层建筑结构中采用了钢管混凝土外框架和钢筋混凝土剪力墙（芯筒）共同组成一种钢—混凝土混合结构体系。该体系是由钢筋混凝土筒体或剪力墙以及钢管混凝土框架柱组成抗侧力体系，以刚度很大的钢筋混凝土墙承担风力和地震作用；框架柱主要承担竖向荷载，以充分发挥钢和混凝土两种结构材料各自的优势，达到良好的技术经济效果。实践证明，钢—混凝土混合结构兼有钢结构施工速度快和混凝土结构刚度大、成本低等优点，被认为是一种符合我国国情的较好的建筑结构形式，值得推广应用（陈富生等，2000）。

钢—混凝土混合结构具有发展前景广阔，近年来已陆续被一些国家用于高层和超高层建筑中。但长期以来，国外在地震区却很少采用钢—混凝土混合结构。其原因就在于人们对钢—混凝土混合结构有一个重要的疑虑就是其抗震性能。因为在抗震概念设计时人们通常要求建筑尤其是高层建筑要有多道抗震防线，而目前人们对钢—混凝土混合结构的一般认识是其抗震性能的好坏主要取决于其内部钢筋混凝土剪力墙或核心筒，外部钢框架（包括钢管混凝土框架和型钢混凝土框架）主要承担重力荷载，仅承担较小的水平剪力，因而在往复地震动的持续作用下，结构进入弹塑性阶段时，墙体一旦产生裂缝，混凝土墙或筒的抗侧刚度大幅度降低，承载能力下降很快，容易导致结构产生整体或局部的破坏（陈富生等，2000）。

对于钢管混凝土混合结构建筑，为了有效提高钢筋混凝土剪力墙或筒体的延性，在不少工程实践中都将普通的钢筋混凝土剪力墙设计成带钢管混凝土边框柱的组合剪力墙，以使钢管混凝土混合结构中的框架柱和剪力墙在变形过程中能更好地达到变形协调。以往国内外的研究者针对型钢混凝土剪力墙的力学性能已开展了不少研究，但针对上述采用钢管混凝土边框柱的组合剪力墙的研究尚少见（廖飞宇等，2006），本文称其为钢管混凝土剪力墙。

为此，课题组共制作了10个钢管混凝土剪力墙试件，同时还分别进行3个带边框柱钢筋混凝土剪力墙和3个型钢混凝土剪力墙的对比试验。所设计的钢管混凝土剪力墙及其对比试件其墙板的设计完全相同；不同边框柱的设计原则是在轴压比相同的情况下，其抗弯承载力基本接近，由此确定边框柱的尺寸和配钢以及钢管和型钢的尺寸。剪力墙的试验方法和上一节进行钢管混凝土框架试验的方法类似，先在边框柱上施加恒定的轴压力，然后对试件施加往复水平荷载。以下简单介绍柱轴压比均为0.3的钢管混凝土和钢筋混凝土剪力墙的试验结果，以了解钢管混凝土剪力墙的工作性能。

试验结果表明，钢管混凝土和钢筋混凝土剪力墙在达到极限承载力前其工作特性均基本相似。即在初始加载阶段，墙板和边框均能较好的共同工作，随着位移的继续加大，在墙板一个对角线方向开始出现45°的初始剪切斜裂缝，反向加载后出现另一方向的交叉斜裂缝。此后，随着位移的继续增大，不断有新的斜裂缝产生，逐渐布满墙面，且原有斜裂缝不断扩展，裂缝宽度增大，最初产生的两条对角斜裂缝逐渐发展为主斜裂缝。但钢筋混凝土剪力墙在加载到$3\Delta_y$（屈服位移）时，就在柱底发现水平弯曲裂缝，而钢管混凝土剪力墙在达到极限承载力前钢管混凝土柱未见有明显变形产生。随着位移的继续增大，墙板水平钢筋和竖向钢筋陆续屈服，裂缝间墙板开始出现压碎现象，试件达到最大承载力。此后，随位移继续增大试件墙板承载力逐渐降低，不同边框及其对墙板的约束作用效果开始显现出来。由于钢管混凝土边框柱其抗震性能以及整体性好，因而相应的剪力墙整体延性较好。钢管混凝土边框柱在加载到 $10\Delta_y$ 时柱脚处才稍有屈曲，但直至试验结束（承载力下降到极限承载力的 85% 左右），屈曲现象也并不严重，边框柱仍具有较好的承载能力。而钢筋混凝土柱在试件达到极限荷载后，裂缝宽度迅速增大，混凝土出现严重剥落。图 33 所示为试验结束后钢管混凝土和钢筋混凝土剪力墙试件破坏形态的对比。

（a）带钢管混凝土边柱　　　　　　　　（b）带钢筋混凝土边柱

图 33　剪力墙试件的破坏形态对比

图 34 所示为上述墙体实测的水平荷载（P）—位移（Δ）滞回关系曲线，图 35 所示为其骨架线的对比情况。从中可见，由于两种墙体均产生剪切破坏，因而滞回曲线也都表现出一定的捏缩现象。在试验初期，两种试件的侧向刚度基本一致，但随着水平位移的增大，试件刚度逐渐降低，但钢管混凝土剪力墙的刚度降低更慢。在进入下降段后，由于钢筋混凝土边框柱的破坏最为严重，因而钢筋混凝土剪力墙的下降段明显，钢管混凝土剪力墙的延性明显好于钢筋混凝土剪力墙。目前，对上述混合结构剪力墙的非线性分析工作正在进行中。

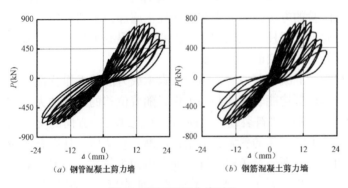 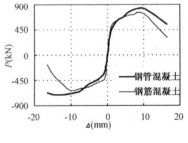

图 34　滞回关系曲线对比　　　　　　　图 35　滞回关系曲线骨架线对比

7. 钢管混凝土混合结构体系的抗震性能

钢—混凝土混合结构体系在应用过程中，其抗震性能一直为大家所关注（陈富生等，2000；徐培福等，2005）。但以往有关钢管混凝土混合结构的抗震性能的报道尚少见（李威，2006）。随着钢管混凝土混合结构建筑的推广应用，为确保其结构在地震作用下的安全可靠，减轻地震危害，有必要对其抗震性能进行深入系统的研究。

为此，课题组委托广州大学抗震研究中心进行了两个钢管混凝土混合结构的地震模拟振动台模型试验，分别采用圆钢管混凝土和方钢管混凝土（广州大学工程抗震研究中心研究报告，2005）。试验模型由外围钢管混凝土框架和位于模型中央的钢筋混凝土剪力墙组成，共30层，结构高度6.3m，标准层平面尺寸为2.2m×2.2m；方形混凝土芯筒位于模型中央，平面尺寸为1.21m×1.21m，如图36（a）所示。楼面主梁与混凝土剪力墙刚性连接。模型框架柱截面分为圆截面（直径30mm）和方形截面（30mm×30mm），钢管壁厚均为1mm，外框架及楼面钢梁均为Ⅰ-40mm×15mm×1mm×1.5 mm。

模型设计时，用设置配重的方法来模拟密度相似关系，同时考虑结构一定的隔墙荷载及恒荷载和活荷载的组合，将这两部分荷载用楼面附加质量来模拟。

试验模型的总质量为16.5t（包括楼板），楼板重1.458t。根据振动台的承载能力和模型的实际重量，可计算出所需配重：每层楼面附加质量（配重块）为320kg，模型总附加质量为0.32×30=9.6t。制作完成的圆钢管混凝土模型如图36（b）所示。

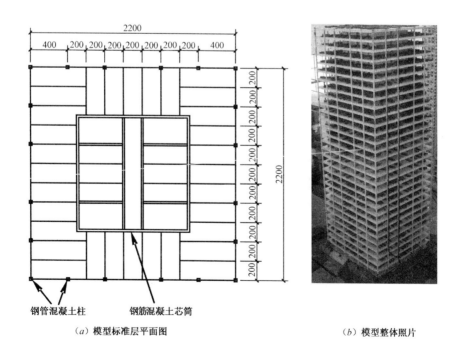

(a) 模型标准层平面图　　　　　(b) 模型整体照片

图36　钢管混凝土混合结构的地震模拟振动台试验模型

试验采用三条真实强震记录，分别为：Taft 波、El Centro 波和天津波，水平加速度峰值采用 0.2g（小震），0.4g（中震），0.6g（大震）和 0.8g（超大震），时间压缩比为 4.47。

通过对两个高层钢管混凝土混合结构模型地震模拟振动台试验，在大量试验数据分析和试验现象观测的基础上，发现两个模型结构一阶和二阶自振频率在震前、小震后和中震后逐渐降低，但降幅不大，表明结构的刚度有所下降，但仍基本处于弹性工作状态；大震和超大震后自振频率下降幅度稍大，方钢管混凝土模型的自振频率最大下降 5.23%，圆钢管混凝土模型的自振频率最大下降 15.75%。

随着地震强度的增加，模型结构的损伤加剧、阻尼逐渐增大，且圆钢管混凝土模型的阻尼比增大程度要高于方钢管混凝土模型，即随着地震强度的增大，圆钢管混凝土模型的损伤程度要高于方钢管混凝土模型。

在同样地震作用下，圆钢管混凝土模型结构的层间位移角大于方钢管混凝土模型结构的层间位移角；天津波作用下模型结构的层间位移角最大，El Centro 波次之，Taft 波引起的层间位移角最小。总体上，在所选地震波及其强度作用下，模型结构的层间位移较小于罕遇地震作用下的层间位移角限值；少数情况下，模型结构的层间位移较小于多遇地震作用下的层间位移角限值。

对模型各部位的应变观测显示，钢管混凝土角柱的应变较大，混凝土芯筒的应变次之，而钢梁应变较小。芯筒各部位中角部的应变反应较大。在大震阶段，芯筒某些部位混凝土已开始出现裂缝，观测得到的裂缝基本上出现在底层芯筒的角部，在个别钢管混凝土柱与基础底板相交处出现裂缝，楼板与梁和柱相交处出现裂缝。但总体来看，即使在超大震作用下，由钢管混凝土柱与钢梁组成的外框架尚处于弹性工作阶段。

从模型结构反应及破坏形态来看，结构无明显薄弱部位，各部分连接较好。只要合理设计，高层钢管混凝土混合结构体系能够满足"小震不坏，中震可修，大震不倒"的抗震要求。

根据上述试验结果，李威（2006）采用 ANSYS 软件建立了两种钢管混凝土框架—混凝土核心筒有限元模型，并对其进行了弹性阶段的动力特性分析、谱分析和时程分析。根据数值计算结果，发现两个试验结构的频率都在 8Hz 左右，说明该模型结构的整体刚度较大。根据时程分析的计算包络图，可见两个模型结构的层间剪力、层间侧移曲线光滑，沿高度变化基本均匀，说明在理想弹性阶段，结构各层刚度基本没有突变。在 0.2g 级别地震作用下，结构最大侧移值为 4 mm 左右，层间位移角在 1/1200 左右，小于《建筑抗震设计规范》GB 50011—2001 所规定的弹性阶段下混凝土框架剪力墙结构的最大层间位移角 1/1000，说明该类结构刚度较大，抗震性能较好。阻尼取值对结果影响较大，需要经过试算得到相对准确的结果。两个模型的计算时程分析结果相比，各对应图形曲线形状相似，但方钢管混凝土模型的各项数值要小于圆钢管混凝土模型，这是由于方钢管混凝土模型的刚度和阻尼要大于圆钢管混凝土模型。实际试验还给出了两个模型在大震和超大震情况下的反应，从破坏形式上看，二者的破坏形态类似，都是核心筒先出现裂缝，然后底层柱子剪坏拔出。这说明模型在大震和超大震情况下的抗震性能良好，能够充分发挥整体结构各个构件的性能。

8. 钢管混凝土曲杆结构

钢管混凝土多被用作直线形构件，但在实际工程中还有不少应用曲线形钢管混凝土构件（在曲线形空钢管内浇筑混凝土而形成的构件）的情况，如大跨度空间结构中的曲线形桁架压杆或者拱桥桁架式拱肋的弦杆等。图 37 所示为典型曲线形钢管混凝土桁架结构体系的示意图。

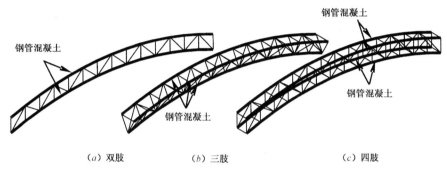

(a) 双肢　　　　(b) 三肢　　　　(c) 四肢

图 37　钢管混凝土曲桁架示意图

以往多研究直线形钢管混凝土，而曲线形钢管混凝土的相关研究工作开展较少（Zheng 等，2006）。和传统的直线形钢管混凝土类似，钢管混凝土曲杆同样可利用钢管和混凝土在受力过程中的相互作用，从而具有较高强度以及抗弯和抗变形能力。为了深入研究钢管混凝土曲杆的力学性能，课题组开展了相关的试验研究和理论分析工作（Zheng 等，2006）。

在钢管混凝土曲杆试验研究方面，进行了 12 个单肢曲杆受竖向荷载（如图 38 所示）和 6 个钢管混凝土直杆在偏心荷载作用下的试验。所进行的曲杆及其对比直杆二者截面尺寸和构件两端直线距离相同，同时在跨中截面施加竖向力到其截面中心的绝对偏心距也相同。试验结果表明，在其他条件相同的情况下，两种构件的破坏形态和力学性能总体基本接近。

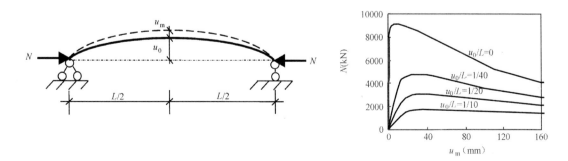

图 38　钢管混凝土受压单肢曲杆示意图　　　　图 39　不同初弯曲的曲杆 N-u_m 曲线

为进一步深入认识钢管混凝土曲杆的受压力学性能，Zheng 等（2006）采用有限元分析软件 ABAQUS 建立了分析模型，对钢管混凝土曲杆进行了计算分析，并和有关试验结果进行了对比，结果表明计算结果和试验结果吻合良好。

图 39 给出了不同初弯曲 u_0 情况下受压钢管混凝土的荷载—变形关系曲线。可见，随着初弯曲的增大，钢管混凝土曲杆的抗弯刚度和极限承载力下降，峰值荷载对应的跨中挠度增大，下降段也趋于平缓。

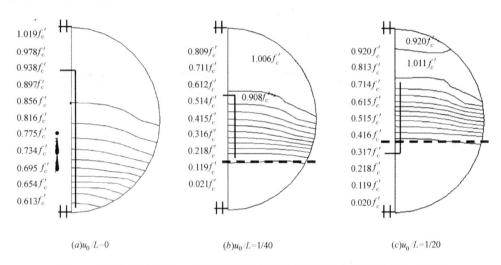

图 40　不同初弯曲的曲杆峰值荷载对应的跨中截面核心混凝土纵向应力分布

图 40 给出了不同初弯曲情况下圆钢管混凝土曲杆达极限承载力时，钢管混凝土曲杆跨中截面核心混凝土的纵向应力（取混凝土圆柱体抗压强度 f'_c 的倍数，$f'_c = 51\text{MPa}$）等值线分布。可以看出，钢管混凝土曲杆的核心混凝土在钢管的约束下强度均有所提高，且随着初弯曲的增大，核心混凝土截面纵向最大应力数值也呈增加趋势。

从图 40 还可以看出，当初弯曲较小时，构件在达到极限承载力时，核心混凝土全截面受压；随着初弯曲的增大，二阶弯矩也随之增大，构件达极限承载力时，核心混凝土截面开始出现受拉区，且混凝土受拉面积随初弯曲的增大而增大，曲杆的极限承载力降低。

图 41 比较了不同初弯曲（u_0/L，L 为曲杆弦长，见图 38）对钢管与受压侧混凝土最外边缘处的相互作用力的影响。从图 41 可见，随着初弯曲的增大，钢管对压区混凝土的约束力逐渐减小。这是因为随着初弯曲的增大，混凝土压区最外纤维到中和轴的距离逐渐减小，其横向变形也随之减小，钢管对其约束作用也相应减小。

有关格构式钢管混凝土曲杆力学性能和设计理论的实验研究及理论分析工作正在进行中。

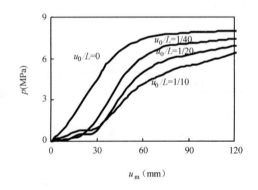

图 41　不同初弯曲下钢管混凝土曲杆 $p\text{-}u_m$ 关系曲线对比（跨中截面混凝土压区最外边缘处）

9 结束语

本文首先简要论述了钢管混凝土结构构件力学性能的系列关键问题，如一次加载和长期荷载作用下的性能、滞回性能、抗火设计和火灾后的力学性能、钢管初应力影响的验算以及混凝土浇筑质量的影响等方面的基本原理和方法；然后介绍了一些新型组合结构，如薄壁钢管混凝土、钢管高性能混凝土、中空夹层钢管混凝土和FRP约束钢管混凝土等的基本原理；本文最后介绍了钢管混凝土结构节点、采用钢管混凝土的框架结构和混合结构的力学性能及一些研究新进展等。

由于篇幅所限，本文主要介绍了作者及其课题组取得的一些研究结果，以期为同类工作起到“抛砖引玉”之作用。

致谢

作者的研究工作先后得到国家自然科学基金（No. 59508007、No. 50425823 和 No. 50608019）、霍英东教育基金、国家教育部优秀青年教师资助计划项目、国家教育部科学技术研究重点项目（No. 204071 和 No. 205083）、国家地震科学联合基金、辽宁省自然科学基金、福建省自然科学基金重点项目、福建省科技计划重点和重大项目、福建省自然科学基金、澳大利亚ARC基金重点项目、福建省引进高层次人才科研启动费、清华大学“百名人才引进计划”基金、“福建建工”青年教师研究基金及其他各类科研项目的资助，另外还得到过不少来自工业界的支持和帮助，特此致谢！作者借此机会感谢公安部天津消防科学研究所和广州大学抗震研究中心等单位在实验方面给予的协作和帮助！

本文第一作者感谢他的学生和合作者们在钢管混凝土结构研究与应用方面的创造性工作和贡献！本文作者借此机会向所有曾给过我们关心和帮助的人们致以诚挚的谢意！

参考文献

[1] 蔡绍怀，2003. 现代钢管混凝土结构［M］. 北京：人民交通出版社

[2] 陈富生，邱国华，范重. 高层建钢结构设计［M］. 北京：中国建筑工业出版社，2000

[3] 陈宗弼，陈星，叶群英，罗赤宇. 广州新中国大厦结构设计［J］. 建筑结构学报，2000，21(3):2-9

[4] 陈周熠. 钢管高强混凝土核心柱设计计算方法研究［D］. 大连：大连理工大学博士学位论文，2002

[5] 福建省工程建设标准 DBJ13—51—2003. 钢管混凝土结构技术规程［S］，福州，2003

[6] 福建省工程建设标准 DBJ13—61—2004. 钢—混凝土混合结构技术规程［S］，福州，2004

[7] 广州大学工程抗震研究中心研究报告. 钢管混凝土框架—混凝土剪力墙混合结构模型模拟地震振动台试验［R］. 广州：广州大学，2005

[8] 韩林海. 钢管混凝土结构［M］. 北京：科学出版社，2000

[9] 韩林海. 钢管混凝土结构—理论与实践（第一版）［M］. 北京：科学出版社，2004

[10] 韩林海. 钢管混凝土结构—理论与实践（第二版）［M］. 北京：科学出版社（即将出版），2006

[11] 韩林海，陶忠. 钢管混凝土结构节点力学性能研究若干问题探讨［J］. 哈尔滨工业大学学报增刊（中国钢结构协会钢—混凝土组合结构分会第十次年会论文集），2005:303-306

[12] 韩林海, 徐蕾, 冯九斌, 杨有福. 钢管混凝土柱耐火极限和防火设计实用方法研究 [J]. 土木工程学报, 2002, 35 (6): 6-13
[13] 韩林海, 杨有福. 现代钢管混凝土结构技术 [M]. 北京: 中国建筑工业出版社, 2004
[14] 洪哲, 陶忠. 劲性钢管混凝土组合柱在我国的研究进展 [J]. 福州大学学报增刊（自然科学版）, 2005, 33: 316-320
[15] 黄宏. 中空夹层钢管混凝土压弯构件的力学性能研究 [D]. 福州: 福州大学博士学位论文, 2006
[16] 黄宏, 韩林海, 陶忠. 方中空夹层钢管混凝土柱滞回性能研究 [J]. 建筑结构学报. 2006, 27 (2): 64-74
[17] 霍静思. 火灾作用后钢管混凝土柱—钢梁节点力学性能研究 [D]. 福州大学博士学位论文, 福州, 2005
[18] 李惠, 王震宇, 吴波. 钢管高强混凝土叠合柱抗震性能与受力机理的试验研究 [J]. 地震工程与工程振动, 1999, 19 (3): 27-33
[19] 李威. 高层建筑钢管混凝土混合结构抗震性能分析 [D]. 北京: 清华大学本科生综合训练论文, 2006
[20] 廖飞宇, 陶忠, 韩林海. 钢—混凝土组合剪力墙抗震性能研究现状简述 [J]. 地震工程与工程振动, 2006, 26 (5). （即将发表）
[21] 林拥军, 程文瀼, 徐明, 左江. 配有圆钢管的钢骨混凝土柱轴压比限值的试验研究 [J]. 土木工程学报, 2001, 34 (6): 23-28
[22] 刘大海, 杨翠如. 型钢、钢管混凝土高楼计算和构造 [M]. 北京: 中国建筑工业出版社, 2003
[23] 刘威, 韩林海. 钢管混凝土受轴向局压荷载时的工作机理研究 [J]. 土木工程学报, 2006, 39 (6): 19-27
[24] 陶忠, 韩林海, 王永昌. 火灾下钢管混凝土梁柱节点性能研究的若干问题探讨 [J]. 钢结构, 2005, 20 (4): 91-94
[25] 陶忠, 于清. 新型组合结构柱—试验、理论与方法 [M]. 北京: 科学出版社, 2006
[26] 天津市工程建设标准 DB29—57—2003. 天津市钢结构住宅设计规程 [S], 天津, 2003
[27] 王文达. 钢管混凝土柱—钢梁平面框架的力学性能研究 [D]. 福州: 福州大学博士学位论文, 2006
[28] 王文达, 韩林海, 陶忠. 钢管混凝土柱—钢梁平面框架抗震性能的试验研究 [J]. 建筑结构学报, 2006a, 27 (3): 48-58
[29] 王文达, 韩林海, 游经团. 方钢管混凝土柱—钢梁外加强环节点滞回性能的实验研究 [J]. 土木工程学报, 2006b, 39 (9) (即将发表)
[30] 徐培福, 傅学怡, 王翠坤和肖丛真. 复杂高层建筑结构设计 [M]. 北京: 中国建筑工业出版社, 2005
[31] 徐毅, 陶忠. 新型 FRP—混凝土—钢管组合柱抗震性能研究 [J]. 福州大学学报增刊（自然科学版）, 2005, 33: 309-315
[32] 杨有福. 新型钢管混凝土柱耐火性能研究 [R]. 福州: 福州大学博士后研究工作报告, 2005
[33] 尧国皇. 钢管混凝土构件在复杂受力状态下的工作机理研究 [D]. 福州: 福州大学博士学位论文, 2006
[34] 浙江省工程建设标准. 建筑钢结构防火技术规范（送审稿）[S]. 杭州, 2003
[35] 中国工程建设标准化协会标准, CECS 159: 2004. 矩形钢管混凝土结构技术规程 [S]. 北京: 中国计划出版社, 2004
[36] 中国工程建设标准化协会标准 CECS 28: 90. 钢管混凝土结构设计与施工规程 [S]. 北京: 中国计

划出版社，1992

[37] 中国工程建设标准化协会标准 CECS 200:2006. 建筑钢结构防火技术规范 [S]. 北京：中国计划出版社，2006

[38] 中国建筑科学研究院. 混凝土结构研究报告选集（3）[M]. 北京：中国建筑工业出版社，1994

[39] 中华人民共和国电力行业标准 DL/T 5085—1999. 钢—混凝土组合结构设计规程 [S]. 北京：中国电力出版社，1999

[40] 中华人民共和国国家标准 GB 50011—2001. 建筑抗震设计规范 [S]. 北京：中国计划出版社，2002

[41] 中华人民共和国国家标准 GB 50045—95. 高层民用建筑设计防火规范 [S]. 北京：中国计划出版社，2001

[42] 中华人民共和国国家建筑材料工业局标准 JCJ 01—89. 钢管混凝土结构设计与施工规程 [S]. 上海：同济大学出版社，1989

[43] 中华人民共和国国家军用标准 GJB 4142—2000. 战时军港抢修早强型组合结构技术规程 [S]. 北京：中国人民解放军总后勤部，2001

[44] 钟善桐. 钢管混凝土结构 [M]. 哈尔滨：黑龙江科学技术出版社，1994

[45] 钟善桐. 高层钢管混凝土结构 [M]. 哈尔滨：黑龙江科学技术出版社，1999

[46] 钟善桐. 钢管混凝土结构 [M]. 北京：清华大学出版社，2003

[47] 周起敬，姜维山，潘泰华. 钢与混凝土组合结构设计施工手册 [M]. 北京：中国建筑工业出版社，1991

[48] ACI 318-02. Building code requirements for reinforced concrete and commentary [S]. Farmington Hills (MI), American Concrete Institute, Detroit, USA, 2002

[49] ACI Committee 209. Prediction of creep, shrinkage and temperature effects in concrete structures [S]. American Concrete Institute, Farmington Hills, Mich., USA, 1992

[50] AIJ. Recommendations for design and construction of concrete filled steel tubular structures [S]. Architectural Institute of Japan (AIJ), Tokyo, Japan, 1997

[51] AISC. Specification for structural steel buildings. ANSI/AISC Standard 360-05, AISC, Chicago, USA, 2005

[52] ASCCS. Concrete filled steel tubes-A comparison of international codes and practices [C]. ASCCS Seminar, Innsbruck, Austria, 1997

[53] British Standards Institutions. BS5400, Part 5, Concrete and composite bridges [S]. London, U.K., 1979

[54] Eurocode 4 (EC4). Design of composite steel and concrete structures-Part1-1: General rules and rules for buildings [S]. EN 1994-1-1:2004, Brussels, CEN, 2004

[55] Ge, H. B. and Usami, T. Strength of concrete-filled thin-walled steel box column: experiment [J]. Journal of Structural Engineering, ASCE, 1992, 118 (11): 3036-3054

[56] Gourley, B. C., Tort, C., Hajjar, J. F. and Schiller, A. A synopsis of studies of the monotonic and cyclic behaviour of concrete-filled steel tube beam-columns [R]. Report No. ST1-01-4 (Version 3.0), Department of Civil Engineering, University of Minnesota, 2001

[57] Han, L. H. Tests on concrete filled steel tubular columns with high slenderness ratio [J]. Advances in Structural Engineering-An International Journal, 2000a, 3 (4): 337-344

[58] Han, L. H. The influence of concrete compaction on the strength of concrete filled steel tubes [J]. Advances in Structural Engineering-An International Journal, 2000b, 3 (2): 131-1137

[59] Han, L. H. Fire performance of concrete filled steel tubular beam-columns [J]. Journal of Constructional Steel Research, 2001, 57 (6):695-709

[60] Han, L. H. Tests on stub columns of concrete-filled RHS sections [J]. Journal of Constructional Steel Research, 2002, 58 (3):353-372

[61] Han, L. H. Flexural behaviour of concrete filled steel tubes [J]. Journal of Constructional Steel Research, 2004, 60 (2):313-337

[62] Han, L. H. and Huo, J. S. Concrete-filled hollow structural steel columns after exposure to ISO-834 fire standard [J]. Journal of Structural Engineering, ASCE, 2003, 129 (1):68-78

[63] Han, L. H., Huo, J. S. and Wang, Y. C. Compressive and flexural behaviour of concrete filled steel tubes after exposure to standard fire [J]. Journal of Constructional Steel Research, 2005a, 61 (7):882-901

[64] Han, L. H. and Lin, X. K. Tests on cyclic behavior of concrete-filled HSS columns after exposure to ISO-834 standard fire [J]. Journal of Structural Engineering, ASCE, 2004, 130 (11):1807-1819

[65] Han, L. H., Lin, X. K. and Wang, Y. C. Cyclic performance of repaired concrete-filled steel tubular columns after exposure to fire [J]. Thin-Walled Structures (Accepted for publication), 2006a

[66] Han, L. H., Lu, H., Yao, G. H. and Liao, F. Y. Further study on the flexural behavior of concrete-filled steel tubes [J]. Journal of Constructional Steel Research, 2006b, 62 (6):554-565

[67] Han, L. H. and Tao Z. Design codes and methods on concrete-filled steel tubular structures in China [C]. Proceedings of the International Symposium on Worldwide Trend and Development in Codified Design of Steel Structures, 2-3 October 2006, Singapore:46-74

[68] Han, L. H., Tao, Z., Huang, H. and Zhao, X. L. Concrete-filled double skin (SHS outer and CHS inner) steel tubular beam-columns [J]. Thin-Walled Structure, 2004a, 42 (9):1329-1355

[69] Han, L. H., Tao, Z. and Liu, W. Effects of sustained load on concrete-filled HSS (hollow structural steel) columns [J]. Journal of Structural Engineering, ASCE, 2004b, 130 (9):1392-1404

[70] Han, L. H., Xu, L. and Zhao, X. L. Temperature field analysis of concrete-filled steel tubes [J]. Advances in Structural Engineering-An International Journal, 2003a, 6 (2):121-133

[71] Han, L. H., Yang, H. and Cheng, S. L. Residual strength of concrete filled RHS stub columns after exposure to high temperatures [J]. Advances in Structural Engineering-An International Journal, 2002a, 5 (2):123-134

[72] Han, L. H. and Yang, Y. F. Influence of concrete compaction on the behavior of concrete filled steel tubes with rectangular sections [J]. Advances in Structural Engineering-An International Journal, 2001, 2 (2):93-100

[73] Han, L. H. and Yang, Y. F. Analysis of thin-walled RHS columns filled with concrete under long-term sustained loads [J]. Thin-Walled Structures, 2003, 41 (9):849-870

[74] Han, L. H. and Yang, Y. F. Cyclic performance of concrete-filled steel CHS columns under flexural loading [J]. Journal of Constructional Steel Research, 2005, 61 (4):423-452

[75] Han, L. H., Yang, Y. F. and Tao, Z. Concrete-filled thin walled steel RHS beam-columns subjected to cyclic loading [J]. Thin-Walled Structures, 2003b, 41 (9):801-833

[76] Han, L. H., Yang, Y. F. and Xu, L. An experimental study and calculation on the fire resistance of concrete-filled SHS and RHS columns [J]. Journal of Constructional Steel Research, 2003c, 59 (4):427-452

[77] Han, L. H., Yang, Y. F., Yang, H. and Huo, J. S. Residual strength of concrete-filled RHS columns after exposure to the ISO-834 standard fire [J]. Thin-Walled Structures, 2002b, 40 (12): 991-1012

[78] Han, L. H. and Yao, G. H. Influence of concrete compaction on the strength of concrete-filled steel RHS columns [J]. Journal of Constructional Steel Research, 2003a, 59 (6): 751-767

[79] Han, L. H. and Yao, G. H. Behaviour of concrete-filled hollow structural steel (HSS) columns with preload on the steel tubes [J]. Journal of Constructional Steel Research, 2003b, 59 (12): 1455-1475

[80] Han, L. H. and Yao, G. H. Experimental behaviour of thin-walled hollow structural steel (HSS) columns filled with self-consolidating concrete (SCC) [J]. Thin-Walled Structures, 2004, 42 (9): 1357-1377

[81] Han, L. H, Yao, G. H, Chen, Z. B. and Yu, Q. Experimental behavior of steel tube confined concrete (STCC) columns [J]. Steel and Composite Structures-An International Journal, 2005b, 5 (6): 459-484

[82] Han, L. H., Yao, G. H. and Zhao, X. L. Behavior and calculation on concrete-filled steel CHS (circular hollow section) beam-columns [J]. Steel and Composite Structures, 2004c, 4 (3): 169-188

[83] Han, L. H., Yao, G. H. and Zhao, X. L. Tests and calculations of hollow structural steel (HSS) stub columns filled with self-consolidating concrete (SCC) [J]. Journal of Constructional Steel Research, 2005c, 61 (9): 1241-1269

[84] Han, L. H., You, J. T. and Lin, X. K. Experiments on the cyclic behavior of self-consolidating concrete-filled HSS columns [J]. Advances in Structual Engineering - An International Journal, 2005d, 8 (5): 497-512

[85] Han, L. H., Zhao, X. L. and Tao, Z. Tests and mechanics model for concrete-filled SHS stub columns, columns and beam-columns [J]. Steel and Composite Structures, 2001, 1 (1): 51-74

[86] Han, L. H., Zhao, X. L., Yang, Y. F. and Feng, J. B. Experimental study and calculation of fire resistance of concrete-filled hollow steel columns [J]. Journal of Structural Engineering, ASCE, 2003d, 129 (3): 346-356

[87] Hibbitt, Karlsson & Sorensen Inc. ABAQUS Version 6.4: Theory manual, users' manual and verification manual [M]. Hibbitt, Karlson and Sorenson Inc., 2003

[88] Huang, C. S., Yeh, Y. K., Liu, G. Y., Hu, H. T., Tsai, K. C., Weng, Y. T., Wang, S. H., Wu, M. H. Axial load behavior of stiffened concrete-filled steel columns [J]. Journal of Structural Engineering, ASCE, 2002, 128 (9): 1222-1230

[89] ISO-834. Fire resistance tests-elements of building construction [S]. Iternational Standard ISO-834, Geneva, 1975

[90] Qu H., Han, L. H. and Tao Z. Seismic performance of concrete-filled steel tubular (CFST) column to reinforced concrete (RC) beam joints [C]. Proceedings of the Ninth International Symposium on Structural Engineering for Young Exports (Editors: Han LH, Ru JP and Tao Z), Science Press, 2006, Vol. 2, 1311-1317

[91] Schneider, S. P. Axially loaded concrete-filled steel tubes [J]. Journal of Structural Engineering, ASCE, 1998, 124 (10): 1125-1138

[92] Shams, M. and Saadeghvaziri, M. A. State of the art of concrete-filled steel tubular columns [J]. ACI Structural Journal, 1997, 94 (5): 558-571

[93] Shanmugam, N. E. and Lakshmi, B. State of the art report on steel-concrete composite columns [J]. Journal of Constructional Steel Research, 2001, 57 (10): 1041-1080

[94] Tao, Z. and Han, L. H. Tests and mechanics model on concrete-filled double skin (RHS inner and RHS outer) steel tubular beam-columns [J], Journal of Constructional Steel Research, 2006, 62 (7) :631-646

[95] Tao, Z, Han, L. H. and Wang, L. L. Compressive and flexural behaviour of CFRP repaired concrete-filled steel tubes after exposure to fire [J] . Journal of Constructional Steel Research (Accepted for publication), 2006a

[96] Tao, Z. , Han, L. H. and Wang, Z. B. Experimental behaviour of stiffened concrete-filled thin-walled hollow steel structural (HSS) stub columns [J] . Journal of Constructional Steel Research, 2005, 61 (7) :962-983

[97] Tao, Z. , Han, L. H. and Zhao, X. L. Behaviour of concrete-filled double skin (CHS inner and CHS outer) steel tubular stub columns and beam-columns [J] . Journal of Constructional Steel Research, 2004, 60 (8) :1129-1158

[98] Tao, Z, Han, L. H. and Zhuang, J. P. Experimental behavior of CFRP strengthened concrete-filled steel tubular stub columns [J] . Advances in Structural Engineering-An International Journal (Accepted for publication), 2006b

[99] Teng J G, Chen J F, Smith S T and Lam L. FRP-strengthened RC structures [M] . John Wiley & Sons, Ltd. , 2002

[100] Teng, J. G. , Yu, T. and Wong, Y. L. Hybrid FRP-concrete-steel double-skin tubular columns: stub column tests [C] . Procedings of the Second International Conference on Steel & Composite Structures, Seoul, Korea, 2004:1390-1400

[101] Yang, Y. F. and Han, L. H. Behaviours and calculations on recycled concrete filled steel tubular (RCFST) columns and beam-columns [J] . Journal of Constructional Steel Research (Article in press), 2006a

[102] Yang, Y. F. and Han, L. H. Compressive and flexural behaviour of recycled aggregate concrete filled steel tubes (RACFST) under Short-Term Loadings [J] . Steel and Composite Structures-An International Journal, 2006b, 6 (3) :257-284

[103] Zhao, X. L. and Han, L. H. Double skin composite construction [J] . Progress in Structural Engineering and Materials, 2006, 8 (3) :93-102

[104] Zheng, L. Q. , Han, L. H. and Tao, Z. Behaviour of curved concrete-filled steel struts [C] . Proceedings of the Ninth International Symposium on Structural Engineering for Young Exports, (Editors: Han LH, Ru JP and Tao Z), Science Press, 2006, Vol. 2, 1539-1545

第八章 | Chapter 8

HIGH PERFORMANCE STEELS AND THEIR USE IN STEEL AND STEEL-CONCRETE STRUCTURES

B. Uy

School of Civil, Mining and Environmental Engineering,
University of Wollongong, Australia. Email: brianuy@uow.edu.au

Abstract: Special steels namely high performance steel (HPS) and stainless steel (SS) generally exhibit improved strength, hardness, and corrosion resistance. When formed into structural elements and juxtaposed with concrete, significant benefits arise in terms of stiffness, strength and potential energy absorption. This chapter examines these benefits and will primarily focus on those which can be exploited with these configurations in columns of major landmark structures. The chapter will firstly focus on the applications of special steels, including past, present and future projects. An overview of the behaviour of these elements and a summary of research which has been conducted over the last decade will then be given. The chapter will then surmise the subtle differences that need to be applied to design approaches in order to ensure both efficient and safe design of these structural elements. In particular the paper will highlight the nuances which are needed to be addressed when designing with HPS and SS in steel and steel-concrete structures using existing international design codes. Finally, the paper will provide a view of future research which is required in order to fully exploit the special properties of these steels.

Keywords: composite columns, high performance steel, high strength steel, stainless steel, tall buildings

INTRODUCTION

High performance steels in this chapter will include high strength quenched and tempered steels and stainless steels. Figure 1 provides an idealised stress-strain curve which illustrates the representative yield and maximum stresses for mild structural steel, high strength steel and stainless steel. Typically mild structural steel (MS) has a nominal yield stress of about 300 MPa, with an elastic modulus of 200

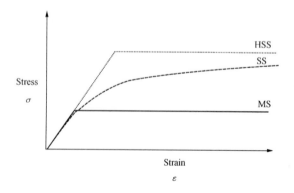

Figure 1 Stress-strain curves for mild structural steel (MS), stainless steel (SS) and high strength steel (HSS)

GPa. High strength steel (HSS) which is produced by a quenching and tempering process (see Figure 2) typically has a nominal yield stress of about 700 MPa with an elastic modulus of 200 GPa. Stainless steel (SS) which is produced with varying amounts of chromium and nickel (see Figure 3) displays much more defined non-linear characteristics, with a 0.2% proof stress in the range of 450 MPa and maximum stress typically about 600 MPa. In the linear range a typical elastic modulus of 200 GPa is also characteristic of stainless steel. Figure 1 also illustrates that SS and HSS exhibit much higher areas under their stress-strain curves which may be able to be harnessed in high temperature situations (such as fire) or in high strain situations (such as blast).

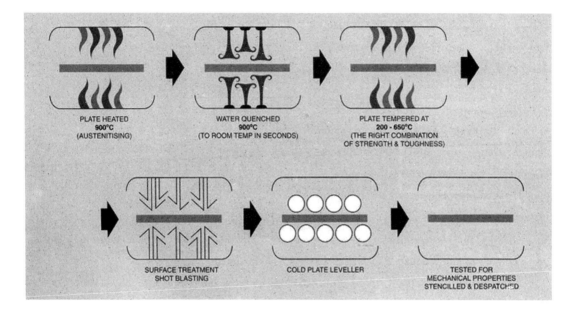

Figure 2 Quenching and tempering process for high strength steel,
(Courtesy of Bisalloy Steels, Unanderra)

PREVIOUS RESEARCH

High strength steel

The majority of previous research using high strength structural steel for columns has been mainly carried out in the regions where it has been applied in practice and this includes research in both Australia and Japan. Firstly Rosier and Croll (1987) considered the benefits of high strength quenched and tempered steel being applied in structures such as bridges, buildings and silos. This study included consideration of the economics of the material over conventional mild structural steel and showed the significant advantages that could be derived from its use.

Rasmussen and Hancock (1992 and 1995) conducted tests on both high strength steel fabricated I-sections and box sections with nominal yield stress of 690 MPa. These tests established local buckling slenderness limits for these high strength steel sections. Furthermore, slender columns were tested and the behaviour of these was compared with the slender column curves of the existing Australian Standard AS 4100 - 1998 (Standards Australia 1998). It was found that providing the local buckling slenderness limits were adhered to, then the slender column behaviour could be described using this standard developed specifically for mild structural steel.

Hagiwara et al. (1995) and Mochizuki et al. (1995) considered the behaviour of high strength structural steel for the application in super high-rise buildings in Japan. These studies considered the reliability inspection and the welding process for heavy gauge steel plate. These studies are pertinent to the application of the use of high strength steel in projects such as the Shimizu Super High Rise in Tokyo, Japan.

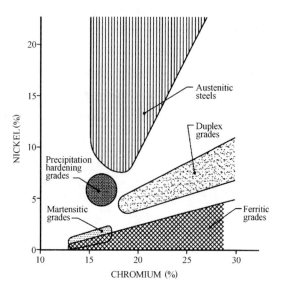

Figure 3 Stainless steel classification according to the percentage of nickel and chromium, AS/NZS 4673:2001, (Standards Australia, 2001)

Sivakumaran and Yuan (1998) considered slenderness limits and ductility of steel sections fabricated with high strength steel with nominal yield stresses between 300 and 700 MPa respectively. The test programme involved testing 12 W shaped stub column sections with the objective being to determine the compression flange strength and strain ductility of sections of different steel grades.

Uy (1996) considered the behaviour of concrete filled steel box columns with a nominal yield stress of 690 MPa. These studies illustrated the advantages derived from filling the sections with concrete to increase the local buckling stresses. Furthermore, the members were considered under combined bending and compression to assess the strength of short columns. The results of these columns were compared with columns designed with normal strength structural steel, to show the reduced cross-sectional dimensions able to be achieved. Furthermore, comparisons of the cross-sectional ductility were made and showed that composite members composed of high strength structural steel still had a large degree of reserve strength after peak loading conditions were attained.

Uy (1999) and Uy (2001a) presented the results of steel and composite sections using high strength structural steel of nominal yield stress 690 MPa. These sections constructed as stubby columns were subjected to concentric axial compression. A theoretical model to predict the axial strength of these columns was provided and shown to be in good agreement with the models suggested by Eurocode 4.

Bergmann and Puthli (2000) conducted an extensive experimental programme on short and slender high strength steel encased sections of 460 MPa grade steel subjected to combined compression and bending. These tests were then compared with the Eurocode 4 approach, which was found to be suitable for predicting the ultimate load for short columns. However, the results of the slender column tests proved to be inconclusive.

Uy (2001b) conducted an extensive experimental programme on short concrete filled steel box columns, which incorporated high strength structural steel of Grade 690 MPa. The experiments were then used to calibrate a refined cross-sectional analysis method, which considered both the non-linear material properties of the steel and concrete coupled with the measured residual stress distributions in the steel. The model and experiments were then compared with the existing approach of Eurocode 4 and it was found that certain modifications were necessary. The Eurocode 4 approach, which employs the rigid plastic analysis method, was found to over predict the strength of the cross-sections. A modified technique known as a mixed analysis was therefore developed and found to be in good agreement with both the test results and the refined analysis procedure. This model considers the concrete to be plastic and the steel to be elastic-plastic and provides a much more realistic design approach for sections utilizing high strength structural steel, particularly when large flexural loads are present.

Mursi and Uy (2004) conducted further research on high strength steel box columns filled with concrete. This study consisted of three short columns and three slender columns to consider both the strength and stability aspects of steel-concrete composite high strength columns. The results of this study, showed that further refinement or adjustments need to be made to the Eurocode 4 approach, to allow for the effects of high strength steel particularly when large flexural loads are present. More recently, Mursi, Haedir and Uy (2004) completed an experimental program on short steel sections and concrete filled high strength steel columns subjected to biaxial bending. The experiments were compared with the existing Australian, European and American Standards, (Standards Australia, 1998; British Standards Institution 2004 and 2005; and American Institute of Steel Construction, 1993) and found to be in good agreement.

Mursi and Uy (2005a and 2005b) have conducted further research on composite columns utilising high strength steel which were subjected to biaxial bending. An experimental program was conducted on short and slender columns with and without concrete infill. A non-linear analysis model was also developed to simulate the experimental results and this was compared with the American, Australian and European standards. The models for each of these standards were found to be extremely conservative when estimating the capacity of hollow and composite columns.

More recent research has also been conducted by Chen, Young and Uy (2006) in relation to the elevated temperature behaviour of high strength steel. These results have illustrated that the degradation in both yield stress and elastic modulus versus temperature is very similar for high strength steel and mild structural steel.

Stainless steel

In Australia, significant research has been carried out by Rasmussen and his research group at The University of Sydney. The work of this group has mainly dealt with thin walled closed cross-sections and has also been extended to consider the design of columns using the column curve approach, (Rasmussen 2000a and Rasmussen 2000b). Much of the research work conducted on stainless steel structures is encompassed in the Australian/New Zealand Standard AS/NZS 4673: 2001, Cold-formed stainless steel structures, (Australian Standards 2001).

In Europe significant research has also been carried out by various groupings into the behaviour of unfilled tubular sections. Kouhi et al. (2000) summarised the significant research being conducted in Finland on stainless steel in construction and suggested that its use would be increased as life cycle costing was considered in building projects. Burgan et al. (2000) summarised UK findings for circular hollow sections made of stainless steel and suggested that the diameter to thickness slenderness limits used for normal carbon steel were far too conservative for stainless steel sections. Nethercot and Gardner (2002) have looked at methods for exploiting the special features of stainless steel in structural design. The premise of the approach is to employ an ultimate stress for design based on the slenderness of the plate elements in tubular sections. For example in sections where very small plate slenderness is present, stresses much larger than the 0.2% proof stress may be employed in the design of these members. Di Sarno et al. (2003) and Di Sarno and Elnashai (2003) have also recently shown that the use of stainless steel can provide up to 3 times the ultimate strain than carbon steels and they can exhibit improved post-local buckling giving them excellent application for seismic regions both for new structures and as braces for rehabilitating structures.

It is now widely accepted throughout the world that stainless steel as a structural material is now able to be designed by various international standards, including ANSI/ASCE-8-90 (American Society of Civil Engineers 1991), Eurocode 3 (British Standards Institution 2005) and AS/NZS 4673:2001 (Standards Australia 2001). The major innovation of this proposal is to conduct experimental and theoretical research and develop design guidelines for the composite behaviour of stainless steel sections filled with concrete. It is believed that there will be significant applications in buildings, bridges, offshore and specialty structures utilising stainless steel in this form as evidenced by the applications already outlined in this proposal. The major benefit of the use of concrete infill is that due to restrained local buckling a concept developed by Uy and Bradford (1996), the thickness of the steel sections can be reduced considerably. Since stainless steel is approximately five times the cost of mild structural steel, this benefit is quite important in ensuring that wider application of the material is made possible. Initial development of a numerical model has been carried out by Roufegarinejad et al. (2004) and the applicant has also successfully supervised two theses on the behaviour of concrete filled stainless steel sections under axial compression at The University of New South Wales in 2004 (Dajanovic 2004 and Gjerding-Smith 2004) and at the University of

Wollongong (Asgar 2005).

APPLICATIONS

High strength steel

Grosvenor Place, Sydney

The Grosvenor Place building is a landmark building on George Street, Sydney which is on the border of the financial and rocks districts. The building has a total height of 180 metres with 44 storeys above ground. The building was designed by the structural engineers, Ove Arup and Partners and construction was completed in 1988 by Concrete Constructions, (Gillett and Watson, 1987). The innovative use of structural steel in this building included quite a few firsts in Australia. The building involved the use of high strength cold formed profiled steel sheets for the decking. The beams which span approximately 16 metres from reinforced concrete core to perimeter frame were designed for serviceability as semi-continuous, with a semi-rigid joint assumed between the beam and core. The columns in the lower levels of the buildings are quite unique and involve three perimeter columns being grouped at the ground level in a single column, with the key objective being the savings in space made for car parking in the basement. This involved the use of high strength quenched and tempered structural steel being used for encased sections in this zone.

Star City, Sydney

The Star City Casino building in Sydney was completed in 1995, (see Figure 4). Upon completion it was the largest building project in Sydney since the construction of the Sydney Opera House. The building was designed by structural engineers Ove Arup and Partners and built by Leighton Contractors. The project entailed a massive podium structure of casino, theatres, shopping and other facilities with a total floor area in a single plate exceeding seven football fields, (King, 1999). The podium was then capped with two twenty storey towers. One tower being for a hotel and the other being for a residential apartment facility. The building comprised a number of innovative composite construction and high strength steel applications. Firstly, in the main gaming areas of the casino, large span composite beams of approximately 16 metres were designed and constructed. In the basement levels of the building, high strength steel fabricated sections were used to miminise the cross-sections of the columns, (Davie, 1997). Another very inno-

Figure 4   Star City, Sydney (1995)

vative structural steel application concerned the roofs of the Lyric and Showroom theatres. Due to constraints with site access for craneage, the trusses for the roofs had to be of minimal weight. This required the design of 36 metre spanning trusses made composite with a topping slab and using high strength structural steel for the sections. Post-tensioning of the trusses helped to alleviate long-term serviceability concerns.

Latitude Tower, Sydney

The Latitude building in Sydney has been completed in 2005 and exists on George Street on the World Square Site and is directly adjacent to Sydney's Chinatown at Haymarket, (see Figure 5). The building is a landmark building which was designed by Hyder Consulting and constructed by Multiplex. The building has a total height of 222 m over 45 floors and has some very innovative features in it's design. The beams in the floor system span a total of 14 metres from core to perimeter frame and in order

Figure 5 Latitude Tower, Sydney (2005)

to achieve this the beam's were pre-cambered by 40 mm to overcome 60 mm long term deflections. The building also uses twin composite columns on the perimeter frame, using 508 mm diameter steel tubes filled with 80 MPa concrete. The building has required the design of 7 metre deep transfer trusses using large diameter steel tubes filled with concrete and large high strength steel boxes filled with concrete, (Chaseling, 2004).

Stainless steel

Gateway Arch, St Louis

The gateway arch constructed in St Louis, Missouri was completed in 1966 and this was constructed of stainless steel. This is a 300 metre tall arch and is composed of a triangular annular cross-section which involved a double skin construction technique. The inner and outer stainless steel skins were filled with reinforced concrete, thus forming a composite steel-concrete composite structure which would provide benefits for durability, as well as stiffness, strength and stability.

Parliament House, Canberra

The new federal parliament house was completed in time for the bicentennial celebrations in Canberra Australia. The 81 metre tall flag masts of the parliament house, shown in Figure 6 were constructed using stainless steel closed sections.

Stonecutters Bridge, Hong Kong

The Stonecutters Bridge in Hong Kong will be the longest cable stayed bridge in the world upon its completion, (1018 metres). The bridge consists of two 290 metre tall masts with their upper third (approximately 100 metres) comprising of a stainless steel section that will be filled with concrete. The major reason for the concrete infill is to ensure that minimal maintenance has to be conducted on this section as it would be difficult to access during the design life. The bridge is due for completion in 2007.

Figure 6   Parliament House, Canberra (1988)

Hearst Tower, New York

Hearst Tower at 959 Eight Avenue, New York City is a 46 storey building completed in 2006. The major lateral load resisting system for this building includes concrete filled steel columns in megacolumns of the diagrid exoskeleton, (Fortner, 2006). These columns incorporate a stainless steel skin as illustrated in Figure 7.

BEHAVIOUR

High strength steel

One of the primary considerations in using high strength steel in a landmark building or structure, is its axial compression capacity. Uy (2001a)

Figure 7   Hearst Tower, New York (2006)

showed that for pure compression concrete filled steel columns utilising high strength steel can be designed using the principles of superposition, (see Figure 8). However, Uy (2001b) illustrated that when large amounts of flexure are present, there is a certain degree of non-conservatism associated with using existing rigid plastic analysis approaches. These will be discussed in the following section.

Some typical load-deflection results are illustrated in Figure 5 for both hollow and concrete filled steel box sections. The columns considered in this test series were constructed with a box section of 110mm × 110mm with a 5mm nominal plate thickness. This gave a plate to thickness ratio of 20 and this is just on the verge of being compact for both the hollow sections and definitely compact for the concrete filled sections, when a nominal yield stress of 690 MPa was chosen. The results in Figure 9 show that both the hollow sections HSSH1 and HSSH2 exhibit quite ductile behaviour, by reaching a peak load and maintaining this load from a deflection of about 3 mm to over 7 mm. The concrete filled sections HSCB1 and HSCB2, reached a peak load and gradually experienced a load reduction, which was as a result of the internal concrete crushing. For high strength steel it seems less likely that confinement will be able to take place as the strains at which yield are achieved are often in the vicinity of the crushing strains of most normal strength concrete.

Figure 8 Axial compression tests of HSS concrete filled steel columns

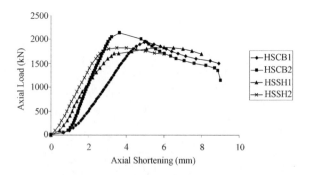

Figure 9 High strength steel axial load experiments

Stainless steel

Tests and analyses have illustrated that when using stainless steel in a concrete filled steel column, (see Figure 10), the prediction of the axial compressive strength using the 0.2% proof stress will result in a conservative estimate of strength. Thus, there is a certain degree of reserve of strength which may be able to be utilised when using stainless steel in columns.

Stainless steel square hollow sections with and without concrete infill have been considered under pure axial compression. Figure 11 illustrates the results of four columns with a nominal dimension of 100 ×100mm with a 5mm nominal wall thickness. The tensile coupon tests revealed a mean 0. 2 % proof stress of this material to be about 340 MPa and the mean ultimate stress to be about 650 MPa, thus a significant opportunit for strength increase is possible . Figure illustrates the load-deflection results for two stainless steel square hollow sections, SSH1 and SSH2. The results for these tests reveal that the columns achieve a maximum load just in excess of 800 kN and then the load begins to stabilize. The reasons for this are that any increase in stress is negated by gradual local buckling of the section. Thus redistribution allows the load to be maintained, but no significant strain hardening appears to develop. The results for two concrete filled steel sections SSCF1 and SSCF2 reveal that the presence of the concrete infill has allowed local buckling to be considerably delayed and a gradual increase in the steel has been allowed. Furthermore, the steel section has appeared to significantly confine the concrete in these sections.

Figure 10 Axial compression tests of SS concrete filled steel columns

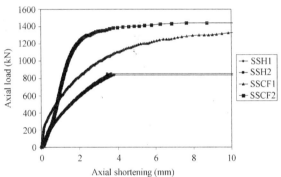

Figure 11 Stainless steel axial load experiments

DESIGN

High strength steel

The use of high strength steel with normal strength concrete generally provides a good mixture as strain compatibility of the two materials ensures that the two materials reach their ultimate stress simultaneously. Typically high strength steel of nominal yield stress of 690 MPa has a yield strain of approximately 3500$\mu\varepsilon$ Normal strength concretes by most international codes of practice are assumed to crush at a strain of about 3000 $\mu\varepsilon$. Thus, the use of high strength steel and high strength concrete in combination may not always provide a suitable solution as high strength concretes if unconfined can have crushing strains which are much lower than 3000 $\mu\varepsilon$. and thus when concrete crushing is initiated the steel may not have fully yielded and may render it inefficient for strength purposes.

Axial strength

The axial strength of composite columns utilising high strength steel can thus be represented as:

$$N_u = N_{uc} + N_{uhss} = f_c A_c + f_{hss} A_s \quad (1)$$

Now the concrete maximum compressive strength, f_c depends on the concrete strength chosen and the steel plate slenderness. For steel sections with compact slenderness the concrete is generally assumed to be adequately confined and thus the concrete strength is assumed to be higher than the mean cylinder strength. The steel compressive yield stress, f_{hss} is also dependent on the concrete strength and steel plate slenderness. Generally if the concrete strength is low, then it is expected that the yield stress will be achieved. However, if the concrete strength is high and the steel section is non-compact then the yield stress may be lower than the nominal yield stress.

Combined strength

When using high strength steel in a concrete filled steel column, Uy (2001a) illustrated as shown in Figure 12, that the strength of columns when subjected to large amounts of flexure will not achieve their full rigid plastic strength. The Eurocode 4 model which utilises a rigid plastic analysis approach is therefore non-conservative when using this for design. Uy (2001a) have therefore suggested a mixed analysis approach which can be used to conservatively predict the interaction between axial force and bending moment.

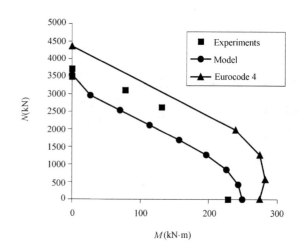

Figure 12 Interaction of axial force and bending moment of HSS short columns, (Uy, 2001a)

Local buckling

Slenderness limits for high strength steel in composite columns can benefit from the effects of the restraint offered by the concrete in increasing the local buckling stress, (Uy and Bradford, 1996). However, these slenderness limits need to be properly adjusted to allow for the non-linear effects of residual stresses and geometric imperfections as outlined by Uy, (2001). With this is mind the following slenderness limits are suggested for plates which are supported on four edges, which is typical of each of the face plates in a box column.

$$\frac{b}{t}\sqrt{\frac{f_y}{250}} \leq 50 \quad (2)$$

where b is the width, t is the plate thickness and f_y is the nominal yield stress of the steel in question.

Global buckling

Mursi and Uy (2004) carried out an extensive experimental and theoretical program of research to determine the global stability behavior of concrete filled columns utilizing high strength steel. The results of the experiments were compared with the model and the suggested model for slender column behavior presently existent in Eurocode 4, (British Standards Institution, 2005). This model which is illustrated in Figure 13 allows one to determine the overall capacity of a column, knowing the plastic resistance of the section, N_{plRd} as well as the critical load of the column N_{cr}. The study by Mursi and Uy (2004) showed that curve a is the most suitable approach to be used for high strength steel composite columns as the level of residual stress as a function of the yield stress, means that the residual stress neutral curve is appropriate for design. Furthermore, the presence of concrete infill ensures that the growth of imperfections is minimized. A generalized expression for the capacity of the slender column, N_c can be given as

$$N_c = \chi N_u \tag{3}$$

where χ is a function of slenderness, residual stresses and geometric imperfections.

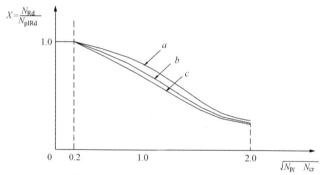

Figure 13 Column curves for slender composite columns, Uy and Liew (2002)

Stainless steel

The use of stainless steel with concrete also provides a very good combination of materials. When the stainless steel component is restrained from local buckling it will then be able to achieve stresses which are significantly greater than the 0.2% proof stress which is generally used in design codes. This has the potential for the steel to significantly exceed the design stress and the steel will also subsequently confine the concrete. To date there has been limited research to consider these effects.

Axial strength

The axial strength of composite columns utilising stainless steel can thus be represented as:

$$N_u = N_{uc} + N_{uss} = f_c A_c + f_{ss} A_s \tag{4}$$

Now the concrete maximum compressive strength, f_c depends on the concrete strength chosen and the steel plate slenderness. For steel sections with compact slenderness the concrete is generally assumed to be adequately confined and thus the concrete strength is assumed to be higher than the mean cylinder strength. The steel compressive yield stress, f_{ss} is also dependent on the concrete strength and steel plate slenderness. Generally if the concrete strength is low, then it is expected that the proof stress will be exceeded. However, if the concrete strength is high and the steel section is non-compact then the maximum stress may be lower than the 0.2% proof stress.

Combined strength

Initial numerical studies by Roufegarinejad, Uy and Bradford (2004) showed that when considering the interaction between axial force and bending moment, the strength of a stainless steel concrete filled column will surpass the rigid plastic strength based on the 0.2% proof stress, (see Figure 14). This is due to the fact that stainless steel continues to increase in stress after the 0.2% proof stress. This is a strength issue which may prove useful in harnessing in future studies. As the strain increases, the stainless steel tube experiences significant strain hardening before the concrete crushes and this explains why the model gives a larger envelope to the Eurocode 4 method, (British Standards Institution, 2004). This is advantageous and should be included in design.

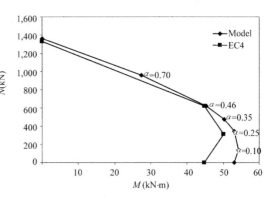

Figure 14 Interaction of axial force and bending of SS short columns, (Roufegarinejad et al. 2004)

Local buckling

Local buckling in stainless steel sections is a far more complex problem than in mild structural steel and high strength structural steel sections which typically adopt a linear elastic, perfectly plastic stress-strain curve. Because of the highly non-linear behaviour of stainless steel it is difficult to actually categorise plate slenderness limits. This has been previously alluded to by Nethercot and Gardner (2002) for hollow steel sections and it is also expected that this issue will be important for concrete filled steel sections. In fact, there will also exist the added complexity associated with concrete filled sections, as the concrete stiffness and strength properties will play an important role in determining the failure strain and local buckling strain of the steel.

Global buckling

Rasmussen and Rondal (2000) examined the strength curves in Eurocode 3 (British Standards Institution, 2004) and found that when using the real material properties which incorporate increased proof stresses in the cold worked corners, that a certain degree of unconservatism was evident.

EXPLOITING THE BENEFITS OF HIGH PERFORMANCE STEELS FOR ABNORNMAL LOADS

The previous sections have illustrated that both high strength steel (HSS) and stainless steel (SS) have significant benefits over mild structural steel (MS) which for static effects in structures could well lead to significant weight savings and reductions in the size of structural elements required in design. This is due to the increased yield stress that may exist for HSS and SS over MS structural elements.

Figure 15 does attempt to provide some strain limitations on these steels. It is generally accepted that HSS has a lower fracture strain than MS, however it has been suggested that SS can have a fracture strain up to 3 times that of MS, (Di Sarno and Elnashai, 2003). If the design objective is to absorb a particlular load with a minimum cross-section size, the benefits of HSS and SS may be in the ability to absorb the energy more efficiently than MS.

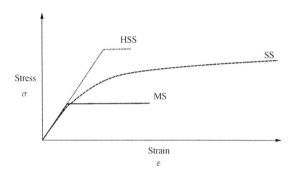

Figure 15 Idealised stress-strain curves for MS, SS and HSS with typical fracture strain limitations

The energy absorption capability of concrete filled columns has been the subject of investigation by Xiao et al. (2005). They carried out experiments on concrete filled columns under axial effects and found that there were significant increases in the capacity associated with the high strain rates. These tests showed promising results and further analysis and experimentation would be necessary to fully exploit the benefits which concrete infill and HPS can provide for critical infrastructure protection (CIP) elements. In analysing such elements the steel materials would need to consider the effects of strain rates and the results of recent studies could be used to incorporate the effects of strain rates on the steel constitutive behaviour, (Rohr, Nahme and Thoma, 2005). Furthermore, the effects of strain rates are also necessary to be considered for the concrete and recent research by Malvar, Morrill and Crawford (2004) has provided some guidance on the effects of strain rates and blast loading on the concrete constitutive behaviour in a triaxial state, and this would need to be utilised in further studies.

CONCLUSIONS AND FURTHER RESEARCH

This chapter has introduced the concept of using high performance steels (namely high strength and stainless steel) in composite construction forms. The paper has initially provided a significant background to the research that has already been carried out for these materials in structural engineering applications. The chapter has also provided significant background on applications of landmark buildings and structures using high performance steel in the past, present and future context. The paper has then provided some commentary on the major differences in the behaviour of these types of materials when using them for short composite columns. Further comments on the subtleties and nuances that need to be considered when applying existing methods in codes of practice have also been provided. Further research will focus on the different types of structural forms and loading in which both high strength steel and stainless steel may provide advantages in composite construction applications and this will be reported on in future.

ACKNOWLEDGEMENTS

The experiments in this chapter were sponsored by various Australian Research Council Small and Large Grants and Faculty and University Research Grants. In kind support by Bisalloy Steels from 1997-2006 has been provided and this is gratefully acknowledged. The author would like to thank all the staff and students at the University of New South Wales and University of Wollonong for their untiring assistance in the conduct of the experiments reported in this chapter.

REFERENCES

American Society of Civil Engineers (1991) *Specification for cold-formed stainless steel structural members*, ANSI/ASCE-8. New York.

American Institution of Steel Construction (1993) *Load and Resistance Factor Design Specification for Structural Steel Buildings*, Chicago.

Asgar, T. (2005) *Behaviour and design of concrete filled columns utilising stainless steel under combined loads*, Thesis submitted in partial fulfilment of the requirements for the award of the degree of Bachelor of Engineering in Civil Engineering.

Bergmann, R. and Puthli, R. (2000) Behaviour of composite columns using high strength steel sections, *Proceedings of Composite Construction IV*, Engineering Foundation Conferences, May-June, Banff, Alberta.

British Standards Institution (2004) Eurocode 3, EN 1994-1-1. *Design of composite steel and concrete structures*, Part 1.1, General Rules and Rules for Buildings.

British Standards Institution (2005) Eurocode 4, ENV 1993-1-1. *Design of steel structures*, Part 1.1, General Rules and Rules for Buildings.

Burgan, B. A., Baddoo, N. R. and Gilsenan, K. A. (2000) Structural design of stainless steel members-comparison between Eurocode 3, Part 1.4 and test results. *Journal of Constructional Steel Research, An International Journal*, 54 (1), pp. 51-73.

Chaseling, C. (2004) Star attraction, *Modern Steel Construction*, 37, 36-42.

Chen, J. Young, B. and Uy, B. (2006) Behaviour of high strength steel at elevated temperatures, *Journal of Structural Engineering*, ASCE, (In press).

Dajanovic, N. (2004) *Strength and deformation characteristics of short stainless steel concrete filled tubes*, Master of Engineering Science Thesis, The University of New South Wales.

Davie, J. C. (1995) The Sydney Casino Project, Concrete 97, *Proceedings of the 18th Biennial Conference of the*

Concrete Institute of Australia, Adelaide, pp. 651-657.

Di Sarno, L. and Elnashai, A. S. (2003) Special metals for seismic retrofitting of steel buildings, *Progress in Structural Engineering and Materials*, 5 (2), pp. 60-76.

Di Sarno, L., Elnashai, A. S. and Nethercot, D. A. (2003) Seismic performance assessment of stainless steel frames, *Journal of Constructional Steel Research, An International Journal*, 59, pp. 1289-1319.

Fortner, B. (2006) Landmark reinvented, **Civil Engineering, Magazine of the American Society of Civil Engineers**, 76 (4), pp. 42-47.

Gillett, D. G. and Watson, K. B. (1987) Developments in steel high rise construction in Australia, *Steel Construction, Journal of the Australian Institute of Steel Construction*, 21 (1), 2-8.

Gjerding-Smith, A. (2004) *Strength and deformation behaviour of short concrete filled stainless steel columns*, Bachelor of Engineering Thesis, The University of New South Wales.

Hagiwara, Y., Kadono, A., Suzuki, T. Kubodera, I. Fukasawa, T. and Tanuma, Y. (1995) Application of HT780 high strength steel plate to structural member of super high rise building: Part 2 Reliability inspection of the structure. *Proceedings of the Fifth East Asia-Pacific Conference on Structural Engineering and Construction, Building for the 21st Century*, Gold Coast, pp. 2289-2294.

King, S. (1999) *Leighton: 50 Years of achievement: 1949-1999*, Technical Resources, St Leonards, Australia.

Kouhi, J., Talja, A, Salmi, P. and Ala-Outinen, T. (2000) Current R&D work on the use of stainless steel in construction in Finland, *Journal of Constructional Steel Research, An International Journal*, 54 (1), pp. 31-50.

Standards Australia (1998) *Australian Standard, Steel Structures, AS4100-1998*, Sydney, Australia.

Mochuziki, H., Yamashita, T., Kanaya, K., and Fukasawa, T. (1995) Application of HT780 high strength steel plate to structural member of super high-rise building: Part 1 Development of high strength steel with heavy gauge and welding process. *Proceedings of the Fifth East Asia-Pacific Conference on Structural Engineering and Construction, Building for the 21st Century*, Gold Coast, pp. 2283-2288.

Mursi, M and Uy, B. (2004) Strength of slender concrete filled high strength steel box columns, *Journal of Constructional Steel Research, An International Journal*, 6 (12), pp. 1825-1848.

Mursi, M and Uy, B. (2005a) Behaviour and design of fabricated high strength steel columns subjected to biaxial bending, Part 1: Experiments, *International Journal of Advanced Steel Construction, Hong Kong Institute of Steel Construction*, (In press).

Mursi, M and Uy, B. (2005b) Behaviour and design of fabricated high strength steel columns subjected to biaxial bending, Part 2: Analysis and design codes, *International Journal of Advanced Steel Construction, Hong Kong Institute of Steel Construction*, (In press).

Mursi, M., Haedir, J. and Uy, B. (2004) Strength of short concrete filled columns using high strength steel under biaxial loading, 2nd International Conference on Steel and Composite Structures, ICSCS04, Seoul, Korea, September.

Nethercot, D. A. and Gardner, L. (2002) Exploiting the special features of stainless steel in structural design, *ICASS'02, 3rd International Conference on Advances in Steel Structures*, Hong Kong, pp. 43-56, (Keynote paper).

Rasmussen, K. J. R., and Hancock, G. J. (1992) Plate slenderness limits for high strength steel sections. *Journal of Constructional Steel Research*, 23, pp. 73-96.

Rasmussen, K. J. R., and Hancock, G. J. (1995) Tests of high strength steel columns. *Journal of Constructional*

Steel Research, 34, pp. 27-52.

Rasmussen, K. J. R. (2000a) Recent research on stainless steel tubular structures, *Journal of Constructional Steel Research, An International Journal*, 54, pp. 75-88.

Rasmussen, K. J. R. and Rondal, J. (2000b) Column curves for stainless steel alloys, *Journal of Constructional Steel Research, An International Journal*, 54, pp. 89-107.

Rosier, G. A. and Croll, J. E. (1987) High strength quenched and tempered steels in structures. *Seminar Papers of Association of Consulting Structural Engineers of New South Wales, Steel in Structures*, Sydney.

Roufegarinejad, A., Uy, B. and Bradford, M. A. (2004) Behaviour and design of concrete filled steel columns utilising stainless steel cross-sections under combined actions, 18th *Australasian Conference on Mechanics of Structures and Materials*, Perth, Australia, December.

Sivakumaran, K. S. and Yuan, B. (1998) Slenderness limits and ductility of high strength steel sections. *Journal of Constructional Steel Research, Special Issue: 2nd World Conference on Steel Construction*, 46, (1-3), pp. 149-151, Full paper on CD-ROM.

Standards Australia (1998) *Australian Standard, Steel Structures, AS4100-1998*, Sydney, Australia.

Standards Australia (2001) *AS/NZS4673: Cold-formed stainless steel structures*, Sydney.

Uy, B. (1996) Behaviour and design of high strength steel-concrete filled box columns. *Proceedings of the International Conference on Advances in Steel Structures*, Hong Kong, pp. 455-460.

Uy, B. (1999) Axial compressive strength of steel and composite columns fabricated with high strength steel plate. *Proceedings of the Second International Conference on Advances in Steel Structures*, Hong Kong, pp. 421-428.

Uy, B. (2001a) Axial compressive strength of short steel and composite columns fabricated with high strength steel plate, *Steel and Composite Structures*, 1 (2), pp. 113-134.

Uy, B. (2001b) Strength of short concrete filled high strength steel box columns, *Journal of Constructional Steel Research*, 57, pp. 113-134.

Uy, B. (2001c) Local and post-local buckling of fabricated thin-walled steel and steel-concrete composite sections, *Journal of Structural Engineering, ASCE*, 127 (6), 666-677.

Uy, B. and Bradford, M. A., (1996) Elastic local buckling of steel plates in composite steel-concrete members, *Engineering Structures, An International Journal*, 18 (3), pp. 193-200.

Uy, B. and Liew, J. Y. R., (2002) Composite steel-concrete structures, *Chapter 51 Civil Engineering Handbook*, (edited by W. F. Chen and J. Y. Richard Liew), CRC Press, Boca Raton.

第九章 | Chapter 9

ISIS TECHNOLOGIES FOR CIVIL ENGINEERING SMART INFRASTRUCTURE

A. Mufti[1] and C. Klowak[2]

[1] Scientific Director and President, ISIS Canada Research Network & Professor, University of Manitoba, Agricultural & Civil Engineering Building, A250-96 Dafoe Road, Winnipeg, Manitoba, R3T 2N2, Canada. E-mail: muftia@cc.umanitoba.ca

[2] Civionics Engineer, ISIS Canada Research Network, A250-96 Dafoe Road, Agricultural & Civil Engineering Bldg, University of Manitoba, Winnipeg, MB, R3T 2N2, Canada

Abstract: ISIS Canada intends to significantly change the design and construction of civil engineering structures by developing innovative new structures. For these new structures to be accepted by the engineering community, it is mandatory that they be monitored and the results reported to the engineering community as well as being incorporated into civil engineering codes. ISIS Canada has been developing such structures and monitoring them through an innovative concept, which involves the development of the new discipline of Civionics where Civil engineering and electrophotonics are being integrated. In this paper, some of the innovations that have been implemented will be described.

Keywords: Innovative structures, corrosion free bridge decks, civionics, fibre reinforced polymers (FRPs), fibre optic sensors (FOSs), structural health monitoring (SHM).

INTELLIGENT SENSING

For many years, engineers have been searching for ways to obtain information on how a structure is behaving in service by incorporating, at the time of construction or subsequently, sensing devices which can provide information about conditions such as strain, temperature, and humidity (Tennyson, 2001). The development of such structurally integrated fibre optic sensors (FOSs) has led to the concept of smart structures.

APPLICATION OF INNOVATIVE TECHNOLOGIES IN THE FIELD

For infrastructure owners, one of the greatest values of ISIS Canada research lies in its practical application. Over the past year, there have been many new opportunities for applying ISIS Canada technology as is evidenced in the growing number of field demonstration projects under-

The Beddington Trail Bridge is a two-span, continuous skew bridge of 22.83 and 19.23-m ⟨s⟩, each consisting of 13 bulb-Tee section, pre-cast, prestressed concrete girders. Two different types of FRP tendons were used to pretension six precast concrete girders. Carbon fibre composite cables produced by Tokyo Rope of Japan were used to pretension four girders while the other girders were pretensioned using Leadline rod tendons produced by Mitsubishi Kasei.

Fibre optic Bragg grating strain and temperature sensors were used to monitor structural behaviour during construction and under serviceability conditions. The four-channel Bragg grating fibre laser sensing system was developed for this purpose at the University of Toronto Institute for Aerospace Studies.

Before constructing the bridge, an experimental program was conducted at the University of Manitoba's W. R. McQuade Laboratory to examine the behaviour of scale model beams pretensioned by the same type, size, and anchorage of the two different tendons used for the bridge girders.

Prestressing of carbon FRP was adapted by coupling the carbon fibre composite cables and Leadline rods to conventional steel strands. Couplers helped to minimize the length of carbon FRP tendons, and were staggered to allow use of the same spacing for the conventional steel reinforcing tendons.

The Leadline rods were cut at the site and two rods were used for each tendon. The carbon fibre composite cables were delivered precut to the specified length with 300-mm die cast at each end to distribute the stresses at the anchoring zone. Construction of the bridge and handling of the girders at the site was typical.

A four-channel Bragg grating fibre optic sensor system was used at different locations along the bridge girders that were pretensioned by the carbon FRP. Each fibre optic sensor was attached to the surface of the tendon after pretension to serve as a sensor. The sensors were connected, through a modular system, to a laptop computer used at the construction site to record the measurements at different stages of construction and after completion of the bridge. The optic sensor system measures the absolute strain rather than a strain relative to an initial calibration value similar to electric resistance strain gauges and mechanical gauges.

In 1999, the bridge was tested statically and dynamically to assess the durability of fibre optic sensors. After six years, all FOSs were functioning Figure 1 (a) and (b). This finding validates the view that FOSs are durable and reliable for long term monitoring.

In November 2004, the bridge was tested again with the same vehicle and weight. Figure 1 (c) and (d) indicate that the FBG sensors are durable and are providing accurate results and the CFRP is performing as designed in 1993.

Hall's Harbour Wharf, Nova Scotia (Hall's Harbour)

Hall's Harbour Wharf in Nova Scotia is a 96-year-old combination wharf/breakwater, shown in Figure 2. It is the world's first marine structure with fibre optic sensors embedded in a steel-free way. The projects range from a concrete steel-free bridge deck for the Salmon va Scotia (Newhook and Mufti, 2000), to the strengthening of a nuclear conta Quebec (Demers et al., 2003).

At least fifty projects are currently being monitored for health in Canada. S described in the following sectizons.

Beddington Trail Bridge, Calgary, Alberta

In 1992, the Beddington Trail Bridge in Calgary (Rizkalla et al., 1994 shown in Figure 1 (a), was the first bridge in Canada to be outfitted with polymer (FRP) tendons and a system of structurally integrated optical sensors nitoring. The bridge opened in 1993, before ISIS Canada was formed. It is sig ISIS network because, for the group of researchers involved, it confirmed the n ganization, like ISIS Canada, that could spearhead transferring this new techno try.

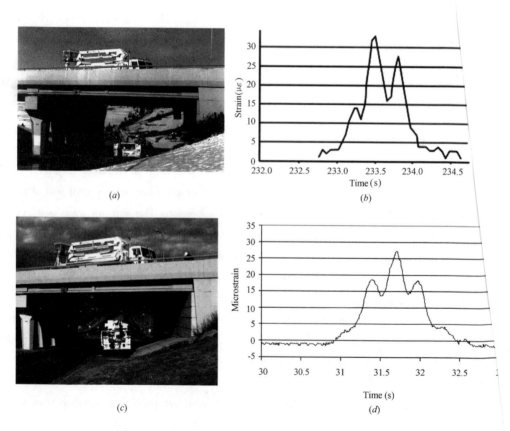

Figure 1   On-site monitoring
(a) accessing fibre optic junction box -1999, (b) Dynamic FBG response to three-axle truck load-1999,
(c) accessing fibre optic junction box-2004, (d) Dynamic FBG response to three-axle truck load-2004

concrete deck for remote monitoring (Figure 3) (Newhook and Mufti, 2000). It is designed to last 80 years - three times longer than traditional construction methods. This design received the "Award of Excellence" from the Canadian Consulting Engineer Association.

Figure 2　Hall's Harbour Wharf

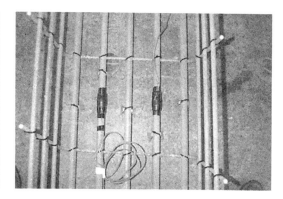

Figure 3　Fibre optic sensors embedded in a steel-free concrete deck for remote monitoring

The new wharf consists of piles supporting concrete beams and deck elements above an armour stone and timber crib breakwater (Figure 4). The concrete beams are designed with a hybrid reinforcement scheme of steel and glass FRP rods. The outer durable layer of FRP protects the inner core of steel reinforcement, which was included for code considerations (Figure 5). The deck consists of precast steel-free concrete bridge deck panels modified to meet the durability needs of this structure. The design is based on draft versions of both the Canadian Highway Bridge Design Code (CHBDC 2000) and the American Concrete Institute Code.

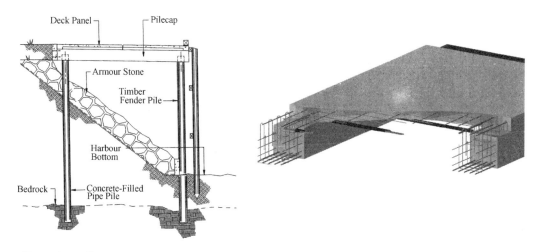

Figure 4　Hall's Habour- Cut-away view

Figure 5　Hybrid reinforcement scheme of steel and glass FRP rods

It is anticipated that rehabilitation of marine structures using FRPs will become standard practice through time, with the inner core of steel reinforcement considered unnecessary.

Portage Creek Bridge-Strengthening Against Earthquakes & Field Assessment, British Columbia

The Portage Creek Bridge (Mufti et al., 2003) (Figure 6) in Victoria, British Columbia, was designed in 1982 by the British Columbia Department of Highways Bridge Engineering Branch. It crosses Interurban Road and the Colquitz River at McKenzie Avenue.

It is a 125 m (410 ft) long 3-span steel structure with a reinforced concrete deck supported on two reinforced concrete piers with abutments on steel H piles. The deck has a roadway width of 16 m (52 ft) with two 1.5 m (5 ft) sidewalks and aluminum railings. The super structure is supported at the ends and has two intermediate supports along the length of the bridge called Pier No. 1 and Pier No. 2.

Figure 6 Portage Creek Bridge

The Portage Creek Bridge is a relatively high profile bridge that has been classified a Disaster-Route bridge. However, it was built prior to current seismic design codes and construction practices and could not resist potential earthquake forces as required by today's standards. Although some consideration was given to seismic aspects as evidenced in the original drawings, it required retrofitting to prevent collapse during a seismic event. The service life of the bridge can thus be increased to 475 years.

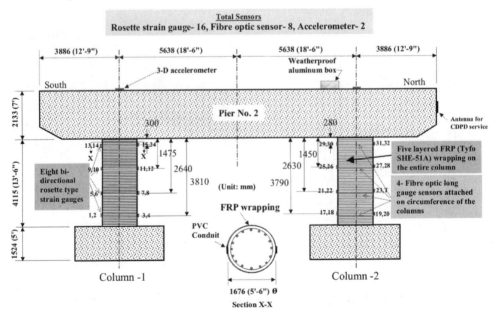

Figure 7 Portage Creek FOS locations

Most of the bridge is being strengthened with conventional materials and methods. The dynamic analysis of the bridge predicts the two tall columns of Pier No. 1 will form plastic hinges during an earthquake. Once these hinges form, additional shear will be attracted by the short columns of Pier No. 2. A nonlinear static pushover analysis indicates that the short columns will not be able to form plastic hinges prior to failure in shear. Therefore, it was decided that FRP wraps should be used to strengthen the short columns for shear without increasing the moment capacity. The bridge is instrumented with 16 foil gauges, 8 fibre optic sensors and 2 accelerometers (Figure7). The bridge is being remotely monitored and data collected with the data logger and monitoring system shown in Figure 8 (a) and (b). ISIS Canada is assisting with the structural health monitoring of this bridge.

(a)   (b)

Figure 8   Data logger and monitoring system

FIRST GENERATION CORROSION FREE BRIDGE DECKS

In the quest for lighter, stronger and corrosion-resistant structures, the replacement of ferrous materials by high-strength fibrous ones is being actively pursued in several countries around the world, both with respect to the design of new structures as well as for the rehabilitation and strengthening of existing ones. In the design of new highway bridges in Canada, active research is focused on a number of specialty areas, including the replacement of steel reinforcing bars in concrete deck slabs by randomly distributed low-modulus fibres, and the replacement of steel prestressing cables for concrete components by tendons comprised of super-strong fibres. Research is also being conducted to repair and strengthen existing structures by the use of FRPs.

The FRPs have perceived disadvantages compared to steel. These are ductility and low thermal compatibility between FRP reinforcement and concrete. The majority of our construction projects in Canada are in non-seismic zones. The ductility is an important characteristic of steel as it allows large deformations and dissipation of energy. Concrete structures reinforced with FRPs at ultimate loads give large deformations. Therefore, reinforced concrete structures, whether reinforced with steel bars or FRPs give the same order of deformability. Research to show that concrete structures with FRPs, if properly designed can dissipate energy is in progress. The design of the proper concrete cover eliminates low thermal compatibility between FRP reinforcement and concrete. It should be noted that Glass Fibre Reinforced Polymer (GFRP) material has a modulus of elasticity comparable to concrete. Therefore, concrete does not feel any intrusion into it and performs well in resisting fatigue under dynamic loading.

These concepts have been implemented to develop corrosion free bridge decks. Several bridge decks have been constructed in Canada and one in Iowa, USA. Some of these bridsge decks are described in the next section.

Salmon River Highway Bridge, Nova Scotia

The first generation steel-free deck-slab in Canada was cast on the Salmon River Bridge, part of the Trans Canada 104 Highway near Kemptown, Nova Scotia (Newhook et al. 2000). Construction of the bridge, which consists of two 31m spans, includes a steel-free deck over one span and a conventional steel reinforced deck over the other. Internal arching in the slabs helps transfer the loads to the girders. Although the cost of the steel-free side was six percent more than the steel-reinforced side, the overall design tends to be less expensive than conventional decks. This is because steel-free decks do not suffer from corrosion, so traditional maintenance costs are greatly reduced. This concept has won six national and international awards including the prestigious NOVA award from the Construction Innovation Forum (CIF) of the United States.

The deck contains no rebar. Instead, longitudinal beams or girders support it. The load is transferred from the deck to the supporting girders in the same way that an arch transfers loads to supporting columns. Although steel straps are applied to tie the girders together, because they are not embedded in the concrete they can be easily monitored and inexpensively replaced.

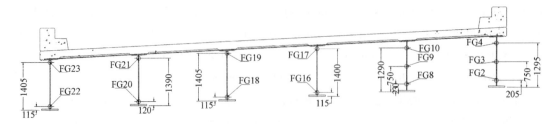

Figure 9 Sensor locations

The Structural Health Monitoring (SHM) of the steel-free bridge deck was conducted by installing sensors, as shown in Figure 9. SHM indicates that the load sharing of the Salmon River Highway Bridge is similar to conventional decks, as shown in Figure 10.

With no steel inside the concrete (Figure 11), no unnecessary weight is added, meaning thinner deck designs. The steel straps are welded to the top flanges of the girders thereby resisting any lateral movement. The Salmon River steel-free bridge deck has withstood a number of Canadian winters, and it appears to be defying the conventional approach to building steel-reinforced bridge decks. There are seven steel free bridge decks across Canada.

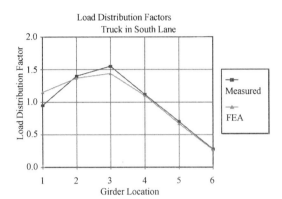

Figure 10 Load sharing of the Salmon River Highway Bridge

Figure 11 Casting of the steel free deck

Crowchild Trail Bridge, Alberta (Calgary)

The original Crowchild Trail Bridge in Calgary, Alberta was a two-lane, three-span prestressed concrete box-girder bridge. The bridge was found to be under-strength as a result of deterioration over 20 years and increased traffic load. As a result, the bridge superstructure was replaced in June 1997 (Afhami & Cheng, 1999).

The new superstructure is the first continuous span steel-free bridge deck in the world. The removal of internal steel reinforcement is made possible by providing lateral restraint to the supporting steel girders through evenly spaced transverse steel straps placed across the tops of the adjacent girders. Glass fibre reinforcements are used at the regions of interior supports and overhanging cantilevers. Prefabricated glass fibre reinforcing grid, NEFMAC, or New Fiber Composite Material for Reinforcing Concrete, is used for the reinforcement of side barriers. A total of 103 strain gauges, two fibre optic strain sensors, and five thermisters were used in the monitoring program.

The first tests (1997) consisted of a static truckload test, an ambient vibration test, an effect of temperature test, and dynamic measurements under passing trucks. The second tests (1998) consisted of static and dynamic truckload tests and ambient vibration test.

To monitor strain distribution in the transverse direction of the bridge deck, 17 embedded strain gauges were installed in a total of five precast blocks-three in the positive moment region and two in the negative moment region. Figure 12 shows the location of the embedded strain gauges.

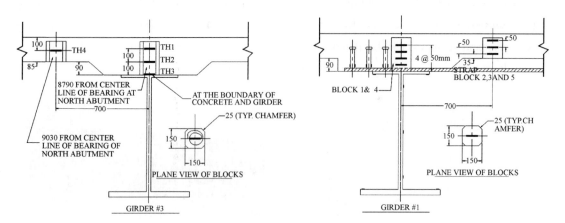

Figure 12 Girder #3 with embedded temperature probes and Girder #1 with embedded strain gauges

Eighteen gauges monitored the performance of the steel straps. Six strain gauges were used to monitor strains in the end shear studs of the strap. Thirty-four strain gauges were used to monitor the steel girders. The webs of all five girders were instrumented with three gauges at both positive and negative moment regions to monitor load sharing among the girders and moment distributions along the girders. Four gauges were also installed on the flanges to measure any warping of the girders. The response of one cross frame was monitored by four strain gauges. At the barriers, two strain gauges were installed on a NEFMAC and two on a stainless steel stud. Six gauges are positioned at the overhanging cantilevers and 14 gauges at the pier monitored glass fibre reinforcement.

To evaluate the use of FOS technology, two commercially available sensors were installed on the glass fibre reinforcement at the same section as the electrical strain gauges in the north span of the bridge. The sensors were of the Fabry-Perot type and non-compensated for temperature. To measure deflections of the bridge under heavy traffic load, a testing program was organized by the City of Calgary before the bridge was open to traffic. Two trucks, each loaded nominally to 355 kN, were used to produce nine different load cases.

Temperature profiles were recorded with the thermisters, and strain measurements were taken using the strain indicator and the manual switching box. As the test took several hours, it was necessary to account for the thermal effects. The results provided preliminary information such as load sharing among the girders, location of the neutral axis, and moment distribution between mid-span and support. Similar information was later obtained from the results of the dynamic measurements. Measured strains were all less than 80 $\mu\varepsilon$ in the girders, and less than 40 $\mu\varepsilon$ in the steel straps. Concrete strains were insignificant.

Thermal effect

Temperature effects produce the most significant strains in the bridge. To evaluate the performance of the two FOS, strains caused by temperature change are plotted in Figure 13. As the sensors were not temperature-compensating gauges, they measured the total strain caused by both free expansion and restrained thermal deformation. The constraint strains for only a 13℃ increase in air temperature are comparable with strain measured during static testing using the truck load shown in Figure 14.

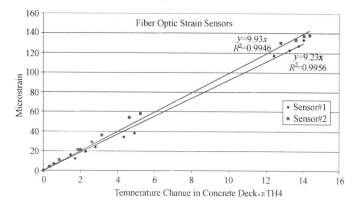

Figure 13　Effect of temperature change on fibre optic strain sensors

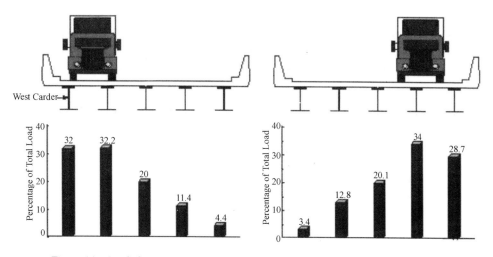

Figure 14　Load sharing among girders based on dynamic strain measurements 1997

Dynamic measurements under passing heavy vehicles

The response of the bridge to car traffic was insignificant. For passing trucks, a 10 s window of data at a scan rate of 1000 readings/s was gathered. Information such as type of truck, number of axles, and passing lane was recorded by videotape. Electronic low pass filters in the data acquisition unit were set at 10 kHz. Figure 14 illustrates the load sharing among girders based on the measured strains and Figure 15 shows the response of one strain gauge to a passing truck.

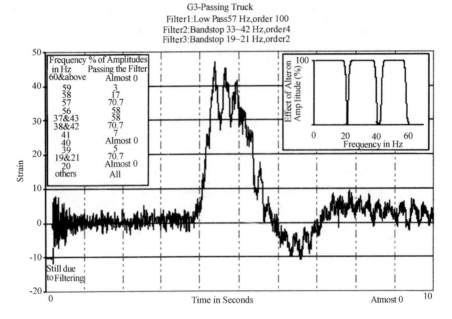

Figure 15　Dynamic measurement on gauge 3 of girder No. 1 and effects of digital filtration

Figure 16 illustrates the vibratory and non-vibratory portions of the response of a strain gauge to the truck. The non-vibratory portion was almost identical to the response in the static test. It can be seen that the vibration of the bridge is quite small in magnitude, resulting in a dynamic amplification factor of less than 1.15.

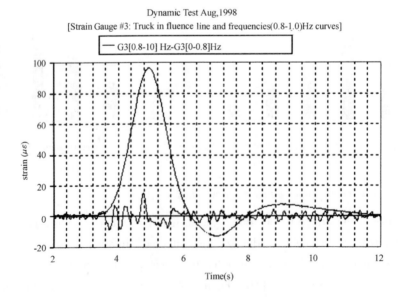

Figure 16　Response of a strain gauge to the truck in 1998 test

SECOND GENERATION CORROSION FREE BRIDGE DECKS

The second-generation steel-free deck slab exhibits the same behavior as the first-generation steel-free deck slab, with the exception of the longitudinal crack at the mid-point between the girders. External steel straps located below the deck provide the structural integrity to the slab. In order to reduce the width of the longitudinal crack that has developed on the five first-generation steel-free decks, researchers at the University of Manitoba (Memon, 2004) have concluded that a bottom mat of GFRP reinforcement with a reinforcement ratio of 0.25 percent is required. In addition, recent fatigue tests have also been undertaken at the University of Manitoba to replicate actual service life conditions for the deck slab. These tests have confirmed that a steel-free deck slab reinforced with a crack-control grid of nominal GFRP reinforcement exhibits a maximum crack width of 0.34 mm, a limit implicitly acceptable by the Canadian Highway Bridge Design Code, CHBDC (2000) (Memon et al, 2003).

Design fundamentals of the second generation steel-free deck slab

As one would expect, two design considerations must be taken into account when designing a second-generation steel-free deck slab (Mufti et al, 2003). The first design parameter that must be investigated is the size and spacing of the external steel straps that provide the transverse confinement for the deck slab. The CHBDC (2000) states that each steel strap must have a minimum cross-sectional area, in millimeters squared, given by

$$A = \frac{F_s S^2 S_1}{Et} 10^9 \qquad (1)$$

where the factor F_s is 6.0 for outer panels and 5.0 for internal panels, S is the spacing of the steel girders that must not exceed 3.0 m, S_1 is the spacing of the steel straps and must not be more than 1.25 m, E is the modulus of elasticity of the straps, and t is the thickness of the deck in millimeters.

After the size and spacing of the steel straps has been calculated, the design engineer can use a computer program called PUNCH to determine the ultimate failure load for punching in order to verify that this load is greater than the factored wheel load from the design truck. The PUNCH Program was developed to predict the behavior of a composite concrete deck slab on laterally restrained girders under the wheel loads of design trucks. Specific geometric and material properties are inputted, and the program then predicts the load deflection behavior and ultimate load. The program was developed for use with the steel-free concrete bridge deck system (Mufti and Bakht, 1996).

The second deign parameter that must be calculated is the allowable stress and strain levels in the GFRP reinforcement under service load conditions. The CBDHC (2000) states that a steel-free deck slab reinforced with a GFRP grid must have a performance or deformability factor, J, greater than 4.0 where

$$J = \frac{M_{ult} \Psi_{ult}}{M_c \Psi_c} \quad (2)$$

in which M_{ult} is the ultimate moment capacity of the slab, M_c is the moment corresponding to a maximum compressive strain in the concrete of 0.001, Ψ_{ult} is the curvature at the moment M_{ult}, Ψ_c is the curvature at the moment M_c.

Parameters that influence crack width are the crack spacing, the quality of bond between the concrete and the reinforcement bars, and above all, the strain in the bars. To control the width of cracks, ACI 318 (1999) limits the stress in the reinforcing steel under service load conditions to sixty percent of the yield stress. For reinforcing steel with a yield stress of 400 MPa, the allowable strain in the steel bars at service, ε_s, is:

$$\varepsilon_s = \frac{0.6 \times 400}{200000} = 1200 \times 10^{-6} \quad (3)$$

When steel reinforcing bars are used, ACI 318 (1999) allows a crack width of 0.3 mm for exterior exposure. When GFRP bars are used there is no risk of corrosion, thus the CHBDC (2000) recommends limiting the crack width to 0.5 mm for exterior exposure. Therefore, the width of cracks allowed for an FRP reinforced steel-free bridge deck is 1.7 times the value allowed for a conventionally reinforced concrete bridge deck. It is assumed that the ratio between crack widths of GFRP and steel reinforcement is 5/3. Thus, the allowable strain in the GFRP reinforcement at service is estimated as:

$$\varepsilon_{frp} = \frac{5}{3} \varepsilon_s = 2000 \times 10^{-6} \quad (4)$$

Strain compatibility can be used to determine the stress and strain levels in the GFRP reinforcement under service load conditions to ensure that the two criteria mentioned above in equations 1 or 2 are satisfied. The second-generation steel-free deck slab design must meet the three criteria outlined above in equations 1, 2 or 4, in order to limit crack widths to those allowed by the CHBDC (2000).

The first application of a second-generation steel-free deck slab was cast in July, 2003 on one span of the Red River Bridge on the North Perimeter Highway in Winnipeg, Manitoba. One of the simply supported spans was designed and cast using this innovative technology as a demonstration project for the Manitoba Department of Transportation and Government Services. This demonstration project was a joint effort between ISIS Canada, Earth Tech (Canada) Inc., JMBT Structures Research Inc. and the Province of Manitoba.

North Perimeter Highway Red River Bridge, Manitoba (Winnipeg)

The Red River Bridge was originally constructed in 1964 to a design loading of HS20 in accordance with Standard Specifications for Highway Bridges, American Association of State and Highway Transportation Officials (AASHTO 1996). This ten span bridge is 347 m long and consists of steel plate girders, spaced at 1.8 m, and a composite, cast-in-place, steel reinforced concrete deck. It is located on the north half of the Perimeter Highway that encircles the City of Winnipeg. Because the Perimeter Highway forms part of the Trans-Canada Highway system, this bridge is subjected to significant daily traffic with approximately twenty percent (20%) truck traffic. The original concrete deck slab began to exhibit signs of deterioration in the early 1980's, and in 1985, the asphalt riding surface was replaced in conjunction with significant deck patching and the installation of a waterproofing membrane. Further deterioration of the deck slab became evident again in the 1990's, and the Province decided to undertake a major rehabilitation of this bridge. The bridge was strengthened for current allowable loading, and the entire concrete deck slab was replaced to improve geometrics and meet current safety requirements. Earth Tech (Canada) Inc. was retained by the Province for the preliminary and detailed design of the major rehabilitation of this bridge.

Nine of the ten spans were designed and cast with a conventional, steel reinforced, cast-in-place concrete deck slab that measured 225 mm in depth. The design of these nine conventional spans was based upon flexural bending of the slab that resulted in 60 kg of steel reinforcement per square meter of deck area. Ternary concrete [a combination of Portland Type 10 cement, silica fume (8%) and fly ash (15%)] was used for these spans in order to provide a dense, durable concrete with a projected service life of 50 years.

The one span utilizing the second-generation steel-free deck technology was designed and cast using a concrete deck slab thickness of 200 mm. GFRP reinforcement was used for both the top and bottom mats in the internal deck panels. The top and bottom transverse and longitudinal reinforcing were comprised of #3 bars spaced at 200 and 600 mm, respectively (Figure 17). CFRP reinforcement was used as main reinforcement in negative moment regions for both the vehicular and pedestrian cantilevers. This transverse reinforcing consisted of 2-#4 bars spaced at 200 mm.

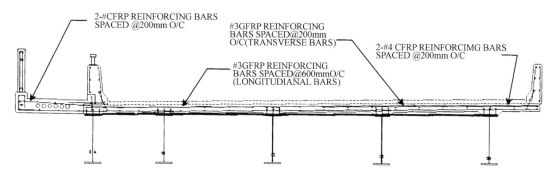

Figure 17 Top and bottom transverse and longitudinal reinforcement

The design of the concrete traffic barriers incorporated stainless steel bars, measuring 19 mm in diameter by 500 mm in length, with a 50 mm diameter by 12 mm in depth stainless steel end disc welded to each end of the bar. These bars were spaced at 300 mm with a minimum embedment length of 175 mm into the concrete deck slab. GFRP reinforcing was used for all straight bars in the barrier wall while galvanized steel reinforcing was used for all bent bars. Currently, the quality of fabrication and cost of bent GFRP reinforcing is not competitive with conventional steel reinforcing.

Transverse confinement of the deck slab was provided by steel straps, measuring 50 mm in width by 30 mm in depth, that have been tack welded to the top flanges of the steel plate girders at a spacing of 1.2 m. To ensure that the steel straps would perform integrally with the deck slab, steel nelson studs were added to the straps in the portion that passed over the girders.

Figure 18 Nelson studs added to steel straps

Because the existing nelson studs welded to the tops of the girders were insufficient in height to ensure composite action between the girders and the new concrete deck, galvanized steel haunch reinforcement was sized and placed in the haunches (Figure 18). This haunch reinforcement will ensure confinement of the concrete deck slab in the longitudinal direction. This reinforcement was galvanized to meet the projected service life and durability consistent with the GFRP reinforced concrete deck slab. Conventional Portland Type 10 cement concrete was used for this span.

The unit prices submitted by the contractor, PCL Constructors Ltd., for the conventional and steel-free deck slabs were Cdn. $993 and $1103 per cubic meter of concrete, respectively. The main reason for the 11% increase in the cost of the steel-free deck slab was that the contractor was unfamiliar with the installation of FRP reinforcement and the amount of work required. They were unsure of potential problems that may have developed during the casting of the conventional deck concrete. It is important to note that the steel-free deck slab becomes a much more economical design when comparing the two alternatives using life-cycle cost analysis because of enhanced durability and extended service life as was demonstrated by Dr. Gordon Sparks in his working paper on Life Cycle Costing (Sparks et al, 2003b).

An integrated structural health monitoring system, or civionic system, allows for continuous, independent, "real-time" monitoring that relates the actual loads on the structure to the performance of monitored components. Typically, a design engineer will specify the type of sensors or gauges to be installed on specific components at specific locations throughout the structure. The choice of the components and location is important to be able to gather meaningful data. The data can then be utilized to assess the performance and condition of the component, specifically, and the bridge as a whole. The strains at the specific locations for each of the monitored components are then relayed to an on-site data acquisition system. The information is processed and recorded before being transmitted through an internet connection to a server that relays the information to a structural health monitoring web page.

Conventional electric strain gauges were developed based upon the principal of the Wheatstone Bridge, where a difference in potential is measured across resistors and this difference can be converted into strain. These strain gauges have been used extensively in laboratory testing; however, the long-term performance of these gauges under field conditions was questionable. ISIS Canada has been researching and developing Fiber Optic Bragg Sensors that are believed to provide consistent readings, without drift, over a long period of time. In addition, because the monitored components will only be subject to strains under service loading conditions, the installed gauges must be very accurate and sensitive to very small changes. Fiber Optic Bragg Sensors are based on the principal of transmitting a fixed wavelength of light through the fiber. At discreet locations, where the strain is to be measured, grooves are cut into the fiber. As the light passes through these grooves, the wavelength is altered and then reflected back down the fiber to a measuring device. The shift in the wavelength is measured, and a strain value can then be calculated based upon the magnitude of the shift.

For the Red River Bridge, the integrated structural health monitoring system, or civionic system, was designed and installed to monitor the components of the steel-free deck slab and provide data on the stresses in the GFRP reinforcement and the transverse steel straps. Stresses in the steel plate girders and the CFRP reinforcement in the negative moment regions for the cantilever sections are also monitored. The system is comprised of a combination of various types of sensors, namely, conventional electric strain gauges, Fiber Optic Bragg Sensors, accelerometers, and thermocouples. A portion of the civionic system for the Red River North Perimeter is shown in Figure 19. Figures 20 (a) and (b) show some of the civionic instrumentation installed by Vector Construction prior to casting of the bridge deck.

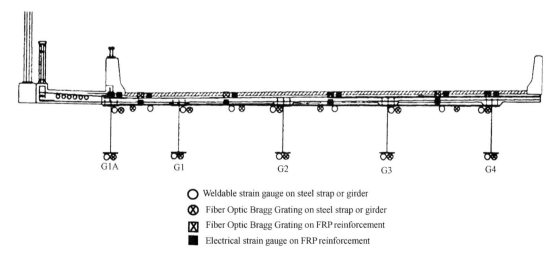

Figure 19 Civionics system for the Red River Bridge North Perimeter

(a)   (b)

Figure 20 Installed civionic instrumentation

Each sensor is connected to an on-site data acquisition system capable of storing the data. In addition, a video camera and weigh-in-motion device are also being installed in order to gather additional information regarding the frequency, configuration and axle loading of truck traffic on this specific bridge.

Thermocouples have been installed at numerous locations throughout the span in order to be able to calibrate the strain information for temperature. ISIS Canada has discovered, from monitoring numerous structures throughout Canada that bridges experience large stresses and strains as a result of temperature variation. It is very important to be able to differentiate the strains caused by temperature from the strains caused by vehicles under service loading conditions.

At the time when the civionic system was being designed for the Red River Bridge, strain measuring devices capable of reading Fiber Optic Bragg Sensor signals were only portable, single channel systems. In recognition of this problem, ISIS Canada had been working with a local electronic engineering design and manufacturing firm, IDERS Incorporated, to design a self-contained SHM System. It is designed specifically to be modular, and capable of reading and logging the data transmitted from the Fiber Optic Bragg Sensors. Additional modular units capable of reading electric strain gauges, accelerometers, thermocouples and real time video are also being designed and manufactured. In addition, the system contains a modem for immediate or timed transfer of data to a specified site. The system can be programmed to send e-mail messages directly to the design engineer, in real time, if pre-determined loads are experienced on the bridge. The system was successfully tested on the Taylor Bridge in Headingley, Manitoba in the summer of 2003 (Mufti, 2003).

One major concern of a civionic system is the enormous quantity of data that can be generated and must be stored in a short period of time. Sensors typically can take up to 100 readings per second, resulting in 8.64 million readings in a single day per sensor. The Red River Bridge contains 64 sensors, which translates into 0.5 billion readings per day. ISIS Canada, in conjunction with IDERS Incorporated, is currently developing an automated system that can be incorporated in the SHM unit. The readings will be scanned for pre-determined strain readings that will initiate a "red flag" notification to the design engineer. The automated system will greatly reduce the time and cost required to review the entire load history of the deck span.

A number of design and installation issues became evident during the course of this project. Fiber Optic Bragg Sensors need to be designed, manufactured and packaged with the rigors of field installation and long-term durability taken into account. The Fiber Optic Bragg Sensors, for example, need to be jacketed in a robust and durable material that will withstand the forces exerted during the pulling of the cable through conduits. Standard specifications need to be developed stipulating: size of conduits, size of junction boxes, capabilities for field splicing of the fiber optic cable, method of attaching conduits to the structure, requirements for heating and cooling an enclosure for the data acquisition system, and electrical requirements for the civionic system. In general, the owners of the structures that are willing to incorporate an integrated SHM system, or civionic system, must have confidence in the reliability and long-term performance of the various components. In order to assist consultants and contractors, ISIS Canada is in the process of developing a Civionic Manual (Rivera and Mufti, 2003). These specifications will provide the information necessary to achieve a reliable and durable SHM, or civionic system that is capable of long-term performance.

South Perimeter Highway Red River Bridge (Winnipeg)

The Red River Bridge is located on the South Perimeter Highway that encircles the city of Winnipeg, Manitoba. It is a 250 m long, 7-span bridge, consisting of three simply supported spans and one continuous span. Over a period of two years, commencing in the spring of 2006, the entire bridge deck will be replaced with a second generation steel-free GFRP hybrid bridge deck. The installation of a civionics structural health monitoring system in this bridge will follow the requirements as outlined in the civionics specifications mentioned later in this paper. Preliminary plans involve a minimal number of sensors including fibre optic sensors, accelerometers, conventional sensors and weigh in motion sensors (Figure 21).

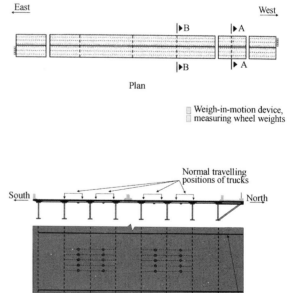

Figure 21 Location of weigh-in-motion sensors and electronic strain gauges on steel straps

STATIC AND FATIGUE STUDIES OF SECOND GENERATION CORROSION FREE BRIDGE DECKS

This section describes the static and fatigue behaviour of two different cast-in-place second generation steel-free bridge decks. Although cast monolithically, the first bridge deck was divided into three segments (A, B and C). Segment A was reinforced according to conventional design with steel reinforcement. Segments B and C were reinforced internally with a carbon fibre reinforced polymer (CFRP) crack control grid and a glass fibre reinforced polymer (GFRP) crack control grid, respectively, and externally with steel straps. The hybrid CFRP/GFRP and steel strap design is called a second generation steel-free concrete bridge deck. All three segments were designed with an almost equal ultimate capacity so that a direct comparison between the segments under fatigue loading conditions could be made. A performance comparison of all three segments for the first bridge deck under 25-ton (222 kN) and 60-ton (588 kN) cyclic loads is reported in this paper. The paper also outlines the details of a larger second generation steel-free bridge deck and briefly outlines some of the preliminary results from the internal panel static test and three static tests conducted on the cantilever. The static behaviour of the bridge deck plays an important role in understanding the fatigue theory developed from the series of fatigue tests outlined in this paper. The deck will also be subjected to a series of fatigue tests that will help confirm fatigue theory derived from the first bridge deck.

Fatigue Testing

Bridge deck details

The overall dimensions of the full-scale bridge deck were 3000 mm wide by 9000 mm long with a deck thickness of 175 mm. It was comprised of an internal panel and two cantilevers. The spacing of the girders was 2000 mm center to center. Although cast monolithically, the slab was conceptually divided into three segments (Figure 22). Segment A of the bridge deck slab was reinforced with steel reinforcement and designed using the empirical design method described in the Canadian Highway Bridge Design Code (CHBDC) (2000). It contained two layers of steel reinforcement with 15M bars spaced at 300 mm in both directions in each layer, providing a total reinforcement ratio of 1.2%.

Segment B of the bridge deck was a steel-free/CFRP hybrid design, reinforced with a CFRP crack control grid internally and external steel straps. The CFRP crack control grid was comprised of #10 CFRP bars spaced at 200 mm in the transverse direction and 300 mm in the longitudinal direction, providing a reinforcing ratio of 0.19% and 0.13% in the transverse and longitudinal directions, respectively (Memon et al. 2004). A steel strap with the dimensions 25 mm by 38 mm spaced at 1000 mm in the transverse direction provided a reinforcing ratio of 0.55% (Figure 22).

Segment C of the bridge deck slab was a steel free/GFRP hybrid design, reinforced with a

GFRP crack control grid internally and external steel straps. The GFRP crack control grid was comprised of #13 GFRP bars spaced at 150 mm in the transverse direction and 200 mm in the longitudinal direction, providing a reinforcing ratio of 0.48 % and 0.36 % in the transverse and longitudinal directions, respectively. A steel strap with the dimensions 25 mm by 38 mm spaced at 1000 mm in the transverse direction provided a reinforcing ratio of 0.55 % (Figure 22).

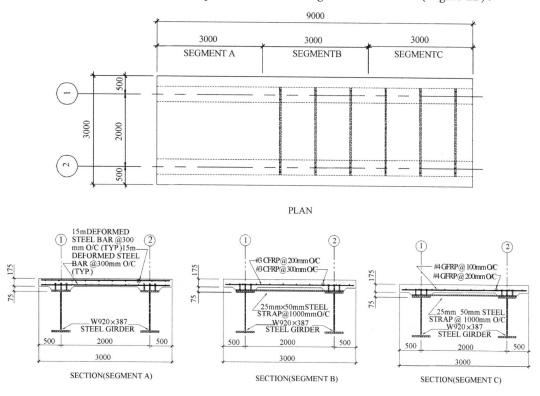

Figure 22　Bridge deck reinforcement details

Test results for 25-ton (222 kN) cyclic loading

Test results show that for a 25-ton or 222 kN cyclic loading, all three segments of the bridge deck completed 1,000,000 cycles without significant damage (Mufti et al. 2005). Vertical deflection of all three segments of the bridge deck were measured by displacement transducers and the maximum deflection was found to be approximately 2 mm after completing 1,000,000 cycles (Figure 23).

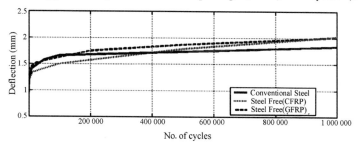

Figure 23　Plot of deflection versus number of cycles at 25 tons (222 kN)

In order to measure the internal and external strain response of each segment of the bridge deck, strain gauges were installed on some of the steel, CFRP and GFRP bars. Maximum strain response was measured under the applied load in transverse direction (Figure 24). All strain values are lower than 1200 micro-strain, therefore serviceability criteria were satisfied. The serviceability limit for either CFRP or GFRP is 2000 micro-strain. Therefore, the test results suggest that there could be a reduction of up to 40% in the area of CFRP or GFRP.

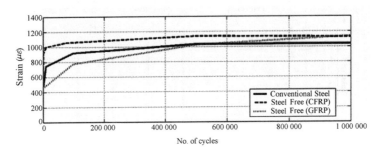

Figure 24 Plot of reinforcement strain versus number of cycles at 25 tons (222 kN)

In order to determine the effect of CFRP or GFRP crack control grids on the growth of crack width, a longitudinal crack was measured with an increasing number of cycles under the applied load (Figure 25). It can be seen that the maximum crack width for all three segments was approximately 0.4 mm, after completing 1,000,000 cycles for a 25-ton or 222 kN cyclic load. The ACI Committee 440 (2001) reported that, for steel reinforced structures, the allowable crack width limits are 0.3 mm for exterior exposure. However, because FRP rods are corrosion resistant, the maximum crack width limitation can be relaxed.

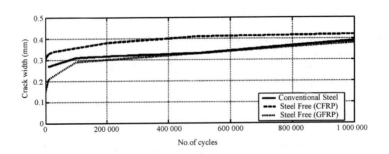

Figure 25 Plot of longitudinal crack width versus number of cycles

Results for 60-ton (588 kN) cyclic loading

Segment C was the first section of the bridge to be tested at a load level above 25 tons (222 kN). After completing 1,000,000 cycles without significant damage the load level was increased to 50 tons or 490 kN. After achieving an additional 1,000,000 cycles without failure, the load level was then increased to 60 tons or 588 kN. Segment C failed at 420,648 cycles. The deflection behaviour of Segment C is plotted against increasing number of cycles in Figure 26.

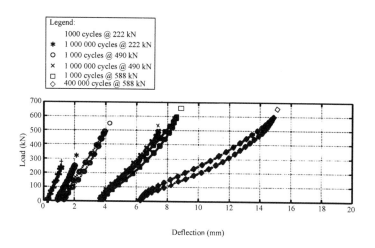

Figure 26　Plot of load versus deflection for steel-free deck with internal GFRP reinforcement

Due to the fact that Segment C did not fail at a load level of 50 tons or 490 kN after completing 1,000,000 cycles, it was determined that Segment A and Segment B would be subjected to 60 tons or 588 kN immediately after completing 1,000,000 cycles at 25 tons or 222 kN. Test results indicate that deck Segment A and deck Segment B failed at 23,162 cycles and 198,863 cycles, respectively, when subjected to a cyclic load of 60 tons or 588 kN. The deflection behaviour for deck Segment A and deck Segment B are shown in Figure 27 and Figure 28.

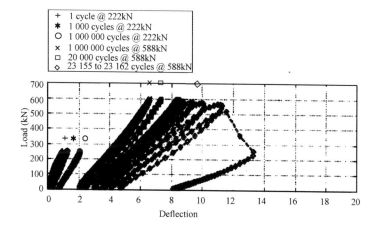

Figure 27　Plot of load versus deflection for deck with conventional steel reinforcement

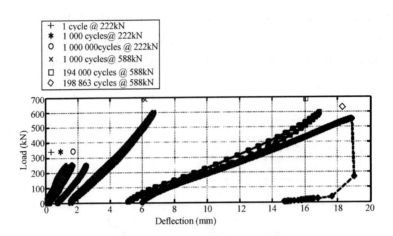

Figure 28　Plot of load versus deflection for steel-free deck with internal CFRP reinforcement

　　Bridge deck Segment C was tested under a 50 ton or 490 kN load level, and Segment A and Segment B were not. Therefore, in order to make a comparison between all three segments, the net deflection behaviour for deck Segment C at 50 tons or 490 kN was deducted or subtracted from the net deflection behaviour recorded under the 60-ton or 588 kN load level (Figure 29). Figure 30 and Figure 31 illustrate the deflection and crack width behaviour for all three bridge deck segments under the 60-ton or 588 kN load level. The results show that deck Segment A fatigued approximately twenty times as fast as deck Segment C, and deck Segment B fatigued approximately twice as fast as deck Segment C. All three segments failed in fatigue and via a punching shear failure (Figure 32).

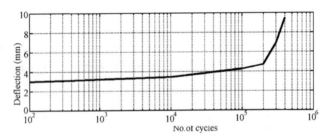

Figure 29　Net Deflection of steel-free deck with internal GFRP reinforcement at 60 tons

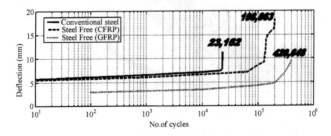

Figure 30　Plot of deflection versus number of cycles at 60 tons

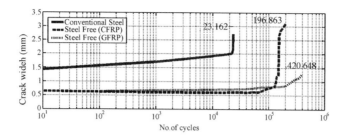

Figure 31　Plot of crack width verses number of cycles at 60 tons

Figure 32　Punching modes of fatigue failure for Segment A, Segment B and Segment C

Static Investigation

Bridge deck details

Construction of a full-scale second generation steel-free bridge deck including the implementation of a complete civionics and structural health monitoring system was completed in March 2004. The bridge deck is a slab on steel girder superstructure comprised of a continuous deck supported by two steel girders. It measures 9000 mm in length and 5000 mm in width (Figure 33). The thickness of the deck is 200 mm, and it contains haunches over the girders that measure 75 mm in depth. The spacing of the girders is 2500 mm center to center. The internal panel is a second generation steel-free design and utilizes 25 mm by 50 mm steel straps spaced at 1200 mm center to center. A bottom mat of GFRP is incorporated in the internal panel to control cracking at service loads. The transverse and longitudinal crack control reinforcement consists of #3 Pultrall V-Rod spaced at 200 mm and 600 mm center to center, respectively. A GFRP crack control grid was chosen for the bridge deck because GFRP had the highest fatigue resistance when compared to CFRP and conventional steel reinforcement (Memon et al. 2004). The top mat of reinforcement required for the cantilevers is divided into three 3000 mm sections in order to provide a comparison between GFRP, CFRP and steel when the deck is subjected to destructive static and fatigue testing. Each of the three sections was designed to have the same nominal moment capacity. The three sections use #10 GFRP Pultrall V-Rod for longitudinal reinforcement spaced at 600 mm center to center. The transverse negative moment reinforcement used two #19 GFRP and two #13 CFRP Pultrall V-Rod spaced at 200 mm center to center in the steel-free portion of the bridge deck, and one 20M steel reinforcing bar spaced at 200 mm center to center for the section of the deck with conventional cantilever reinforcement. Occasionally on bridge deck rehabilitation projects, such as the Red River Bridge, existing studs on the steel supporting girders have insufficient length to fully ensure composite action between the deck and the girders. For such cases, it may be more economical to extend the depth of stud penetration into the deck through the use of additional steel reinforcement called top hats, rather than completely removing existing studs and replacing them with new ones. As a means of investigating such design alternatives, short steel studs were welded to the girders and additional top hat reinforcement comprised of 10M steel bars at 200 mm center to center with alternate top hats doubled up were placed in the haunches over the girders (Figure 34).

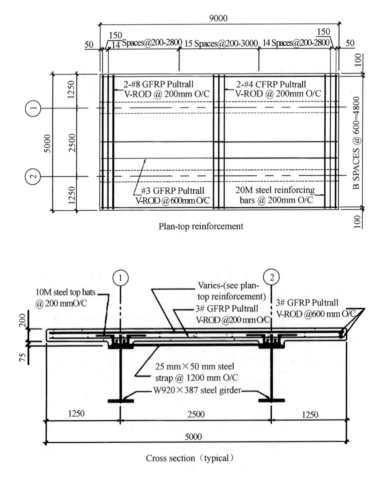

Figure 33 Bridge deck details

Figure 34　Steel top-hat reinforcement

Internal panel load deflection behaviour

The static ultimate capacity of the deck's internal panel when loaded monotonically until failure was 980 kN and the maximum deflection at the time of punching failure was recorded to be 16. 2 mm. The theoretical ultimate load and deflection determined using a computer program called PUNCH were 1093 kN and 19. 8 mm, respectively. Researchers were able to successfully predict the ultimate load of the deck slab to within 11. 5 percent. However, the predicted maximum deflection was 3. 6 mm or 22. 2 percent greater than that observed during the test (Figure 35).

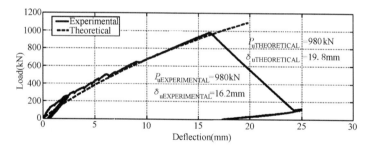

Figure 35　Plot of load versus deflection for internal panel static test

Internal panel load strain behaviour

The strain in the straps at failure was determined to be 1483 micro-strain and 1495 micro-strain in straps one and two, respectively (Figure 36). The strains in the straps were approximately 74 percent of their yielding strength at the time of failure. A computer program called PUNCH, developed by Newhook and Mufti (1998), was used to determine the theoretical strains in the straps. The theoretical ultimate strain in the straps was found to be 1540 micro-strain, approximately 3. 4 percent greater that that observed during the test.

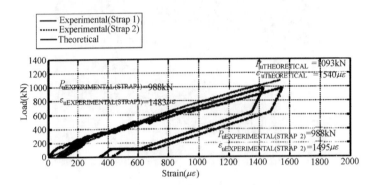

Figure 36 Plot of load versus strain of steel straps
for internal panel static test

Internal panel mode of failure

Test Location 1 exhibited a typical punching failure. Failure occurred at an ultimate load of 980 kN and a maximum deflection of 16.2 mm. A longitudinal crack was first apparent at a load of 250 kN and developed in length until failure. Radial cracks were visible at a load of 300 kN and continued to move outwards until failure. Punch angles along the north, east, and south axis were 18°, 20°, and 25°, respectively (Figure 37).

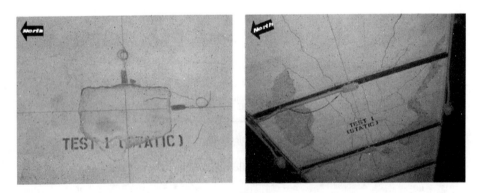

Figure 37 Top and bottom view of internal panel punch cones

Cantilever load deflection behaviour

The static results for the steel, CFRP and GFRP reinforced cantilevers are presented in this section of the paper. The results are showing that cantilevers may not be behaving the way in which they were designed. If we assume that the wheel load is distributed at a 45° (angle from the loading plate to the center line of the girder, and compute the moment resistance based on strains in the top transverse reinforcing bars, we can calculate an ultimate load that is approximately only two thirds of the ultimate load determined from static testing (CHBDC 2000). The preliminary results indicate that there may be some arching action taking place in the cantilevers. However, researchers at the University of Manitoba are currently working with a finite element program developed for steel-free bridge decks as well as commercially available finite element software to gain a better understanding into the mode of failure of the three cantilever sections.

The static ultimate capacity of the cantilever section reinforced CFRP when loaded monotonically to failure was 287 kN and the maximum deflection at the time of failure was 35.9 mm. The second test on the cantilever was carried out on April 4, 2005, and consisted of testing the cantilever section with steel reinforcement. The ultimate capacity of the cantilever was found to be 301 kN and the maximum deflection was recorded to be 25.8 mm. The cantilever reinforced with GFRP was tested on June 21, 2005. The ultimate load was 294 kN and the maximum deflection was 29.3 mm (Figure 38).

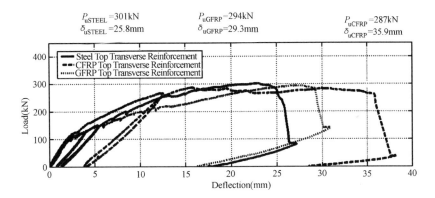

Figure 38 Plot of load versus deflection for steel, CFRP, and GFRP cantilever static tests

Cantilever load strain behaviour

Strain gauges were placed at various locations along the length of the top transverse reinforcing bars for all three sections of the cantilever to measure tensile strains throughout the test. The results discussed in this section only deal with the average of the gauges placed on the two transverse bars that were directly under the loading plate over the centerline of the girder. The average strain in the 20M deformed steel reinforcing bars at the time of failure was 1478 micro-strain, approximately 74% of their yielding strength (Figure 39). The average strain in the CFRP and GFRP reinforcing bars at the time of failure was 1863 and 3377 micro-strain, respectively. The strain in the CFRP bars was approximately 20% of their ultimate strain and the strain in the GFRP bars was approximately 21% of their ultimate strain.

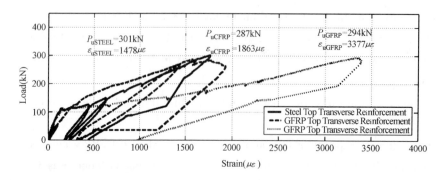

Figure 39   Plot of load versus strain for steel, CFRP, and GFRP cantilever static tests

Cantilever modes of failure

All three cantilevers exhibited a punching type failure similar to that of an internal panel steel free bridge deck. The shape of the punch cone was approximately a half circle compared to a full circle punch cone typical of an internal panel. Failure of the cantilever reinforced with traditional steel top transverse reinforcing occurred at an ultimate load of 301 kN. An ultimate load 287 kN and 294 kN was measured for the cantilevers reinforced with CFRP and GFRP, respectively. Longitudinal cracks over the girder were apparent at a load of 100 kN and continued to grow in length and width until failure. A transverse crack, extending from the steel girder to the free edge of the cantilever, developed on the underside of all three cantilevers at approximately 130 kN and grew in width until failure. Circumferential cracks started to develop at a load of approximately 200 kN and additional circumferential cracks developed as the load was increased moving inwards towards the loading plate. Punch angles along the east and south axis of the steel reinforced cantilever were 14° and 28°, respectively (Figure 40). The punch angle along the east axis was determined to be 7°, and 27° along the south axis for the cantilever section reinforced with CFRP (Figure 41). Punch angles along the east and south axis of the GFRP reinforced cantilever were 9° and 35°, respectively (Figure 42).

Figure 40   Top and bottom view of steel cantilever punch cones

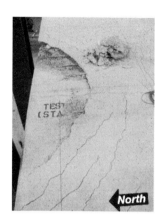 

Figure 41 Top and bottom view of CFRP cantilever punch cones

 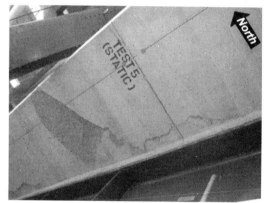

Figure 42 Top and bottom view of GFRP cantilever punch cones

Static and Fatigue Relationship

Due to the limitations of performing full-scale tests in a laboratory, only the bridge deck outlined earlier in this paper was subjected to cyclic loading conditions. In the absence of static loading tests, all three segments of the deck were analyzed using a computer program called PUNCH (Newhook and Mufti, 1998). The ultimate capacity of the steel-free deck with internal GFRP reinforcement was determined to be 1145 kN or 117 tons and the static load deflection behaviour predicted by the program is shown in Figure 43. The figure illustrates how the deflection under cyclic loading conditions is much higher than that predicted by the static monotonic loading conditions. Cyclic loading of variable amplitudes results in cumulative damage that reduces the capacity of the deck for further load distribution. Similar results were found by Oh et al. (2005). In order to verify deflections under monotonic and cyclic loading conditions, the bridge deck outlined in the previous section of this paper was constructed; however, fatigue data for the bridge deck is not yet available.

Based on the limited fatigue tests outlined in section 2 of this paper, a method for estimating the fatigue strength is suggested by Memon (2005). According to this method, the number of cycles to fatigue failure can be predicted by:

$$n = 10^5 \left[\frac{1}{P/P_u} - 1 \right]^{0.5} \qquad (5)$$

where n is the number of cycles to failure, P is the magnitude of the applied load, and P_u is the ultimate static capacity of the bridge deck. Further refinement is expected as additional fatigue data become available.

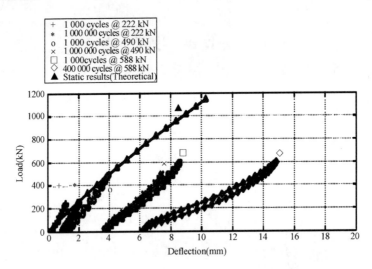

Figure 43　Plot of static (theoretical) and fatigue load versus deflection for steel-free deck with internal GFRP reinforcement

A bridge deck containing an internal CFRP or GFRP crack control grid and external steel straps prevents the growth of longitudinal crack widths and eliminates corrosion completely. Experimental results suggest that the area of CFRP or GFRP can be reduced by up to 40% (Memon and Mufti 2004). Experimental results indicate that fatigue damage induced under a 25-ton cyclic load is within permissible limits, and that all three bridge deck segments were subjected to 1,000,000 cycles under a 25-ton cyclic load without significant damage. The results show that all three segments of the bridge deck failed in fatigue under a 60-ton or 588 kN cyclic load. Conventional steel reinforcement failed after completing 23,162 cycles. The steel-free deck with CFRP internal reinforcement failed after completing 198,863 cycles and the steel-free deck with internal GFRP reinforcement failed after completing 420,648 cycles at 60 tons. The second generation steel-free bridge deck with external steel straps and internal GFRP reinforcement for crack control provides the best fatigue performance and is an efficient, economical and corrosion free system for bridge deck superstructures. The ability to predict ultimate loads to within an accuracy of 11.5 percent indicates that the static behaviour of a second generation steel-free bridge deck is well known, however, the ability to predict maximum deflections is less accurate. The preliminary test results from the static testing of the different cantilever sections suggests that cantilevers are not behaving totally the way in which they were designed. Flexural theory predicts an ultimate load of approximately two thirds of the ultimate load observed, indicating that there is some arching action taking place in the cantilevers. Researchers at the University of Manitoba are currently looking at finite element methods to gain a better understanding into the mode of failure of the different cantilever sections.

CIVIONICS

CIVIONICS (Mufti, 2003b) is a new term coined from the integration of Civil-Electronics, which is derived from the application of electronics to civil structures. It is similar to the term Avionics used in the Aerospace Industry. For structural health monitoring to become part of civil structural engineering, it should include the new discipline of Civionics, which requires the development of formal definitions and guidelines to encourage its use as an academic discipline in the curriculum of Canadian universities and technical colleges.

Realistically, it is true that consulting engineers and contractors will only invest in the development of the expertise created by graduates of the Civionics discipline when they can be assured that the prospects for business are good in this field. The ISIS Canada experience of integrating FOSs and FRPs into innovative structures that have been built across Canada demonstrates that these opportunities do exist. Following are some business opportunities that have been identified, as well as some recommendations for the Civil Structural Engineering Industries to develop the opportunities that currently exist.

Figure 44 Civionics Specifications Manual

In Canadian Universities and Technical Colleges, research, academic and training courses are established or are in process of being developed. It is intent of ISIS Canada that future generations of structural engineers and technologists will be familiar with the requirements of industries related to Civionics. To that end, ISIS Canada has developed Civionics Specifications (Rivera et al., 2004), which is being utilized to assist in the training of highly qualified personnel (Figure 44).

Civionics Application in a Second Generation Steel Free Bridge Deck

In the spring of 2004 research engineers at the University of Manitoba constructed a full-scale second generation steel free bridge deck. The bridge deck is the first of its kind to fully incorporate a complete civionics structural health monitoring system to monitor the deck's behaviour during destructive testing.

Throughout the construction of the bridge deck the entire installation of the civionics system was carried out by research engineers to simulate the actual implementation of such a system in a large scale construction environment. One major concern that consulting engineers have raised is the impact that a civionics system that uses conduit, junction boxes, and other electrical ancillary protection, will have when embedded and installed externally on full-scale infrastructure. The full-scale destructive testing of the second generation steel free bridge deck that utilizes a civionics system designed and implemented using the Civionics Specifications Manual at the University of Manitoba will provide engineers with the information necessary to address the constructability and structural integrity issues.

The complete civionics system for the test bridge deck at the University of Manitoba includes rigid PVC conduit and junction boxes to provide protection for the network of sensors embedded within the deck and all of the sensors installed externally on the steel straps and steel bridge girders. The placement of the system of conduits and junction boxes is important because it should minimize its effect on the structural integrity of the bridge deck, as well as minimize the amount of conduit and exposed cable within the deck. In this case, the conduits were placed within the haunches of the deck (Figure 45).

Figure 45 Rigid PVC conduit embedded with bridge deck haunches

The sizing of the conduit was determined using the Civionics Specifications. The conduit size and junction boxes required for this project were 38 mm in diameter. It should be noted that for research purposes this bridge deck was equipped with several additional sensors and that the conduit was sized accordingly. A civionics system for an actual steel free bridge deck would contain a minimal number of embedded sensors, thus reducing the amount and size of conduit required. For a typical installation on an actual bridge the conduit would be placed by the local electrical contractor and all of the necessary wires would be terminated at the junction boxes. A short time prior to the casting of the bridge deck, and after all major placement of reinforcement has

Figure 46 Protected sensors complete with properly located and secured lead wires

been completed, qualified personnel would be required to install the specified sensors and make the appropriate connections at the junction box to the wires already supplied and installed by the electrical contractor. Field installation of the sensors requires the application of several protective coatings and the exposed lead wires should be located in such a manner as to prevent damage during placement of concrete (i. e. wires should be placed on the underside of the reinforcement and properly secured - Figure 46). The sensors embedded within the bridge deck on the reinforcing bars allow engineers to monitor live load stresses during and after construction.

Surface mounted sensors located on the external steel straps and on the steel girders also provide engineers with pertinent information relating to the overall performance of a bridge. However, these sensors must also be properly protected from vandalism and local environmental conditions. Therefore, a system of rigid PVC conduits and junction boxes were permanently installed on the steel girders and steel straps. The conduit and lead wires would be installed by an electrical contractor after all of the formwork has been removed. The wires must be properly labelled to ensure that the correct locations of all sensors are known at a later date. Once again, qualified personnel perform the installation of the necessary gauges within the PVC boxes and splice all lead wires.

The size of the conduit and junction boxes on the straps were 25 mm in diameter to provide enough space in the junction box to install the sensor. The back of the PVC junction box had to be removed in order to facilitate surface mounting of the sensors (Figure 47). Strain gauges installed on the straps will allow engineers to monitor live load strains. Flexible conduit measuring 19 mm in diameter was used with the junction boxes on the top and bottom flanges of the steel girder (Figure 47). The purpose of the gauges installed on the girders is to determine the live load stress ranges and to determine the transverse load distribution pattern. Also, the gauges will allow engineers at the University of Manitoba to determine exactly if placement of the conduit within the haunches has any significant effect on the composite action between the girders and the deck.

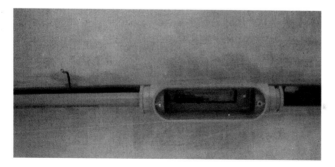 

Figure 47   Rigid PVC Conduit and Junction Boxes for Externally Installed Gauges

STRUCTURAL HEALTH MONITORING DATA ON WEB SITE

For the structures that have been instrumented with remote monitoring capabilities, data is constantly collected at varying frequencies. As the data grows daily, so does the need for a sophisticated archival management system. It becomes one of the most vital aspects of structural monitoring. ISIS is developing an on-line archiving system whereby authorized researchers submit raw data that will be accessible to users. In a user-friendly, world wide web interface, the site will offer access to sensor characteristics and locations, and response measurements from static and dynamic load tests. The archive will enable interested parties to browse the content, view the relevant documentation and down load data for analysis.

Several different strategies are being used for data collection. In some cases, the sensors are triggered manually at selected intervals of time, and the sensor data is stored on computer disks. In other cases sensor data is collected continuously at a specified scan rate and automatically transmitted to a remote location, where it is stored on a computer.

STUDIES OF CONCRETE REINFORCED WITH GFRP SPECIMENS FROM FIELD DEMONSTRATION PROJECTS

The methods used in this study (Mufti et al., 2005) to investigate the degradation of GFRP reinforced concrete are Scanning Electron Microscopy (SEM) and Energy Dispersive X-ray (EDX), Light Microscopy (LM), Differential Scanning Calorimetry (DSC) and Infrared Spectroscopy. To obtain reliable information from the results/observations using such methods, special attention was given to sample preparation. During specimen preparation, the glass fiber can be scratched and microcracks can be induced into the matrix and concrete; the glass fiber can be debonded and the glass and matrix polished surfaces can be contaminated with elements from each other and with elements from the concrete. When such events take place the interpretation of the results/observations become laborious.

Details on sample preparation, presentation of the results obtained on the entire set of GFRP reinforced concrete specimens removed from the five structures and an extended discussion of the results will be presented in the ISIS Monograph on GFRP durability and at the 3rd International Conference on Composites in Construction 2005 in Lyon, France (Mufti et al., 2005).

The following is a brief excerpt of the main findings suggested by the results obtained to date from the analysis performed by Onofrei (2005) at the University of Manitoba on the randomly selected core specimens from the field demonstration projects using the SEM and EDX analyses. Examples of SEM micrographs and EDX spectra on GFRP specimens cored from the Joffre Bridge and Hall's Harbor Wharf as well as for a set of control (unexposed specimens) GFRP specimens are presented in Figure 48 (a) and (b). It should be noted that, although the entire surface of each specimen was examined, the attention was focused on areas close to the concrete/GFRP rod interface, where any degradation could be expected to take place, from which the micrograph presented in Figure 48 (a) and (b) has been taken.

Data from SEM analyses suggests that that there are no visible signs of any degradation taking place on the reinforced bars/grids. The individual fibers are intact with no gaps between fibers and the matrix (Figure 48 (a)). There is also good contact between the concrete and the matrix (Figure 48 (b)).

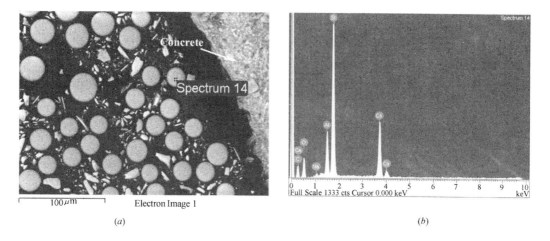

Figure 48   SEM photomicrograph
(a) the cross section of core specimen from Joffre Bridge. EDX spectrum,
(b) shows the chemical composition of the glass fiber- magnification factor is 500X

It can be concluded that for the range of conditions of field demonstration projects included in this study there is no degradation of the GFRP reinforcement. Since the pH of the concrete pore water solution is expected to decrease further with time, it is quite probable that for all practical problems the degradation of GFRP can be considered to be insignificant.

The results to date obtained in this study do not support the results obtained in the laboratory studies that report GFRP degradation in various degrees. There are several reasons for this disagreement. First, in the present study the specimens under investigation were from real life engineering structures, whereas in laboratory studies that observed GFRP degradation the conditions were very different from the real systems. The lab studies are usually conducted in alkaline solutions that are kept at constant high pH values and/or high temperatures, the GFRP is generally kept in contact with infinite leachant volume or a very large effective surface area of the road is exposed to the solution. With the exception of special projects, such conditions do not really exist in real life. For most practical engineering projects, it is quite probable that the degradation of GFRP reinforced concrete in most cases is not significant. The full details will be published in an ISIS Canada Technical Report in 2005.

CONCLUSIONS AND RECOMMENDATIONS

As mentioned earlier, ISIS Canada intends to significantly change the design and construction of civil engineering structures. For changes in design and construction to be accepted, it is necessary that innovative structures be monitored for their health. To assist in achieving this goal, ISIS Canada is developing a new discipline, which integrates Civil Engineering and Electrophotonics under the combined banner of Civionics.

The new discipline of Civionics must be developed by Civil Structural Engineers and Electrophotonics Engineers to lend validity and integrity to the process. Civionics will produce engineers with the knowledge to build "smart" structures containing the SHM equipment to provide much needed information related to the health of structures before things go wrong. This discipline will, thereby, assist engineers and others to realize the full benefits of monitoring civil engineering structures.

ACKNOWLEDGMENTS

The authors acknowledge financial support of ISIS Canada, NSERC and NCE for the research worked conducted. In particular we wish to acknowledge collaboration of Dr. Baidar Bakht, Dr. Gamil Tadros, Dr. Amjad Memon and Dr. John Newhook on the development of the Corrosion Free Bridge Decks concept that is incorporated in the Winnipeg Principles.

In addition, the assistance of Ms. Nancy Fehr, Executive Assistant to the President of the I-SIS Canada Research Network for editing and improving the structure of the paper is gratefully acknowledged.

REFERENCES

ACI 440.1R-01. 2001. Guide for the design and construction of concrete reinforced with FRP bars. Reported by ACI Committee 440, USA.

Canadian Highway Bridge Design Code (CHBDC). 2000. Canadian Standards Association International, Toronto, Ontario, Canada

Demers, M., Popovic, A., Neale, K., Rizkalla, S. and Tadros, G. (2003). FRP Retrofit of the Ring-Beam of a Nuclear Reactor Containment Structure", ACI Special Publication, SP-215: 303-316.

Klowak, C. (2005). The Design, Construction, and Structural Health Monitoring of an Innovative Second Generation Steel-Free/FRP Hybrid Bridge Deck. M.Sc. thesis to be submitted to the University of Manitoba-in progress. Winnipeg, Manitoba, Canada.

Memon A. H., Mufti A. A. and Bakht, B. (2003). Crack Control with GFRP Bars in Steel-Free Concrete Deck Slab, Proceedings of the Canadian Society for Civil Engineering Annual Conference. Moncton, New Brunswick, Canada.

Memon, A. H. (2005). Fatigue behaviour of steel-free concrete bridge deck slabs under cyclic loading. Thesis, University of Manitoba, Winnipeg, Manitoba, Canada.

Memon, A. H. and Mufti, A. A. (2004). Fatigue behavior of second generation steel-free concrete bridge deck slab. Proceedings of the Second International Conference on FRP Composites in Civil Engineering (CICE), Adelaide, Australia.

Memon, A. H., Mufti A. A. and Bakht, B. (2004). Crack control with GFRP bars in steel-free concrete deck slab. Proceedings of the 2003 Canadian Society for Civil Engineering Annual Conference, Moncton, New Brunswick, Canada.

Mufti, A. A. (2003). Structural Health Monitoring of Innovative Bridge Decks, Keynote Lecture, Proceedings for the International Workshop on Structural Health Monitoring of Bridges / Colloquium on Bridge Vibration. pp. 55-62. Kitami, Japan.

Mufti, A. A. and Bakht, B. (1996). Steel Free Fibre Reinforced Concrete Bridge Decks: New Construction and Replacement, APWA International Public Works Congress, NRCC/CPWA Seminar Series-Innovations in Urban Infrastructure, pp. 63-71.

Mufti, A. A., (2003). "Structural health Monitoring of Manitoba's Golden Boy" Canadian Journal of Civil engineering.

Mufti, A. A, Tadros, and Bakht, B. 2003. Design Calculations for Steel free Deck for Red River Bridge. Report Submitted to EarthTech Consultants. Winnipeg, Manitoba, Canada.

Mufti, A. A., (2003b). "Integration of Sensing in Civil Structures: Development of the New Discipline of Civionics", Proceedings for the First International Conference on Structural Health Monitoring and Intelligent Infrastructure (SHMII-1 2003), pp. 119-129, Tokyo, Japan.

Mufti, A. A., Memon, A. H. and Klowak, C. (2005). Study of static and fatigue behaviour of second generation bridge decks. Proceedings of the International Workshop on Innovative Bridge Deck Technologies, Winnipeg, Manitoba, Canada.

Mufti, A. A., Neale, K., Rahman, S. and Huffman, S. (2003). "GFRP Seismic Strengthening and Structural Health Monitoring of Portage Creek Bridge Concrete Columns", Proceedings for the fib2003 Symposium-Concrete Structures in Seismic Regions, Athens, Greece.

Mufti, A. A., Onofrei, M., Benmokrane, B., Banthia, N., Boulfiza, M., Newhook, J., Tadros, G., Bakht, B. and Brett, P. (2005). "Studies of Concrete Reinforced with GFRP Specimens from Field Demonstration Projects", Proceedings for the Third International Conference on Composites in Construction (CCC 2005). Lyon, France.

Newhook, J. P, Bakht, B., Tadros, G. and Mufti, A. A. 2000. Design and construction of a concrete marine structure using innovative technologies. ACMBS-III Conference Proceeding, pp. 777-784. Ottawa, Ontario, Canada.

Newhook, J. P. and Mufti, A. A. (1998). Punch Program User Manual. Nova Scotia CAD/CAM Centre, Dalhousie University. Halifax, Nova Scotia.

Newhook, J. P. and Mufti, A. A. 2000. A reinforcing steel free concrete bridge deck for the Salmon River Bridge. Concrete International. Vol. 18 (6): 30-34.

Newhook, J. P. and Mufti, A. A., (2000). "A Reinforcing Steel Free Concrete Bridge Deck Slab for the Salmon River Bridge", Concrete International, Vol. 18 (6): 81-90.

Newhook, J. P., Bakht, B., Mufti, A. A. and Tadros, G. (2001). "Structural Health Monitoring of Innovative Bridge Decks", Proceedings, Annual Conference of the International Society for Optical Engineering, held in Newport, California, USA.

Oh, H., Sim, J. and Meyer, C. 2005. Fatigue life of damaged bridge deck panels strengthened with carbon fibre sheets. ACI Structural Journal, Vol. 102 (1): 85-92.

Rivera, E., Mufti A. A. and Thomson, D. (2004). Civionics Specifications, Design Manual, ISIS Canada Research Network, Winnipeg, Manitoba, Canada.

Rizkalla, S. H. and Tadros, G., (1994): "A Smart Highway Bridge in Canada", Concrete International, pp. 42-44.

Tennyson, R., (2001). "Installation, Use and Repair of Fibre Optic Sensors", ISIS Canada Research Network, Winnipeg, Manitoba, Canada.

www.isiscanada.com. (2003). ISIS Canada Research Network Web Site, Winnipeg, Manitoba, Canada.

第十章 | Chapter 10

STRUCTURAL APPLICATIONS OF FRP COMPOSITES IN CONSTRUCTION

J. G. TENG

Department of Civil and Structural Engineering
The Hong Kong Polytechnic University, Hong Kong, China.
E-mail: cejgteng@polyu.edu.hk

Abstract: Fibre-reinforced polymer (FRP) composites have found increasingly wide applications in construction. The most popular area for the application of FRP composites has been the strengthening and retrofit of deficient structures, but the new construction area has also attracted increasing attention. The main aim of this chapter is to provide a summary of recent advances in three major areas: (a) strengthening of concrete structures, (b) strengthening of steel structures, and (c) FRP-confined columns, with particular attention to recent research undertaken at The Hong Kong Polytechnic University (PolyU). The first area is discussed with a particular emphasis on the research underpinning some of the key design provisions in the draft Chinese Code for the Structural Use of FRP Composites in Construction. The second area is a more recent area compared to the first, but intensive research is now being conducted around the world. The third area is of great importance to the application of FRP in new construction.

Keywords: Fibre-reinforced polymer, strengthening, new construction, state of the art, design methods, research needs

INTRODUCTION

Structural engineering innovations have been strongly influenced by the introduction of new structural materials. The development and application of high-performance, durable, and intelligent structural materials form an important part of the emerging technologies for third-generation structures (Teng et al. 2003a). Over the past decade, fibre-reinforced polymer (FRP) composites have gradually become accepted as a new structural material, due to their superior properties. FRP composites are formed by embedding continuous fibres in a polymeric resin matrix which binds the fibres together. Common fibres used in FRP composites include carbon, glass, and aramid fibres while common resins are epoxy, polyester, and vinyl ester resins. The most widely used FRP composites are glass fibre-reinforced polymer (GFRP) composites, carbon fibre-reinforced polymer (CFRP) composites, and aramid fibre-reinforced polymer

(AFRP) composites. More recently, the application of basalt fibre-reinforced polymer (BFRP) composite has attracted increasing attention, and BFRP is expected to become an important addition to the FRP family in the near future. A useful general background on the composition of these materials and their mechanical properties can be found in ACI-440 (1996) and Hollaway and Head (2001).

Compared to traditional structural materials, FRP composites possess the following advantages: (1) high strength-to-weight ratios. The tensile strength of CFRP can be more than 10 times that of common carbon steel, but its density is only approximately one quarter of that of steel; (2) excellent resistance to corrosion. In particular, CFRP shows excellent resistance to environmental attacks of all types and to fatigue loading; (3) flexibility in design. Desirable macroscopic properties and geometric shapes can be achieved through suitable designs of fibre orientations, layout and volume ratios; (4) ease for structural health monitoring. Embedding sensors such as fibre optical sensors in FRP composites is both convenient and reliable. Due to these and other advantages, FRP composites have a great potential for applications both in the retrofit of structures and in the construction of new structures.

FRP composites also have some significant weaknesses: (1) their stress-strain behaviour is linear elastic, followed by brittle failure (Figure 1). As a result, structures incorporating FRP tend to fail in a brittle manner. The current theoretical framework for structural design based on the availability of a sufficient degree of material ductility is no longer applicable, and the development of a new theoretical framework for structural design is urgently needed; (2) FRP composites possess very limited resistance to high temperatures. Although this drawback is not a significant issue of concern for outdoor structures such as bridges, it significantly limits the scope of application of FRP composites in buildings; (3) high material cost. The cost effectiveness of FRP composites in structural applications depends strongly on whether cost savings can be made through ease of construction, shortened construction time, reduction of structural self-weight, reduced labour cost and reduced maintenance cost. It is also reasonable to expect that as FRP composites become more widely used, the economy of scale will come into play so that the material cost of FRP will come down and their structural use will become more economically competitive.

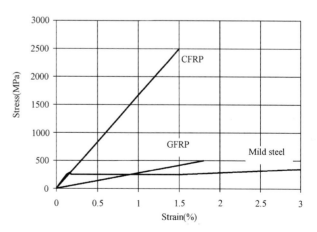

Figure 1 Typical FRP and mild steel stress-strain curves

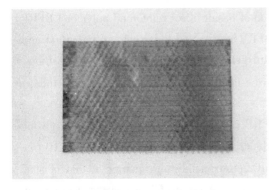

(a) Carbon fibre sheet for wet lay-up application (b) CFRP pultruded plate

Figure 2   Typical composite products for strengthening applications

The structural applications of FRP composites in construction fall into the following areas: (1) strengthening (or retrofit) of existing structures; (2) reinforcement or prestressing of concrete structures; (3) Hybrid structures of FRP and other structural materials; and (4) All-FRP structures. Available FRP products include: (a) laminates (both in-situ wet lay-up and prefabricated) (Figure 2), mainly for use in the strengthening of structures; (b) bars (Figure 3), mainly for the reinforcement and prestressing of concrete structures and as near-surfaced mounted reinforcement in the strengthening of structures; (c) cables, to replace steel cables in prestressing applications and as support cables in large span roofs and bridges (Figure 4). (d) FRP profiles such as FRP tubes (Figure 5) and FRP I-sections, mainly for use in hybrid structures and all-FRP structures. The wet lay-up process involves the in-situ impregnation of dry fibre/fabric sheets with epoxy resin and the subsequent bonding of these impregnated plates/strips. A state-of-the-art survey of the entire field is given in Teng (2006), where key issues for future research are also identified. Readers are referred to Teng (2006) for a general understanding of the current status and future needs of the broad field of FRP composites in construction. For a comprehensive and rigorous treatment of the behaviour and design of FRP strengthening systems, readers are referred to Teng et al. (2002), whose Chinese translation has also been published by the China Architecture and Building Press in 2005.

Figure 3   FRP reinforcing bars

Figure 4   First FRP cable-stayed bridge in China
(Courtesy of Prof. J. W. Zhang, Southeast University)

第十章 Structural Applications of FRP Composites in Construction

Figure 5 FRP tube

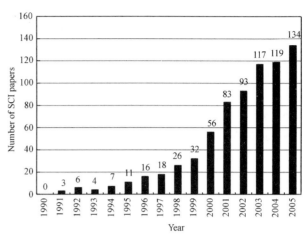

Figure 6 FRP in concrete structures: growth of SCI papers since 1990

This chapter aims to provide a summary of the latest research advances in the following three major areas of the broad field of FRP composites in construction: (a) FRP strengthening of concrete structures, (b) FRP strengthening of steel structures, and (c) FRP-confined columns, with particular attention to recent research at The Hong Kong Polytechnic University (PolyU), much of which was conducted in collaboration with other universities, particularly the University of Edinburgh and the University of Surrey in UK, the University of Lecce in Italy, and Tsinghua University, Jinan University and Zhejiang University in mainland China. The first area is discussed with a particular emphasis on the research underpinning some of the key design provisions in the draft Chinese Code for the Structural Use of FRP Composites in Construction (hereafter referred to as the draft Chinese code or the draft code for brevity). The second area is a more recent area compared to the first, so a detailed discussion of situations where this technique may be attractive is provided, followed by a summary of recent research at PolyU. The third area covers two types of FRP-confined columns: (a) concrete-filled FRP tubes, for which the design provisions in the draft Chinese code are discussed; and (b) FRP-concrete-steel double-skin tubular columns, which have recently been developed at PoyU.

BRIEF HISTORICAL NOTE

The recent surge of interest in the use of FRP in structures dates back to the early 1980s, but there had been occasional use of the material well before 1980s (ACI-440 1996; Hollaway and Head 2001; Bakis et al. 2002; Teng et al. 2002; Hollaway 2003). The activities in the area really took off only about a decade ago. Over the past decade, there has been a dramatic growth of activities in the area, both in terms of practical applications and research activities.

The number of journal papers published each year and indexed in the SCI database of the Institute for Scientific Information on the application of FRP in concrete structures is shown in Figure 6. These numbers were found from searches conducted using "FRP" and "concrete" as the keyword combination. By nature of the key words used, these searches are believed to have excluded most papers on non-concrete applications of FRP; the total number of SCI papers published each year on all types of FRP applications in structures can be expected to be much greater. The exponential growth of SCI papers in the past 10 years illustrates clearly the development of strong research activities.

As a result of this intensive research effort, design guides and codes have been developed by various countries (Teng 2006). At the time of preparing this chapter, the draft Chinese Code for the Structural Use of FRP Composites in Construction is being finalized for public comments. This code is unique in that it covers a wide range of topics, including the strengthening of concrete and masonry structures, FRP-reinforced/prestressed concrete and hybrid FRP structures.

FLEXURAL STRENGTHENING OF CONCRETE BEAMS AND SLABS

General

A large amount of research has been conducted on the flexural strengthening of reinforced concrete (RC) beams and slabs. Much of this research has been conducted for beams such as that shown in Figure 7, where an FRP plate is bonded to the soffit of a simply-supported RC beam to enhance its flexural strength. Nevertheless, the research outcomes can, in principle, be applied to indeterminate beams by treating each segment between two adjacent points of inflection as a simply-supported beam. Therefore, only such beams are explicitly referred to in this section.

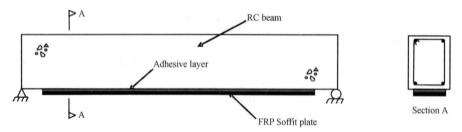

Figure 7 RC beam with an FRP plate bonded to its soffit

The design procedure for the flexural strengthening of RC beams using externally bonded FRP plates is easy to establish if debonding failures do not become critical (Teng et al. 2002). Therefore, an important focus of the existing and current research is on debonding failures of various forms (e.g. Teng et al. 2002, 2003b; Oehlers and Seracino 2004; Chen and Teng 2005), including the development of a better understanding of debonding failure mechanisms and the establishment of more rational and reliable design methods against debonding failures. Existing research has included a large amount of experimental work, but the associated theoretical work has been more limited. Research in this area is currently very active (Chen and Teng 2005) and rapid progress is being made. A more detailed treatment of debonding failures in flexurally-strengthened RC beams is given later in this section.

Near-surface mounted (NSM) FRP reinforcement is a new technique that has emerged over the last few years for the strengthening of concrete and masonry structures (De Lorenzis and Teng 2006). This method involves the grooving of the cover of concrete or masonry members and the subsequent embedding of FRP bars or strips into the grooves using a suitable binding material. The NSM method greatly reduces the likeliness of debonding failures of the FRP reinforcement, and represents an important extension of the method of externally bonded FRP reinforcement. Existing research on this technique is still very limited, so a great deal of further research is required before the method can be widely accepted in practice (De Lorenzis and Teng 2006). Similar to externally bonded FRP reinforcement, the behaviour of the interface between NSM FRP and concrete is a key issue for research (e. g. Teng et al. 2006a).

The FRP material may be prestressed before being bonded to concrete members for strengthening purposes. The strengthening of structures with prestressed FRP is an important technique, as prestressing leads to a fuller exploitation of the high tensile strength of FRP. Prestressing is particularly attractive when improvement to the serviceability of a structure is required. A key obstacle to the practical implementation of the prestressing technique is that it is difficult to prestress and anchor the FRP reinforcement on site. A number of attempts have been made to develop simple prestressing and anchorage methods (e. g. El-Hacha et al. 2001; Meier et al. 2001; Wu et al. 2003; Shawulieti et al. 2004), but the success has so far been limited and further work is needed.

In the remainder of this section, attention is focused on debonding failures in RC beams flexurally-strengthened with externally bonded FRP plates, as these failures are now reasonably well understood and are covered in the draft Chinese code.

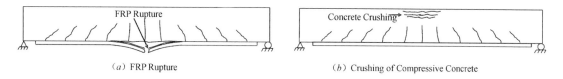

Figure 8 Conventional flexural failure modes of an FRP-plated RC beam

Failure Modes

General

A number of distinct failure modes for RC beams bonded with an FRP soffit plate (i. e. FRP-plated RC beams) have been observed in numerous experimental studies (Teng et al. 2002; Oehlers and Seracino 2004). Schematic representations of these failure modes are shown in Figures 8 and 9. Failure of an FRP-plated RC beam may be by the flexural failure of the critical section (Figure 8) or by debonding of the FRP plate from the RC beam (Figure 9). In the former type of failure, the composite action between the bonded plate and the RC beam is maintained up to failure, while the latter type of failure involves a loss of this composite action. Debonding failures generally occur in the concrete, which is also a common assumption in the existing design guidelines/codes. This is because, with the strong adhesives currently available and with appropriate surface preparation for the concrete substrate, debonding failures along the physical interfaces between the adhesive and the concrete and between the adhesive and the FRP plate are generally not critical.

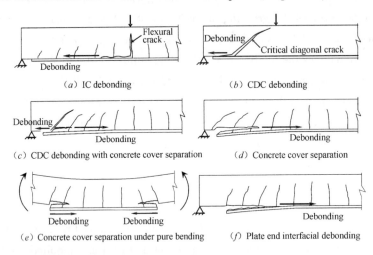

Figure 9 Debonding failure modes of an FRP-plated RC beam

Figure 10 FRP-plated RC beam: FRP rupture

Debonding may initiate at a flexural or flexural-shear crack in the high moment region and then propagates towards one of the plate ends (Figure 9a). This debonding failure mode is commonly referred to as intermediate crack (IC) induced interfacial debonding (or simply IC debonding) (Teng et al. 2002, 2003c; Yao et al. 2005a; Lu et al. 2006). Debonding may also

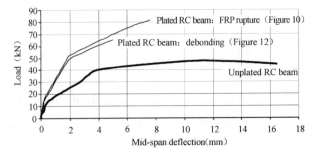

Figure 11 Typical load-deflection curves of plated and unplated RC beams

occur at or near a plate end (i. e. plate end debonding failures) in four different modes: (a) critical diagonal crack (CDC) debonding (Figure 9b) (Oehlers and Seracino 2004), (b) CDC debonding with concrete cover separation (Figure 9c) (Teng and Yao 2005), (c) concrete cover separation (Figures 9d and 9e) (Teng et al. 2002), and (d) plate end interfacial debonding (Figure 9f) (Teng et al. 2002).

Flexural failure modes of an FRP-plated RC beam section

The flexural failure of an FRP-plated RC section can be in one of two modes: tensile rupture of the FRP plate (Figures 8a and 10) or compressive crushing of the concrete (Figure 8b). These modes are very similar to the classical flexural failure modes of RC beams, except for small differences due to the brittleness of the bonded FRP plate. FRP rupture generally occurs following the yielding of the steel tension bars, although steel yielding may not have been reached if the steel bars are located far away from the tension face.

Figure 11 shows a typical load-deflection response of a simply supported RC beam bonded with an FRP plate subjected to four point bending (Hau 1999). In this particular beam, the plate was terminated very close to the support and the beam failed in flexure by FRP rupture. Compared to the corresponding response of a control RC beam, the strengthened beam achieved a strength increase of 76%, but showed much reduced ductility. The strength increase and the ductility decrease are

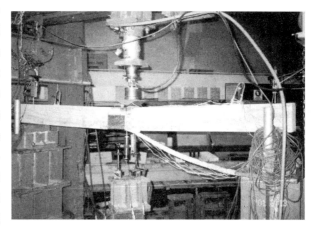

Figure 12 FRP-plated RC beam: IC debonding

the two main consequences of flexural strengthening of RC beams using FRP plates. FRP-strengthened beams which fail by concrete crushing when a large amount of FRP is used also show significantly reduced ductility (Buyukozturk and Hearing 1998).

IC debonding

When a major flexural or flexural-shear crack is formed in the concrete, the need to accommodate the large local strain concentration at the crack leads to immediate but very localized debonding of the FRP plate from the concrete in the close vicinity of the crack, but this localized debonding is not yet able to propagate. The tensile stresses released by the cracked concrete are transferred to the FRP plate, so high local interfacial stresses between the FRP plate and the concrete are induced near the crack. As the applied loading increases further, the tensile stresses in the plate and hence the interfacial stresses between the FRP plate and the concrete near the crack also increase. When these stresses reach critical values, debonding starts to propagate towards one of the plate ends, generally the nearer end.

A typical picture of flexural intermediate crack-induced debonding is shown in Figure 12. A thin layer of concrete remained attached to the plate (Figure 12), which suggests that failure occurred in the concrete, adjacent to the adhesive-to-concrete interface. IC debonding failures are more likely in shallow beams and are, in general, more ductile than plate end debonding failures.

Concrete cover separation

Concrete cover separation involves crack propagation along the level of the steel tension reinforcement. Failure of the concrete cover is initiated by the formation of a crack near the plate end. The crack propagates through the concrete cover and then along the level of the steel tension reinforcement, resulting in the separation of the concrete cover. As the failure occurs away from the bondline, this is not a debonding failure mode in strict terms, although it is closely associated with stress concentration near the ends of the bonded plate (Smith and Teng 2001). A typical picture of a cover separation failure is shown in Figures 13a and 13b shows a close-up view of the detached plate end, where the flexural tension reinforcement of the beam can be clearly seen. The cover separation failure mode is a rather brittle failure mode (Figure 11).

(a) overall view

(b) close-up view

Figure 13 FRP-plated RC beam: concrete cover separation

Plate-end interfacial debonding

A debonding failure of this form is initiated by high interfacial shear and normal stresses near the end of the plate that exceed the strength of the weakest element, generally the concrete. Debonding initiates at the plate end and propagates towards the middle of the beam (Figures 9f and 14). This failure mode is only likely to occur when the plate is significantly narrower than

Figure 14 FRP-plated RC beam: plate end interfacial debonding

the beam section, as otherwise, failure tends to be by concrete cover separation (i.e. the steel bars-concrete interface controls the failure instead).

CDC debonding

This mode of debonding failure occurs in flexurally-strengthened beams where the plate end is located in a zone of high shear force but low moment (e.g. a plate end near the support of a simply-supported beam) and the amount of steel shear reinforcement is limited. In such beams, a major diagonal shear crack (critical diagonal crack, or CDC) forms and intersects the FRP plate, generally near the plate end. As the crack widens, high interfacial stresses between the plate and the concrete are induced, leading to the eventual failure of the beam by debonding of the plate from the concrete; the debonding crack propagates from the crack towards the plate end (Figures 9b and 15).

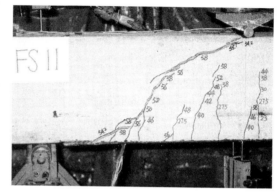

Figure 15   FRP-plated RC beam: CDC debonding

In a beam with a larger amount of steel shear reinforcement, multiple shear cracks of smaller widths instead of a single major shear crack dominate the behavior, so CDC debonding is much less likely. Instead, concrete cover separation takes over as the controlling debonding failure mode. In some cases, particularly when the plate end is very close to the zero-moment location, CDC debonding leads only to the local detachment of the plate end, but the beam is able to resist high loads until cover separation occurs (Figure 9c). The local detachment due to CDC debonding effectively moves the plate end to a new location with a larger moment, and cover separation then starts from this "new end". The CDC failure mode is thus related to the cover separation failure mode. It is surmised that if a flexurally-strengthened beam is also shear-strengthened with U-jackets to ensure that the shear strength remains greater than the flexural strength, the CDC debonding failure mode may be suppressed.

Other aspects of debonding

The risk of debonding is increased by a number of factors associated with the quality of on-site application. These can include poor workmanship and the use of low-quality adhesives. The effects of these factors can be minimized if due care is exercised in the application process to ensure that debonding failure is controlled by the concrete. In addition, small unevenness of the concrete surface may cause localized debonding of the FRP plate but such localized debonding is unlikely to propagate over a large portion of the plate.

Strength Models for Debonding Failures
IC debonding

Over the past few years, a significant number of studies have been concerned with IC debonding in FRP-strengthened RC beams (e. g. Teng et al. 2003c). More recently, intensive research at PolyU has led to a new IC debonding strength model. This recent research included a large number of tests (Yao et al. 2005a, 2005b), new models for the bond behaviour (Yuan et al. 2004; Lu et al. 2005a, 2005b), a new smeared crack approach for the finite element simulation of the IC debonding process (Lu et al. 2006), and an IC debonding strength model based on interfacial shear stress distributions from finite element analysis (Lu et al. 2006). The finite element model was shown to be accurate through comparisons with the results of a large number of beam tests, and the new IC debonding strength model represents significant improvements to existing models. Work is currently underway to develop an alternative strength model for IC debonding failures (Teng et al. 2006b; Chen et al. 2006a). Preliminary assessment of this alterative approach shows it to be promising (Chen et al. 2006b).

Plate end debonding failures

Many strength models have been proposed for plate end debonding failures; Smith and Teng (2002a, 2002b) reviewed and assessed those proposed prior to their work. In the years that followed, additional models have been published, including Colotti et al. 's (2004) model covering debonding as well as other failure modes, Gao et al. 's (2005) model for concrete cover separation failures, and Oehlers et al. 's (2004, 2005) model which was intended for CDC debonding failures. All of these three new models, particularly the first two, are significantly more complex than are desirable for design use. The accuracy of these models also needs improvement (Teng and Yao 2006).

A large number of tests were recently conducted at PolyU on plate end debonding failures (Smith and Teng 2003; Yao 2005; Yao and Teng 2006; Teng and Yao 2006). Based on existing knowledge and experimental observations, key parameters controlling the debonding failure load were identified and a debonding strength model was proposed (Teng and Yao 2005; 2006). In this model, pure flexural debonding of a plate end located in a pure bending region and pure shear debonding of a plate end located in a high-shear zero (or low) -moment region are first dealt with. The general case of a plate end under the combined action of shear and bending is treated as the interaction of these two extreme conditions. The proposed model reflects explicitly the contributions of both the internal steel shear reinforcement and the bonded plate to the debonding failure load and has been shown to be reasonably accurate through comparisons with available test results. Within the general framework of the proposed model, future refinements to improve its accuracy can be easily achieved when more results from tests and/or numerical predictions become available. The model relates the debonding failure load to the shear capacity of the beam and a number

Design Provisions in the Draft Chinese Code

In the draft Chinese code, design against IC debonding is based on the recent research by Lu et al. (2006) with modifications to cater for specific constraints imposed by the context of the draft code, which is explained in detail in Ye et al. (2005). No design calculations are provided for the other debonding failure modes; instead, U jackets are specified to suppress the other failure modes. The installation of U jackets specified in the draft code is based on limited evidence and shall be thoroughly examined in future research. As an alternative to U jackets, the design proposal given by Teng and Yao (2005, 2006) may be used to ascertain whether plate end debonding controls the strength of the beam. If plate end debonding is found to be non-critical, then limited U jacketing at the two ends of the plate shall still be adopted to improve the robustness of the strengthening system. If plate end debonding is found to be a critical mode, then the installation of suitably located U jackets is still needed. Methods for designing such U jackets are not yet available, but it appears that plate end debonding may not control the strength of the beam, if a small amount of U jacketing is employed and the provided U jackets are sufficient to ensure that the shear strength exceeds the flexural strength of the strengthened RC beam.

SHEAR STRENGTHENING OF RC BEAMS

General

FRP shear strengthening of RC beams has commonly been achieved through the external bonding of FRP composites. In addition, it can also be achieved using near-surface mounted FRP bars and strips (De Lorenzis and Teng 2006) or CFRP straps with prestressing (Lees et al. 2002). The latter methods are relatively new, with only limited research (Teng et al. 2004a), so in the remainder of this section, attention is focused on shear strengthening of RC beams with externally bonded FRP.

A variety of possibilities exist with the use of externally bonded FRP in shear strengthening. Common ways of attaching FRP shear reinforcement to a beam include: (a) side bonding in which the FRP is bonded to the sides only, (b) U-jacketing in which FRP U-jackets are bonded on both the sides and the soffit (Figure 16b), and (c) complete wrapping in which the FRP is wrapped around the entire cross section (Figure 16a). Instead of discrete strips, continuous sheets may be used. The fibres in the FRP or the FRP strips may also be oriented at different angles to meet different strengthening requirements. For brevity, "strips" are used in this chapter as a generic term to refer to FRP shear reinforcement including continuous sheets as a special case.

As the FRP strips need to be bent around the corners of the beam in both U-jacketing and complete wrapping, the wet lay-up process (Teng et al. 2002) is commonly used, which involves the in-situ impregnation of dry fibre/fabric strips with epoxy resin and the subsequent bonding of these impregnated strips to the beam with appropriate surface preparation and corner rounding. One notable variation is the use of prefabricated L-shaped CFRP plates, a solution developed by Sika (Basler et al. 2001).

(a) Rupture of complete FRP wraps   (b) Debonding of FRP U-jackets

Figure 16 Shear failure modes of FRP-strengthened RC beams

Failure Modes

The shear failure process of FRP-strengthened RC beams involves the development of either a single major diagonal shear crack or a number of diagonal shear cracks, similar to normal RC beams without FRP strengthening. For ease of description herein, the existence of a single major diagonal shear crack (the critical shear crack) is assumed whenever necessary. Eventual failure of almost all test beams occurred in one of the two main failure modes: tensile rupture of the FRP and debonding of the FRP from the concrete. Generally, both failure modes start with a debonding propagation process from the critical shear crack. Tensile rupture starts in the most highly-stressed FRP strip, followed rapidly by the rupture of other FRP strips intersected by the critical shear crack (Figure 16a). In beams with complete FRP wraps, it is also common that many of the FRP strips intersected by the critical shear crack have debonded from the sides over the full height of the beam before tensile rupture failure occurs. In beams whose failure is by debonding of the FRP from the RC beam, failure involves a process of sequential debonding of FRP strips starting from the most vulnerable strip (Figure 16b).

The FRP rupture failure mode has been observed in almost all tests on beams with complete FRP wraps and in some tests on beams with FRP U-jackets, while the debonding failure mode has been observed in almost all tests on beams with FRP side strips and most tests on beams with FRP U-jackets. Mechanical anchors can be used to prevent debonding and thus change the failure mode from debonding to rupture. However, care needs to be excised to avoid local failure adjacent to the anchors.

Strength Models

Several different approaches have been used to predict the shear strength of FRP-strengthened RC beams (Teng et al. 2004a). These include the modified shear friction method, the compression field theory, various truss models and the traditional design code approach widely adopted by existing codes of practice for concrete structures.

The vast majority of existing research has adopted the design code approach (Teng et al. 2002; 2004a). The total shear resistance of FRP-strengthened RC beams in this approach V_n is commonly assumed to be given in the following form:

$$V_n = V_c + V_s + V_{frp} \tag{1}$$

where V_c is the contribution of concrete, V_s is the contribution of steel shear reinforcement and V_{frp} is the contribution of FRP. V_c and V_s may be calculated according to provisions in existing design codes for concrete structures. The contribution of FRP is found by truss analogy, similar to the determination of the contribution of steel shear reinforcement. Two parameters are important in determining the FRP contribution: The shear crack angle which is generally assumed to be 45° for design use and the average stress (or effective stress) in the FRP strips intersected by the critical shear crack. Different models differ mainly in the definition of this effective stress.

Chen and Teng (2003a, b) proposed two models following the design code approach to overcome the limitations of previous models, in which the effective stress of the FRP strips is rationally defined instead of being given by empirical expressions from curve-fitting of test results (Teng et al. 2004a). They made a clear distinction between rupture failure and debonding failure, and developed two separate models for them. The key contribution of these two models is the realization and the explicit account taken of the fact that the stress distribution in the FRP along the shear crack is likely to be strongly non-uniform at failure. Their debonding failure model has the additional advantage that an accurate bond strength model (Chen and Teng 2001) was employed, leading to accurate predictions. Chen and Teng's (2003a, b) design models have been shown to be more accurate than other similar models when assessed using available test data (Chen 2003).

The design code approach neglects the interactions between the external FRP and internal steel stirrups and concrete. The validity of this assumption has been questioned by several researchers (e. g. Teng et al. 2002, 2004a; Denton et al. 2004, Qu et al. 2005 and Mohamed Ali et al. 2006), but the approach is the least involved for design, most mature and appears to be conservative for design in general.

Design Provisions in the Draft Chinese Code

In the draft Chinese code, shear strengthening design is based on the simple code approach. The design provisions are based on the work of Lu (2004) which follows the conceptual framework of Chen and Teng (2003a, 2003b). In this work, the FRP-to-concrete bond-slip model of Lu et al. (2005b) and idealized shear crack patterns (Lu et al. 2005c, 2005d) based on experimental observations and numerical simulations were employed in a numerical study to arrive at a design equation which is more reliable and relates more closely to the failure mechanism.

STRENGTHENING OF RC COLUMNS

General

An important application of FRP composites in the strengthening/retrofit of RC structures is to provide confinement to columns to enhance either their ultimate loads (strengthening) or seismic resistance (seismic retrofit) (Figure 17). This technique exploits the well-known phenomenon that both the strength and ductility of concrete under axial compression increases with lateral confinement. Typically, such columns fail by the tensile rupture of the FRP jacket (Figure 17).

Various methods have been used to achieve confinement to columns using FRP composites (Teng et al. 2002). In-situ FRP wrapping has been used in the vast majority of cases, in which uni-directional fibre sheets or woven fabric sheets are impregnated with polymeric resins and wrapped around columns in a wet lay-up process, with the main fibres orientated in the hoop direction. In addition, filament winding and prefabricated FRP jackets have also been used.

For the reliable design of FRP jackets for columns, accurate stress-strain models for FRP-confined concrete need to be developed. Once the stress-strain curve of FRP-confined concrete is defined, the behaviour of the column subjected to combined compression and bending can be evaluated using established methods. Therefore, this section is focussed on recent advances in understanding and modelling the behaviour of FRP-confined concrete, followed by a section which summarises the design provisions in the draft Chinese code.

(a) Circular columns: uniform confinement   (b) Rectangular columns: non-uniform confinemnt

Figure 17 Failure of FRP-confined concrete specimens

Stress-Strain Models for FRP-Confined Concrete

Classification of stress-strain models

Many theoretical and experimental studies have been carried out on FRP-confined concrete (Teng and Lam 2004). Existing models can be classified into two categories: (a) design-oriented models, and (b) analysis-oriented models. In the first category, stress-strain models are presented in closed-form expressions, while in the second category, stress-strain curves of FRP-confined concrete are predicted using an incremental/iterative numerical procedure.

Hoop rupture strain

An important phenomenon of FRP-confined concrete is that the FRP jacket fails under hoop tension at a strain much lower than the corresponding ultimate strain from a flat coupon tensile test. As a result, reliable stress-strain models must take this phenomenon into account (Teng and Lam 2004). In an attempt to explain why FRP hoop rupture strains measured in FRP-confined concrete cylinder tests fall substantially below those from flat coupon tensile tests, Lam and Teng (2004) conducted a study on the ultimate condition of FRP jackets in which the tensile rupture strains for two types of FRP (CFRP and GFRP) obtained from three types of tests were presented and compared: flat coupon tensile tests, ring splitting tests and FRP-confined concrete cylinder tests. Based on the comparisons of these test results, it was concluded that the average hoop rupture strains measured in FRP-confined concrete cylinders are affected by at least three factors: (a) the curvature of the FRP jacket; (b) the deformation non-uniformity of cracked concrete; and (c) the existence of an overlapping zone in which the measured strains are much lower than strains measured elsewhere. Another important factor which may be responsible for much of the reduction is the biaxial stress state of the FRP jacket (hoop tension plus axial compression) (Yu and Teng 2006), an issue which is currently being examined at PolyU. These factors combine to produce an average FRP hoop rupture strain in confined cylinders which is much lower than that from flat coupon tests.

Analysis-oriented stress-strain models

Although design-oriented models are simple and convenient for use in everyday design, analysis-oriented models are more versatile and accurate in general, are often the preferred choice for use in more involved analysis than is required in design (e. g. nonlinear finite element analysis), and are applicable/easily extendible to concrete confined with materials other than FRP. They can also be employed to generate numerical results for use in the development of a design-oriented model.

Among the many analysis-oriented stress-strain models, the recent model of Teng et al. (2006c) has been found to be the most accurate (Jiang and Teng 2006a). The key new feature of this model compared to existing models is a more accurate and more widely applicable lateral strain equation based on careful interpretations of test results of unconfined, actively confined and FRP-confined concrete. The model provides accurate predictions not only for FRP-confined concrete, but also for concrete confined with other materials. A refined version of this model has also been proposed, which provides more accurate predictions for weakly confined concrete than the original model (Jiang and Teng 2006a).

Design-oriented stress-strain models

Many different design-oriented stress-strain models have been published. The model proposed by Lam and Teng (2003a) is advantageous over other models due to its simplicity and accuracy. It consists of a parabolic first portion which connects smoothly to a linear second portion. This model, with some modification, has been adopted by a UK design guideline (Concrete Society 2004). This model adopts a simple form which automatically reduces to that for unconfined concrete when no FRP is provided (Figure 18). Its simple form also caters for easy improvements to the definition of the ultimate condition (the ultimate axial strain and compressive strength) of FRP-confined concrete, which are the key to the accurate prediction of stress-strain curves of FRP-confined concrete by this model.

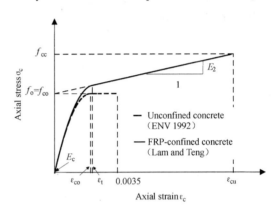

Figure 18 Lam and Teng's stress-strain model for FRP-confined concrete

Although Lam and Teng's (2003a) model was developed on the basis of a large test database, some significant issues could not be readily resolved using the test database available to them at that time. In particular, there was considerable uncertainty with the hoop tensile rupture strain reached by the FRP jacket, which has an important bearing on the definition of the ultimate condition. In addition, the different testing conditions and the limited ranges of the type and the amount of the confining FRP covered by that test database also affected its interpretation. To address the deficiencies of that database and hence those of Lam and Teng's (2003a) stress-strain model based on that database, a large number of additional tests on FRP-confined concrete cylinders were recently conducted at PolyU (Lam and Teng 2004; Jiang and Teng 2006a; Teng et al. 2006d; Lam et al. 2006). Further theoretical modeling work on FRP-confined concrete was also carried out (Teng et al. 2006c; Jiang and Teng 2006a). With these new test results and the new understandings from experimental and theoretical work, refinements of Lam and Teng's design-oriented stress-strain model were recently proposed (Teng et al. 2006e). The refined version of Lam and Teng's model provides significantly closer predictions of test stress-strain curves than the original version.

FRP-confined concrete under cyclic compression

For the safe and economical design of FRP jackets in the seismic retrofit of RC columns, the stress-strain behaviour of FRP-confined concrete under cyclic compression needs to be properly understood and modelled. Only a few studies have been concerned with FRP-confined concrete under cyclic compression, and these studies are reviewed in Lam et al. (2006). Lam and Teng (2006) have recently established a new cyclic stress-strain model. Figure 19 shows a comparison of a test stress-strain curve from Lam et al. (2006) with the prediction of the proposed stress-strain model. The only previous cyclic stress-strain model by Shao (2003) was shown by Lam et al. (2006) to require significant improvement.

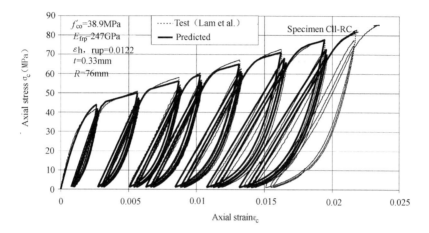

Figure 19 Test stress-strain curve vs proposed model

Non-uniform confinement and finite element modelling

This section has so far been concerned with concrete uniformly confined with FRP as is found in circular columns. In rectangular columns as well as other non-circular columns, the concrete is non-uniformly confined by the FRP jacket. Although some stress-strain models such as that proposed by Lam and Teng (2003b) exist, recent work at PolyU suggests that their accuracy still requires improvement. New monotonic and cyclic compression tests on FRP-confined rectangular columns have recently been conducted at PolyU which provided test data required for a reliable calibration of finite element models. Finite element modelling has also been carried out (Yu 2006). A more accurate stress-strain model is expected to arise from this combined experimental and theoretical approach.

Design Provisions in the Draft Chinese Code

Short versus slender columns

A short RC column, for which the slenderness effect may be neglected, may need to be reclassified as a slender column when provided with FRP confinement, for which the slenderness effect needs to be considered. This is because FRP confinement may lead to large strength increases which are achieved with a tangent modulus of the confined concrete being much smaller than the elastic modulus of the unconfined concrete.

A comprehensive numerical study (Jiang and Teng 2006b) has recently been conducted to develop a reliable definition of the slenderness limit for short FRP-confined RC columns, in which a computer analysis of slender FRP-confined RC columns was developed using the refined version of Lam and Teng's stress-strain model for FRP-confined concrete in compression (Lam and Teng 2003a; Teng et al. 2006e). The expression developed for FRP-confined RC columns consider columns with unequal load eccentricities at the two ends and closely match existing provisions in GB-50010 (2002) and ACI-318 (2005) for conventional RC columns.

Strengthening design of short RC columns

The draft Chinese code includes provisions only for the strengthening of short columns with a slenderness ratio below the limit defined by Jiang and Teng (2006c). Existing research on FRP-confined slender RC column is still very limited (e.g. Tao et al. 2004a). Research is currently underway at PolyU to develop a design method for FRP-confined slender RC columns.

Short column design makes use of section analysis based on the plane section assumption and a stress-strain curve for the concrete. Three different section forms are covered: (a) circular sections, (b) rectangular sections; (c) inflated rectangular sections, which are modified from rectangular sections by expanding each lateral surface into a cylindrical surface with a rise of at least one twentieth of the side length. The refined version of Lam and Teng's design-oriented stress-strain model (Lam and Teng 2003a; Teng et al. 2006e) was adopted for circular sections; for rectangular and inflated rectangular sections, the same equation forms are employed, but the definitions of the ultimate point are different. For rectangular sections, these definitions are based on the work of Lam and Teng (2003b) with some simplifications by Teng and Jiang (2006), and on the work conducted at the Harbin Institute of Technology (Lai et al. 2004). A simplified section analysis procedure including the use of appropriate stress-block factors is included in the draft code (Jiang and Teng 2006c). Further details of the design provisions in the draft Chinese code are available in Teng and Jiang (2006).

STRENGTHENING OF STEEL STRUCTURES

Appropriate Use of FRP in the Strengthening of Steel Structures

General

The use of FRP composites to strengthen steel structures has received much recent attention. Since steel is also a material of high elastic modulus and tensile strength, the use of FRP in strengthening steel structures calls for innovative exploitations of the advantages of FRP, including their high strength-to-weight ratio and flexibility in shape (applicable to FRP laminates formed via the wet lay-up process only). Furthermore, when wet lay-up FRP laminates instead of prefabricated FRP plates/shapes are used, the combination of adhesive bonding with shape flexibility makes bonded FRP laminates an attractive strengthening method for a number of applications. Needless to say, steel plates can also be adhesively-bonded but bonding is less attractive for steel plates due to their heavy weight and inflexibility in shape. Furthermore, for the same tensile capacity, a steel plate has a much larger bending stiffness than an FRP laminate so a steel plate leads to higher peeling stresses at the interface between the steel plate and the steel substrate. It is also easier to anchor FRP laminates by wrapping FRP jackets.

When bonded FRP laminates are used to strengthen concrete, masonry or timber structures, excellent corrosion resistance is another of its advantages over externally bonded steel plates. For example, in the case of a concrete structure, an externally bonded steel plate may corrode easily and thus needs protection against corrosion. The advantage of corrosion resistance of FRP laminates over steel plates is no longer so significant when strengthening a steel structure, as the existing structure, which occupies a much large volume, needs to be protected against corrosion as well, except when a structure made of a corrosion resistant material (e. g. stainless steel) is being strengthened.

The use of both CFRP and GFRP to strengthen steel structures has been explored. For the strength enhancement of steel structures, CFRP is preferred over GFRP due to the much higher elastic modulus of the former. In particular, when the enhancement of buckling resistance is the aim, the use of high or ultra-high modulus CFRP is very attractive. Table 1 shows the properties of pultruded CFRP plates supplied by SIKA. By contrast, for the confinement of steel tubes, particularly when ductility enhancement is the main aim, GFRP is more attractive as it is cheaper and offers a greater strain capacity (>2%). An issue to note is the galvanic corrosion when steel is in direct contact with CFRP (Tavakkolizadeh and Saadatmanesh 2001; Hollaway and Cadei 2002), so a layer of GFRP has been advised to be sandwiched between them by some researchers (e. g. Hollaway and Cadei 2002). A detailed recent discussion of the issue of galvanic corrosion is given in Schnerch et al. (2006a).

Properties of Sika CFRP plate   Table 1

Product	Young's modulus (GPa)	Tensile strength (MPa)	Elongation (%)
Sika CarboDur S	165	2800	1.70
Sika CarboDur M	210	2400	1.20
Sika CarboDur H	300	1300	0.45

* Extracted from manufacturer's product data sheet

Since FRP composites, particularly CFRP composites are an expensive material, in all applications, the amount of FRP material required should be minimised. For this reason, where the amount of FRP material required is small by nature of the problem (e. g. local strengthening under a concentrated force), FRP strengthening is more likely to be attractive.

High-strength-to-weight ratio

The main advantage of FRP over steel in the strengthening of steel structures is its high strength-to-weight ratio, leading to ease and speed of transportation and site operation, reduced disturbance to service and traffic and hence reduced economic losses as a result of suspension of service. The use of FRP to strengthen steel or steel-concrete bridges (e. g. Miller et al. 2001; Sen et al. 2001; Liu et al. 2001; Hollaway and Cadei 2002; Tavakkolizadeh and Saadatmanesh 2003; Schnerch et al. 2006b) exploits this advantage. Pultruded CFRP plates can have a tensile strength around 3000 MPa (Table 1), but a density of only a quarter of steel, so for the same total tensile capacity, the weight of the strengthening material can be greatly reduced. CFRP, which has a much higher elastic modulus and tensile strength, is generally preferred to GFRP in the strength enhancement of flexural members. However, GFRP has also been used in the flexural strengthening of beams in laboratory tests (e. g. El Damatty et al. 2003; El Damatty et al. 2005); the use of GFRP means that a much thicker laminate is needed than if a CFRP laminate is used.

High local stresses from concentrated forces often lead to local buckling or yielding problems, and such problems can be easily addressed by bonding FRP laminates over the local region. Such applications also exploit the high strength-to-weight ratio of FRP, as on-site implementation, often at heights or in congested areas, becomes much easier with lightweight laminates. The local nature of such strengthening also means that the higher material cost of FRP over steel is not a significant concern. An example problem of this type is the strengthening of cold-formed sections against web crippling failure (Zhao et al. 2006).

Flexibility in shape

Another significant advantage of FRP, which applies only to FRP laminates formed via the wet lay-up process, is the ability of such FRP laminates to follow curved and irregular surfaces of structures. This is difficult to achieve using steel plates. As a result, FRP composites can be used to confine steel tubes/shells or concrete-filled steel tubes, to delay or eliminate local instability problems in the steel tube/shell, thereby enhancing the strength and/or seismic resistance of such structures (e. g. Tani et al. 2000; Teng and Hu 2004, 2006a; Xiao 2004; Xiao et al. 2005). The method of FRP confinement is attractive not only in the strengthening of steel tubular structures, but also in the construction of new tubular columns.

Adhesive bonding

Bonding of FRP laminates are more attractive than the welding or bonding of steel plates. Due to the thinness of FRP laminates, bonded FRP laminates are superior to bonded steel plates as the former are less likely to debond and can follow curved or irregular surfaces. Compared to the welding of steel plates, bonding of FRP laminates is attractive for the following situations:

(1) Bonding of FRP laminates for enhanced fatigue resistance has the advantage that the

strengthening process does not introduce new residual stresses, which is thus superior to the method of welding steel plates;

(2) In certain applications (e.g. oil storage tanks and chemical plants) where fire risks must be minimized, welding needs to be avoided when strengthening a structure; bonding of FRP laminates is then a very attractive alternative; and

(3) High-strength steels suffer significant local strength reductions in the heat-affected zones of welds, and bonded FRP laminates offer an ideal strength compensation method (Jiao and Zhao 2004).

In the remainder of this section, a brief summary of research being undertaken at PolyU on the strengthening of steel structures is given.

Flexural Strengthening of Steel Beams

As for an RC beam, a steel beam can be also be strengthened by bonding an FRP (generally CFRP) plate to its soffit. Research has recently been undertaken at PolyU on the behaviour and design of such FRP-plated steel beams. A number of failure modes are possible for these beams: (a) in-plane bending failure; (b) lateral buckling; and (c) debonding at the plate ends; (d) yielding-induced debonding (from the mid-span if it is a simply-supported beam under symmetric loading). Additional but less likely failure modes include: (e) local buckling of the compression flange; and (f) local buckling of the web.

The in-plane bending capacity of an FRP-plated steel beam can be easily determined, provided debonding does not become critical and hence the plane section assumption can still be used (ICE 2001; Cadei et al. 2004). The elastic lateral buckling of an FRP-plated steel beam has recently been addressed by Zhang and Teng (2006), but much more work is needed on this problem.

Debonding in FRP-plated concrete beams has been an issue of major concern and has been widely studied. Research on bond behaviour and debonding failures of FRP-plated steel beams has also attracted considerable attention (e.g. ICE 2001; Cadei et al. 2004; Lenwari et al. 2006; Schnerch 2006a) and is currently being studied at PolyU. Such research should start with the characterisation of the bond behaviour between steel and FRP, and a study of this type has been conducted by Xia and Teng (2005) who presented the first bond-slip model for FRP-to-steel interfaces.

In the strengthening of steel or steel-concrete composite bridges, the speed of strengthening operations is of great importance when closure of traffic needs to be avoided to reduce economic losses. Collaborative research between PolyU and the University of Surrey in UK has recently been conducted on the rapid strengthening of bridges using prepregs and film adhesive (Hollaway et al. 2006; Zhang et al. 2006). Using this new method, a bridge may be strengthened in as short as four hours. The study also examined the effect of traffic induced vibration during the curing of the FRP system on the performance of the strengthened structure. This study established the effectiveness and reliability of this rapid strengthening method for steel structures.

Strengthening of Steel Structures under High Local Stresses

In practice, high stresses in a local zone often arise, due to the need to introduce discrete supports, openings and other local features. Under local high compressive stresses, local buckling failure is likely to control the thickness of a thin-walled steel structure. Such local buckling failure can be easily prevented by bonding CFRP patches. A practically important problem is the web crippling failure of thin-walled sections under a bearing force. Zhao et al. (2006) has shown that bonded CFRP can be an effective solution to this problem. Local high tensile stresses can also be addressed in the same way. At PolyU, the local strengthening of thin-walled steel structures under high local stresses is being investigated, in collaboration with Monash University, Australia, with web crippling being one of the problems being addressed.

FRP Confinement of Hollow Steel Tubes

Hollow and concrete-filled steel tubes are widely used as columns in many structural systems and local buckling can occur when they are subject to axial compression alone or in combination with monotonic/cyclic lateral loading. For example, hollow steel tubes are often used as bridge piers and such bridge piers suffered extensive damage and even collapse during the 1995 Hyogoken-nanbu earthquake (Kitada et al. 2002). Figure 20a shows a local buckling failure mode at the base of a steel bridge pier and the repair of the pier by the addition of welded vertical stiffeners. Such local buckling is often referred to as elephant's foot buckling. In typical circular tubular structures, elephant's foot buckling appears after yielding and the appearance of this inelastic local buckling mode normally signifies the exhaustion of the load carrying capacity and the end of ductile response. The latter is of particular importance in seismic design, as the ductility and energy absorption capacity of the column dictates its seismic resistance. In rectangular (including square) steel tubes, a similar failure mode can occur. Here, the buckling deformation is normally outwards on the flanges and inwards on the webs.

(a) Failure near the base of a steel tube   (b) Failure at the base of a liquid storage tank

Figure 20 Elephant's foot buckling in a steel tube or shell
(Courtesy of Dr. H. B. Ge, Nagoya University & Prof. J. M. Rotter, Edinburgh University)

The enhancement of ductility and hence seismic resistance of hollow tubular columns through confinement by an FRP jacket has recently been explored at PolyU (Hu 2003; Teng and Hu 2004, 2006a). The technique was shown to be highly effective. The failure modes of hollow steel tubes with and without FRP confinement are shown in Figures 21a and 21b, while the axial load-axial shortening curves are shown in Figure 21c. It is clear that through FRP confinement, the elephant's foot mode of buckling failure is prevented and the ductility of the tube is greatly enhanced. Nishino and Furukawa (2004) also explored the same technique for hollow steel tubes independently.

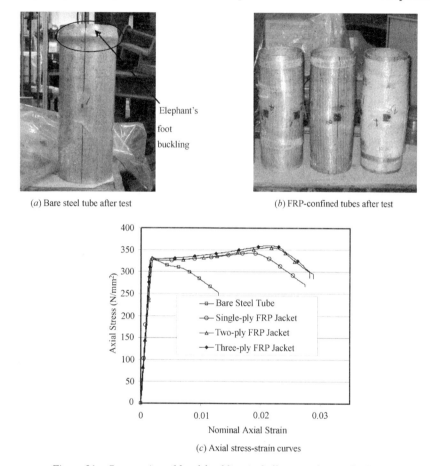

Figure 21 Suppression of local buckling in hollow circular steel tubes

These results also show that when the jacket thickness reaches a threshold value for which inward buckling deformations dominate the behaviour, further increases in the jacket thickness do not lead to significant additional benefits as the jacket provides little resistance to inward buckling deformations. It is significant to note that FRP confinement of steel tubes leads to large increases in ductility but limited increases in the ultimate load, which is often desirable in seismic retrofit of columns which are part of a larger structure, so that the retrofitted tube will not attract forces which are so high that adjacent members may be put in danger.

The elephant's foot buckling mode is not only the critical failure mode in commonly used circular steel tubular columns under axial compression and/or bending, it also occurs in much thinner cylindrical shells in steel storage silos and tanks under combined axial compression and internal pressure (Figure 20b). This failure mode has been commonly observed in earthquakes (Manos et al. 1985) and under static loading (Rotter 1990). The use of FRP jackets to strengthen thin steel cylindrical shells against local elephant's foot buckling failure at the base has also been explored through finite element analyses by Teng and Hu (2006a). The limited numerical results for a thin cylindrical shell with a radius-to-thickness ratio of 1000 and subjected to axial compression in combination with internal pressure indicate that the method leads to significant increases of the ultimate load. The FRP jacketing of steel cylindrical shells can also be used in the construction of new tanks and silos to enhance their performance. A similar and related study on the strengthening of such cylindrical shells has been conducted by Chen et al. (2006c) where an optimally-located ring stiffener is proposed as the strengthening method. This ring stiffener may well be a CFRP cable that provides the same circumferential stiffness and the needed strength.

FRP Confinement of Concrete-Filled Steel Tubes

In concrete-filled steel tubular columns, the concrete and the steel tube interact in a beneficial manner: the steel tube confines the concrete and the concrete delays the occurrence of local buckling in the steel tube. Concrete-filled steel tubes are thus an economic form of structural members, mainly as columns for buildings and bridges, and research on concrete-filled steel tubes is abundant. In concrete-filled steel tubes, although inward buckling deformations of the tube are prevented by the concrete, local outward buckling deformations of the steel tube can still occur and can lead to degradation in the confinement to the concrete provided by the steel tube, and in the overall strength and ductility of the column. When concrete-filled steel tubes are used as columns and are subject to combined axial and lateral loads, the critical regions for the local buckling of the steel tube are at the ends of the column where the moments are the largest. Under seismic loading, plastic hinges form at the column ends and large plastic rotations without significant degradation in stiffness and strength are demanded here.

(a) Concrete-filled steel tube　　(b) FRP-confined concrete-filled steel tube

Figure 22　Failure modes of concrete-filled steel tubes
with and without FRP confinement

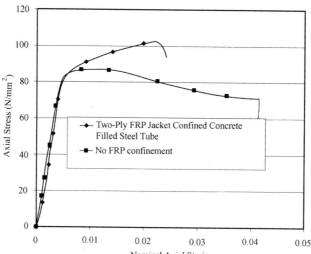

Figure 23 - Axial stress-strain curves of concrete filled steel tubes with and without FRP confinement

Xiao (2004) and Xiao et al. (2005) recently explored the use of FRP jackets for the confinement of the critical regions of circular concrete-filled steel tubes. Although his work was directed at new construction, the same concept can be employed in the retrofit of columns. In such concrete-filled steel tubular columns, the inward buckling deformation of the steel tube is prevented by the concrete core while the outward buckling deformation is prevented by the FRP jacket. FRP jacketing therefore provides an effective means of suppressing local buckling failures at the column ends.

At PolyU, the effect of FRP jacketing on concrete-filled steel tubes has investigated both experimentally and theoretically (Teng and Hu 2005, 2006b; Hu and Teng 2006). GFRP was used in their study, as the larger strain capacity was important in such confinement applications. Two test specimens after failure are shown in Figure 22: one without FRP jacketing and one with FRP jacketing. The failure mode of the specimen without FRP jacketing was local outward buckling (Figure 22a) which has been observed by many other researchers in previous tests on such tubes. The local buckling deformations occurred in the mid-height region on one side and near the end on the other side. The failure mode of the specimen with a thin FRP jacket was rupture of the FRP jacket (Fig. 22b) due to hoop tension. Once rupture of the FRP jacket occurred, the confinement effect of the FRP jacket disappeared and the load carried by the tube reduced immediately and rapidly.

The axial stress-nominal axial strain (i. e. axial shortening divided by specimen length) curves of these two specimens are shown in Figure 23. It is clear that the axial strength is significantly enhanced but the ductility of the tube may be reduced. Hu and Teng (2006) also examined the inclusion of a gap between the steel tube and the FRP, which is a measure proposed by Xiao et al. (2005) to make the FRP-confined steel tube more ductile. While this measure was found to have some beneficial effect on ductility, it was shown to be much less effective than initially expected. A theoretical model for the behaviour of concrete-filled steel tubes confined by an FRP jacket has also been developed (Teng and Hu 2006b).

CONCRETE-FILLED FRP TUBES (CCFTs)

General

Concrete-filled FRP tubes (CFFTs) are one of the most popular examples of hybrid compression members incorporating FRP. When a CFFT is under compression, the concrete is subjected to axial compression and confinement from the outer FRP tube, which is consequently subjected to tension in the circumferential direction. As a result, a very ductile structural member results from the combination of two brittle materials. In addition to good ductility, the advantages of CFFTs include their lightweight and excellent corrosion resistance. With these advantages, CFFTs have been proposed for use as bridge columns and piles. Extensive existing research has been carried out on CFFTs, on their compressive behaviour (e.g. Mirmiran and Shahawy 1996; Fam and Rizkalla 2001a, 2001b; Davol 1998), flexural behaviour (e.g. Mirmiran et al. 1999; Fam and Rizkalla 2002; Davol et al. 2001), and beam-column behaviour (e.g. Mirmiran et al. 1999; Fam et al. 2003).

Design Provisions in the Draft Chinese Code

Basis of the design provisions

Yu and Teng (2006) recently summarized and interpreted the existing research on CFFTs for the development of a design procedure. On the basis of this work, they proposed design provisions for the draft Chinese code in collaboration with researchers of Tongji University. A brief summary of the work by Yu and Teng (2006) together with the key design provisions in the draft Chinese code is given below.

Stress-strain curves of FRP tubes

It is widely acknowledged that the stress-strain behaviour of FRP is predominantly linear elastic, especially when it is loaded in the direction of the fibres. Consequently, most of the previous research on CFFTs adopted a linear elastic constitutive law based on the classical lamination theory of FRP laminates. The classical lamination theory provides a feasible method to evaluate the mechanical properties of FRP tubes.

The classical lamination theory, however, involves complicated calculations for FRP laminates, especially for asymmetric unbalanced laminates. Furthermore, in a number of situations, the stress-strain behaviour becomes significantly nonlinear. Such situations include FRP laminates subjected to in-plane transverse loading (when the laminate is predominantly unidirectional) and in-plane shear, and angle-ply laminates; such nonlinearity cannot be predicted by the lamination theory. Some existing research (e.g. Fam 2000) has shown that the error of predictions of the lamination theory may be up to 40% for the ultimate strength, up to 25% for the elastic modulus and up to 50% for the Poisson's ratio. Therefore, standardized material tests are important for the more accurate evaluation of the mechanical properties of FRP tubes.

In the draft Chinese code, two alternative approaches are available for determining the properties of FRP laminates, formed from a number of laminae with fibres oriented in different directions. The theoretical approach relies on the classical lamination theory and the Tsai-Wu failure criterion which is widely used for one-ply FRP laminae under biaxial stress states. In the experimental approach, a series of tests are specified to determine the material properties of FRP tubes. In particular, a short column test procedure is specified to determine the ultimate state of CFFTs, which requires axial compression tests on at least six CFFTs with the size of 152.5 mm × 305 mm.

When the stress-strain curves of FRP tubes are determined using the experimental approach and the experimental stress-strain curves are nonlinear, the following procedure is recommended by the draft Chinese code to find the equivalent elastic modulus:

1) When the slope of the experimental curve continuously reduces (Figure 24a), the secant modulus at the peak stress point should be taken as the equivalent elastic modulus.

2) When the slope of the experimental curve increases continuously (Figure 24b), the equivalent elastic modulus should be based on energy equivalence: at the ultimate state, both the strain and the stored strain energy of the experimental curve should be the same as those of the equivalent curve. In Figure 24b, this means that the area of the shaded region is equal to the area enclosed by the experimental curve, the horizontal axis and the vertical line at $\varepsilon = \varepsilon_{fu}$.

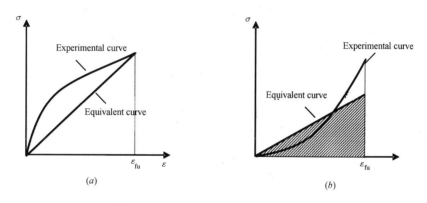

Figure 24 Definition of equivalent elastic modulus

Stress-strain curves of confined concrete in CFFTs

The concrete in CFFTs is subjected to a stress state similar to that in an RC column retrofitted with an FRP jacket. In terms of the behaviour of the confined concrete, the major difference between these two types of structural members is that FRP tubes in CFFTs normally have significant longitudinal stiffness, while FRP jackets used in retrofitting normally have negligible longitudinal stiffness. As a result, the FRP tube is able to resist significant axial tension and compression. In the draft Chinese code, this difference is duly reflected in defining the ultimate state of the FRP tube and that of the confined concrete. Apart from this aspect, the stress-strain model for FRP-confined concrete adopted by the draft code for column strengthening design is also employed in the design of CFFTs, except for a modification to allow the effect of load eccentricity on the stress-strain behaviour of concrete in FRP tubes to be included (Yu and Teng 2006).

Design of CCFTs under eccentric compression

Based on the stress-strain curves for the FRP tube and the concrete defined in the draft code, the design calculations for short CFFTs can be easily executed based on the plane section assumption. The slenderness limit defined for FRP-strengthened columns was also adopted for short concrete-filled FRP tubular columns. Furthermore, the stress block factors used are also the same as those for FRP-confined RC columns.

HYBRID FRP-CONCRETE-STEEL COLUMNS

Concrete-filled FRP tubes have their own disadvantages, particularly when used as building columns. These include poor fire resistance, difficulty for connection to beams, inability to support substantial construction loads, brittle failure in bending, and high cost as the tube needs to be relatively thick in order to resist axial loads. As a bridge column, the fire resistance and connection problems are not significant.

To overcome the disadvantages of concrete-filled FRP tubes, a new form of hybrid columns (Figure 25) has recently been proposed by the author and investigated at PolyU (Teng et al. 2003a; Teng et al. 2004b; Teng et al. 2006d). The new column consists of a steel tube inside, an FRP tube outside and concrete in between. The inner void may be filled with concrete if desired. The FRP tube is provided with fibres which are predominantly oriented in the circumferential direction to provide confinement to the concrete. The new column is an attempt to combine the advantages of all three constituent materials and those of the structural form of double-skin tubular columns (DSTCs), so as to achieve a high-performance structural member. Steel-concrete DSTCs with both skins being made of steel were first reported in late 1980s (Shakir-Khalil and Illouli 1987) and have since received much attention (e. g. Shakir-Khalil 1991; Wei et al. 1995; Han et al. 1995; Zhao et al. 2002; Tao et al. 2004b; Zhao and Han 2006). In DSTCs, the inner void reduces the column weight without significantly affecting the

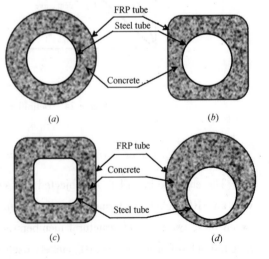

Figure 25 Typical cross-sections of double-skin tubular members

e 26 shows load-displacement curves from three pairs of tests on stub hybrid columns failed eventually by the rupture of the FRP tube as a result of hoop tension (Figure 27). pairs correspond to three FRP tube thicknesses (Teng et al. 2004b). The applied load asing or remained almost constant with the increment of the axial strain, depending on ess of the FRP tube (Figure 26). The stub column tests confirmed some of the expected s of the new hybrid column. The concrete in the new column was found to be very effec- fined, leading to a very ductile response. The inner void in the double-skin column was have almost no effect on the effectiveness of confinement provided by the FRP tube. rge amount of research has recently been undertaken at PolyU on this type of hybrid col- study their compressive, flexural and beam-column behaviour (Yu et al. 2006; Yu A simple design method for such columns under bending or combined bending and com- has been developed.

:LUDING REMARKS

field of FRP composites in construction is wide and rapidly growing. This chapter has ated on recent advances in three major areas: (a) strengthening of concrete structures, ngthening of steel structures, and (c) FRP-confined columns, with particular attention to search undertaken at The Hong Kong Polytechnic University (PolyU). Recent research in area has been summarised with particular attention to work underpinning some of the key rovisions in the draft Chinese code for the use of FRP composites in construction. The sec- is a more recent area compared to the first, and a detailed discussion of situations where nique may be attractive has been examined; a summary of recent research at PolyU in this also been given. The third area covers two types of FRP-confined columns: (a) concrete 'RP tubes, and (b) FRP-concrete-steel double-skin tubular columns. The latter type of has been proposed to overcome the disadvantages of the first type of columns, and existing at PolyU has shown conclusively that the latter type of columns possess several advantages d of them. From the information presented in this chapter, it is clear that while existing re- as led to impressive progresses in these areas, a great deal of further work is required to new horizons for applications and to improve design theories. The field of FRP composites ruction is a field that offers a challenging and rewarding path for a new generation of struc- gineering researchers and practitioners.

NOWLEDGEMENTS

e author is grateful to The Hong Kong Research Grants Council, The Hong Kong Polytech- versity (PolyU), and the Natural Science Foundation of China for their financial support. hor also wishes to thank his students, assistants and collaborators for their contributions to

bending rigidity of the section and allows the easy passage of service ⟨
with both skins made of FRP have also been proposed (Fam and Rizl
The novel feature of the new column form proposed by the auth
DSTCs is that the inner tube is made of steel but the outer tube is mad⟨
mainly in the hoop direction to provide confinement to the concrete f
simple change to the existing DSTC forms offers many advantages, le
easy to construct and highly resistant to corrosion and earthquakes.
made between the new column and the steel-concrete DSTC or betwe
FRP-concrete DSTC. Compared to the steel-concrete DSTC, the ad⟨
include: (a) a more ductile response of concrete as it is well confin
does not buckle; (b) no need for fire protection of the outer tube as the
as a form during construction and as a confining device during earthqu
rosion protection as the steel tube inside is well protected by the concre
pared to the FRP-concrete DSTC, the advantages of the new column i⟨
port construction loads through the use of the inner steel tube; (b) ea
due to the presence of the inner steel tube; (c) savings in fire protecti
required only as a form during construction and as a confining d⟨
(d) better confinement of concrete as a result of the increased rigidity
ly, the new column also has many advantages over other composite/hyb⟨
crete-filled steel tubes, concrete-filled FRP tubes and concrete-encas⟨

The cross-section shown in Figure 25a consists of two circular tube⟨
binations of tubes are possible. Figures 5b and 5c show some of such v⟨
the section form can also be employed in a beam, in which case the inne
towards the tension side (Figure 25d). It should be noted that if the col
centric tubes is deployed in situations where axial compression does n⟨
loading, the column should be provided with some longitudinal reinforce⟨
ment of large tensile cracks.

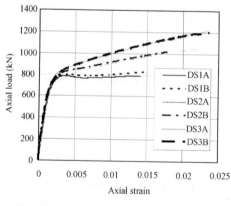

Figure 26 Axial load-strain behaviour of DSTCs

Figure 27 DSTC speci

the FRP composites research program at PolyU. In particular, Dr. J. F. Chen of the University of Edinburgh made many important contributions to this research program over many years.

Thanks are due to members of the committee for the Chinese Code for the Structural Use of FRP in Construction for the opportunities to work with and learn from them, and particularly to Prof. Y. D. Xue for his contributions to the design provisions for concrete-filled FRP tubes and to Profs J. L. Pan and X. N. Jin for their contributions to the design provisions for the strengthening of RC columns in the draft Chinese Code.

REFERENCES

ACI-318 (2005). *Building Code Requirements for Structural Concrete and Commentary*, ACI Committee 318, Detroit, USA.

ACI-440 (1996). *State-of-the-Art Report on Fiber Reinforced Plastic Reinforcement for Concrete Structures*, ACI Committee 440, Detroit, USA.

Bakis, C. E., Bank, L. C., Brown, V. L., Cosenza, E., Davalos, J. F., Lesko, J. J., Machida, A., Rizkalla, S. H. and Triantifilliou, T. C. (2002). "Fiber-reinforced polymer composites for construction: State-of-the-art review", *Journal of Composites for Construction*, ASCE, 6 (2), 73-87.

Basler, M., Mungall, B. and Fan, S. (2001). "Strengthening of structures with the Sika CarboDur composite strengthening systems", *Proceedings, International Conference on FRP Composites in Civil Engineering*, Elsevier Science, Oxford, 253-262.

Buyukozturk, O. and Hearing, B. (1998). "Failure behavior of precracked concrete beams retrofitted with FRP", *Journal of Composites for Construction*, ASCE, 2 (3), 138-144.

Cadei, J. M. C., Stradford, T. J., Hollaway, L. C. and Duckett, W. G. (2004). *Strengthening Metallic Structures Using Externally Bonded Fibre-Reinforced Polymers*, Publication C595, Construction Industry Research and Information Association (CIRIA), London, UK.

Chen, J. F. (2003). "Design guidelines on FRP for shear strengthening of RC beams", *Proceedings (CD-ROM), Tenth International Conference on Structural Faults and Repair*, 1-3 July, London, UK.

Chen, J. F. and Teng, J. G. (2001). "Anchorage strength models for FRP and steel plates bonded to concrete", *Journal of Structural Engineering*, ASCE, 127 (7), 784-791.

Chen, J. F. and Teng, J. G. (2003a). "Shear capacity of FRP strengthened RC beams: fibre reinforced polymer rupture", *Journal of Structural Engineering*, ASCE, 129 (5), 615-625.

Chen, J. F. and Teng, J. G. (2003b). "Shear capacity of FRP-strengthened RC beams: FRP debonding", *Construction and Building Materials*, 17, 27-41.

Chen, J. F. and Teng, J. G. (Eds) (2005). *Proceedings, International Symposium on Bond Behaviour of FRP in Structures*, 7-9 December, Hong Kong, China.

Chen, J. F., Yuan, H. and Teng, J. G. (2006a). "Debonding failure along a softening FRP-to-concrete interface between two adjacent cracks in concrete members", *Engineering Structures*, in press.

Chen, J. F., Teng, J. G. and Yao, J. (2006b). "Strength model for intermediate crack debonding in FRP-strengthened concrete members considering adjacent crack interaction", *Proceedings, Third International Conference on FRP Composites in Civil Engineering*, 13-15 December, Miami, Florida, USA.

Chen, J. F., Rotter, J. M. and Teng, J. G. (2006c). "A simple remedy for elephant's foot buckling in cylindrical silos and tanks", *Advances in Structural Engineering*, 9 (3), 409-420.

Colotti, V., Spadea, G. and Swamy, R. N. (2004). "Structural model to predict the failure behavior of plated reinforced concrete beams", *Journal of Composites for Construction*, ASCE, 8 (2), 104-122.

Concrete Society (2004). *Design Guidance for Strengthening Concrete Structures Using Fibre Composite Materials*, Technical Report No. 55, 2nd Edition, Concrete Society, U. K.

Davol, A. (1998). *Structural Characterization of Concrete Filled Fiber Reinforced Shells*, PhD Thesis, University of California, San Diego, USA.

Davol, A., Burgueno, R. and Seible, F. (2001). "Flexural behavior of circular concrete filled FRP shells", *Journal of Structural Engineering*, ASCE, 127 (7), 810-817.

De Lorenzis, L. and Teng, J. G. (2006). "Near-surface mounted FRP reinforcement: an emerging technique for strengthening of structures", *Composites Part B: Engineering*, in press.

Denton, S. R., Shave, J. D. and Porter, A. D. (2004). "Shear strengthening of reinforced concrete structures using FRP composites", *Proceedings, International Conference on Advanced Polymer Composites for Structural Applications in Construction*, Woodhead Publishing Limited, U. K., 134-143.

El Damatty, A. A., Abushagur, M. and Youssef, M. A. (2003). "Experimental and analytical investigation of steel beams rehabilitated using GFRP sheets", *Steel and Composite Structures*, 3 (6), 421-438.

El Damatty, A. A., Abushagur, M. and Youssef, M. A. (2005). "Rehabilitation of composite steel bridges using GFRP plates", *Applied Composite Materials*, 12, 309-325.

El-Hacha, R., Wight, R. G. and Green, M. F. (2001). "Prestressed fibre-reinforced polymer laminates for strengthening structures", *Progress in Structural Engineering and Materials*, 3 (2), 111-121.

Fam, A. Z. (2000). *Concrete-Filled Fiber-Reinforced Polymer Tubes for Axial and Flexural Structural Members*, PhD Thesis, Department of Civil and Geological Engineering, The University of Manitoba, Winnipeg, Manitoba.

Fam, A. Z. and Rizkalla, S. H. (2001a). "Confinement model for axially loaded concrete confined by circular fiber-reinforced polymer tubes", *ACI Structural Journal*, 98 (4), 451-461.

Fam, A. Z. and Rizkalla, S. H. (2001b). "Behavior of axially loaded concrete-filled circular fiber-reinforced polymer tubes", *ACI Structural Journal*, 98 (3), 280-289.

Fam, A. Z. and Rizkalla, S. H. (2002). "Flexural behavior of concrete-filled fiber-reinforced polymer circular tubes", *Journal of Composites for Construction*, ASCE, 6 (2), 123-132.

Fam, A.., Flisak, B. and Rizkalla, S. (2003). "Experimental and analytical modeling of concrete-filled fiber-reinforced polymer tubes subjected to combined bending and axial loads", *ACI Structural Journal*, 100 (4), 499-509

Gao, B., Leung, C. K. Y. and Kim, J. K. (2005). "Prediction of concrete cover separation failure for RC beams strengthened with CFRP strips", *Engineering Structures*, 27 (2), 177-189.

GB-5 0010 (2002). *Code for Design of Concrete Structures*, China Architecture and Building Press, China.

Han, B., Xia, J., and Zhao, J. (1995). "The strength of concrete filled double-skin steel tubes under axial compression", *Journal of Shijiazhuang Railway Institute*, 8 (3), 75-80.

Hau, K. M. (1999). *Experiments on Concrete Beams Strengthened by Bonding Fibre Reinforced Plastic Sheets*, MSc Thesis, The Hong Kong Polytechnic University, Hong Kong, China.

Hollaway, L. C. (2003). "The evolution of and the way forward for advanced polymer composites in the civil infrastructure", *Construction and Building Materials*, 17 (6&7), 365-378.

Hollaway, L. C. and Cadei, J. (2002). "Progress in the technique of upgrading metallic structures with advanced polymer composites", *Progress in Structural Engineering and Materials*, 14, 131-148.

Holloway, L. C. and Head, P. R. (2001). *Advanced Polymer Composites and Polymers in the Civil Infrastructure*, Elsevier.

Hollaway, L. C., Zhang, L., Photiou, N. K., Teng, J. G. and Zhang, S. S. (2006) "Advances in adhesive joining of carbon fibre/polymer composites to steel members for repair and rehabilitation of bridge structures", *Advances in Structural Engineering*, 9 (6).

Hu, Y. M. (2004). *Behaviour of FRP-Confined Steel Tubes under Axial Compression*, BEng Thesis, Department of Civil and Structural Engineering, The Hong Kong Polytechnic University, April.

Hu, Y. M. and Teng, J. G. (2006). "Behaviour of FRP-jacketed circular concrete-filled steel tubes under axial compression", in preparation.

Moy, S. S. J. (Ed.) (2001). *ICE Design and Practice Guides: FRP Composites-Life Extension and Strengthening of Metallic Structures*, Thomas Telford.

Jiang, T. and Teng, J. G. (2006a). "Analysis-oriented stress-strain models for FRP-confined concrete", submitted for publication.

Jiang, T. and Teng, J. G. (2006b). "Slenderness limit for short FRP-confined RC columns", in preparation.

Jiang, T. and Teng, J. G. (2006c). "Strengthening of short circular RC columns with FRP jackets: a design proposal", *Proceedings, Third International Conference on FRP Composites in Civil Engineering*, 13-15 December, Miami, Florida, USA.

Jiao, H. and Zhao, X. L. (2004). "CFRP strengthened butt-welded very high strength (VHS) circular steel tubes", *Thin-Walled Structures*, 42 (7), 963-978.

Kitada, T., Yamaguchi, T., Matsumura, M., Okada, J., Ono, K. and Ochi, N (2002). "New technologies of steel bridges in Japan", *Journal of Constructional Steel Research*, 58 (1), 21-70.

Lai, W. H., Pan, J. L and Jin, X. N. (2004). "Compressive behavior of concrete confined by fiber reinforced polymer", *Industrial Construction*, 34 (10), 81-84. (in Chinese)

Lam, L. and Teng, J. G. (2003a). "Design-oriented stress-strain model for FRP-confined concrete", *Construction and Building Materials*, 7 (6&7), 471-489.

Lam, L. and Teng, J. G (2003b). "Design-oriented stress-strain model for FRP-confined concrete in rectangular columns", *Journal of Reinforced Plastics and Composites*, 22 (13), 1149-1186.

Lam, L. and Teng, J. G. (2004). "Ultimate condition of fiber reinforced polymer-confined concrete", *Journal of Composites for Construction*, ASCE, 8 (6), 539-548.

Lam, L. and Teng, J. G. (2006). "Stress-strain mode for FRP-confined concrete under cyclic axial compression", in preparation.

Lam, L., Teng, J. G., Cheng, C. H. and Xiao, Y. (2006). "FRP-confined concrete under axial cyclic compression", *Cement and Concrete Composites*, in press.

Lees, J. M., Winistorfer, A. U. and Meier, U. (2002). "External prestressed carbon fiber-reinforced polymer straps for shear enhancement of concrete", *Journal of Composites for Construction*, ASCE, 6 (4), 249-256.

Lenwari, A., Thepchatri, T. and Albrecht, P. (2006). "Debonding strength of steel beams strengthened with CFRP plates", *Journal of Composites for Constructions*, ASCE, 10 (1), 69-78.

Liu, X., Silva, P. F. and Nanni, A. (2001). "Rehabilitation of steel bridges members with FRP composite materials", *Proceedings, International Conference on Composites in Construction*, 10-12 October, Porto, Portugal, 613-617.

Lu, X. Z. (2004). *Studies on FRP-Concrete Interface*, PhD Thesis, Tsinghua University, Beijing, China.

Lu, X. Z., Teng, J. G., Ye, L. P and Jiang, J. J. (2005a). "Bond-slip models for FRP sheets/plates bonded to concrete", *Engineering Structures*, 27 (6), 920-937.

Lu, X. Z., Ye, L. P., Teng, J. G. and Jiang, J. J. (2005b). "Meso-scale finite element model for FRP sheets/plates bonded to concrete", *Engineering Structures*, 27 (4), 564-575.

Lu, X. Z., Chen, J. F., Ye, L. P., Teng, J. G. and Rotter, J. M. (2005c). "Theoretical analysis of FRP stress distribution in U-jacketed RC beams", *Proceedings*, *Third Internationmal Conference on Composites in Construction*, 11-13 July, Lyon, France, 541-548.

Lu, X. Z., Chen, J. F., Ye, L. P., Teng, J. G. and Rotter, J. M. (2005d). "Theoretical analysis of stress distributions in FRP side-bonded to RC beams for shear strengthening", *Proceedings*, *International Symposium on Bond Behaviour of FRP in Structures*, 7-9 December, Hong Kong, China, 363-370.

Lu, X. Z., Teng, J. G., Ye, L. P. and Jiang, J. J. (2006). "Intermediate crack debonding in FRP-strengthened RC beams: FE analysis and strength model", *Journal of Composites for Construction*, ASCE, in press.

Manos, G. C. and Clough, R. W. (1985). "Tank damage during the Coalinga earthquake", *Earthquake Engineering and Structural Dynamics*, 13, 449-466.

Meier, U., Stocklin, I. and Terrasi, G. P. (2001). "Making better use of the strength of advanced materials in structural engineering", *Proceedings*, *International Conference on FRP Composites in Civil Engineering*, 12-15 December, Hong Kong, China, 41-48.

Miller, T. C., Chajes, M. J., Mertz, D. R. and Hastings, J. N. (2001). "Strengthening of a steel bridge girder using CFRP plates", *Journal of Bridge Engineering*, ASCE, 6 (6), 514-522.

Mirmiran, A. (2003). "Stay-in-place FRP form for concrete columns", *Advances in Structural Engineering*, 6 (3), 231-241.

Mirmiran, A. and Shahawy, M. (1996). "A new concrete-filled hollow FRP composite column", *Composites Part B: Engineering*, 27B (3-4), 263-268.

Mirmiran, A., Shahawy, M. and Samaan, M. (1999). "Strength and ductility of hybrid FRP-concrete beam-columns", *Journal of Structural Engineering*, ASCE, 125 (10), 1085-1093

Mohamed Ali, M. S., Oehlers, D. J. and Seracino, R. (2006). "Vertical shear interaction model between external FRP transverse plates and internal steel stirrups", *Engineering Structures*, 28 (3), 381-389.

Nishino, T. and Furukawa, T. (2004). "Strength and deformation capacities of circular hollow section steel member reinforced with carbon fiber", *Proceedings*, *Seventh Pacific Structural Steel Conference*, American Institute of Steel Construction, March.

Oehlers, D. J., Seracino, R. (2004). *Design of FRP and Steel Plated RC Structures-Retrofitting Beams and Slabs for Strength, Stiffness and Ductility*. Elsevier, UK.

Oehlers, D. J., Liu, I. S. T., Seracino, R. and Mohamed Ali, M. S. (2004). "Prestress model for shear deformation debonding of FRP- and steel-plated RC beams", *Magazine of Concrete Research*, 56 (8), 475-486.

Oehlers, D. J., Liu, I. S. T., Seracino, R. (2005). "Shear deformation debonding of adhesively bonded plates", *Proceedings of the Institution of Civil Engineers: Structures and Buildings*, 158 (1), 77-84.

Qu, Z., Lu, X. Z. and Ye, L. P. (2005). "Size effect of shear contribution of externally bonded FRP U-jackets for RC beams", *Proceedings*, *International Symposium on Bond Behaviour of FRP in Structures*, 7-9 December, Hong Kong, China, 371-380.

Rotter, J. M. (1990). "Local collapse of axially compressed pressurized thin steel cylinders", *Journal of Structural Engineering*, ASCE, 1116 (7), 1955-1970.

Schnerch, D., Dawood, M., Rizkalla, S. and Sumner, E. (2006a). "Bond behaviour of CFRP strengthened steel structures", *Advances in Structural Engineering*, 9 (6).

Schnerch, D., Dawood, M., Rizkalla, S. and Sumner, E. (2006b). "Proposed design guidelines for strengthening of steel bridges with FRP materials", *Construction and Building Materials*, in press.

Sen, R., Liby, L. and Mullins, G. (2001). "Strengthening steel bridge sections using CFRP laminates", *Composites Part B: Engineering*, 32 (4), 309-322.

Shakir-Khalil, H. (1991). "Composite columns of double-skinned shells", *Journal of Constructional Steel Research*, 19, 133-152.

Shakir-Khalil, H. and Illouli, S. (1987). "Composite columns of concentric steel tubes", *Proceedings, Conference on the Design and Construction of Non-Conventional Structures* 1987, 8-10 December, London, UK, 73-82.

Shao, Y. (2003). *Behavior of FRP-Confined Concrete Beam-Columns under Cyclic Loading*. PhD Thesis, North Carolina State University, USA.

Shawulieti, B. K. Y., Ye, L. P., Yang, Y. X. and Zhuang, J. B. (2004). "Construction technology of prestress CFRP sheet strengthening reinforcement concrete beams", *Construction Technology*, 33 (6), 23-24. (in Chinese)

Smith, S. T. and Teng, J. G. (2001). "Interfacial stresses in plated beams", *Engineering Structures*, 23 (7), 857-871.

Smith, S. T. and Teng, J. G. (2002a). "FRP-strengthened RC beams-I: Review of debonding strength models", *Engineering Structures*, 24 (4), 385-395.

Smith, S. T. and Teng, J. G. (2002b). "FRP-strengthened RC beams-II: Assessment of debonding strength models", *Engineering Structures*, 24 (4), 397-417.

Smith, S. T. and Teng, J. G. (2003). "Shear-bending interaction in debonding failures of FRP-plated RC beams", *Advances in Structural Engineering*, 6 (3), 183-199.

Tani, K., Matsumura, M., Kitada, T. and Hayashi, H. (2000). "Experimental study on seismic retrofitting method of steel bridge piers by using carbon fiber sheets", *Proceedings, Sixth Korea-Japan Joint Seminar on Steel Bridges*, Tokyo, Japan, 437-445.

Tao, Z., Teng, J. G., Han, L. H. and Lam, L. (2004a) "Experimental behaviour of FRP-confined slender RC columns under eccentric loading", *Proceedings, Second International Conference on Advanced Polymer Composites in Construction*, University of Surrey, UK, 20-22 April, 203-212.

Tao, Z., Han, L. H. and Zhao, X. L. (2004b). "Behaviour of concrete-filled double skin (CHS inner and CHS outer) steel tubular stub columns and beam-columns", *Journal of Constructional Steel Research*, 60 (8), 1129-1158.

Tavakkolizadeh, M. and Saadatmanesh, H. (2001). "Galvanic corrosion of carbon and steel in aggressive environments", *Journal of Composites for Construction*, ASCE, 5 (3), 200-210.

Tavakkolizadeh, M. and Saadatmanesh, H. (2003). "Strengthening of steel-concrete composite girders using carbon fiber reinforced polymer sheets", *Journal of Structural Engineering*, ASCE, 129 (1), 30-40.

Teng, J. G. (2006). "Fibre-reinforced polymer composites in construction: current research and future challenges", *Proceedings, Ninth International Symposium on Structural Engineering for Young Experts*, 18-21 August, Fuzhou & Xiamen, China.

Teng, J. G. and Hu, Y. M. (2004). "Suppression of local buckling in steel tubes by FRP jacketing", *Proceedings, Second International Conference on FRP Composites in Civil Engineering*, 8-10 December, Adelaide, Australia, 749-753.

Teng, J. G. and Hu, Y. M. (2005). "Enhancement of seismic resistance of steel tubular columns by FRP jacketing", *Proceedings, Third International Conference on Composites in Construction*, 11-13 July, Lyon, France, 307-314.

Teng, J. G. and Hu, Y. M. (2006a). "Behaviour of FRP-jacketed circular steel tubes and cylindrical shells under axial compression", *Construction and Building Materials*, in press.

Teng, J. G. and Hu, Y. M. (2006b). "Theoretical model for FRP-confined circular concrete-filled steel tubes under axial compression", *Proceedings, Third International Conference on FRP Composites in Civil Engineering*, 13-15 December, Miami, Florida, USA.

Teng, J. G. and Jiang, T. (2006). "Design of FRP jackets for the strengthening of RC columns", in preparation.

Teng, J. G. and Lam, L. (2004). "Behavior and modeling of fiber reinforced polymer-confined concrete", *Journal of Structural Engineering*, ASCE, 130 (11), 1713-1723.

Teng, J. G. and Yao, J. (2005). "Plate end debonding failures of FRP- or steel-plated RC beams: a new strength model", *Proceedings, International Symposium on Bond Behaviour of FRP in Structures*, 7-9 December, Hong Kong, China, 291-298.

Teng, J. G. and Yao, J. (2006). "Plate end debonding in FRP-plated RC beams-II: Strength model", submitted for publication.

Teng, J. G., Chen, J. F., Smith, S. T. and Lam, L. (2002). *FRP Strengthened RC Structures*. John Wiley & Sons, Ltd., England.

Teng, J. G., Ko, J. M., Chan, T. H. T., Ni, Y. Q., Xu, Y. L., Chan, S. L., Chau, K. T. and Yin, J. H. (2003a). "Third-generation structures: intelligent high-performance structures for sustainable urban systems", *Proceedings, International Symposium on Diagnosis, Treatment and Regeneration for Sustainable Urban Systems*, 13-14 March, Japan, 41-55.

Teng, J. G., Chen, J. F., Smith, S. T. and Lam L. (2003b). "Behaviour and strength of FRP-strengthened RC structures: a state-of-the-art review", *ICE Proceedings: Structures and Buildings*, 156 (1), 51-62.

Teng, J. G., Smith, S. T., Yao, J. and Chen, J. F. (2003c). "Intermediate crack-induced debonding in RC beams and slabs", *Construction and Building Materials*, 17 (6&7), 447-462.

Teng, J. G., Lam, L. and Chen, J. F. (2004a). "Shear strengthening of RC beams using FRP composites", *Progress in Structural Engineering and Materials*, 6, 173-184.

Teng, J. G., Yu, T. and Wong, Y. L. (2004b). "Behaviour of hybrid FRP-concrete-steel double-skin tubular columns", *Proceedings, Second International Conference on FRP Composites in Civil Engineering*, 8-10 December, Adelaide, Australia, 811-818.

Teng, J. G., De Lorenzis, L., Wang, B., Li, R., Wong, T. N. and Lam, L. (2006a). "Debonding failures of RC beams strengthened with near surface mounted CFRP strips", *Journal of Composites for Construction*, ASCE, 10 (2), 92-105.

Teng, J. G., Yuan, H. and Chen, J. F. (2006b). "FRP-to-concrete interfaces between two adjacent cracks: theoretical model for debonding failure", *International Journal of Solids and Structures*, 43 (18&19), 5750-5778.

Teng, J. G., Huang, Y. L., Lam, L., Ye, L. P. (2006c). "Theoretical model for fiber reinforced polymer-confined concrete". *Journal of Composites for Construction*, ASCE, in press.

Teng, J. G., Yu, T., Wong, Y. L. and Dong, S. L. (2006d). "Hybrid FRP-concrete-steel tubular columns: concept and behaviour", *Construction and Building Materials*, in press.

Teng, J. G., Jiang, T., Lam, L. and Luo, Y. Z. (2006e). "Refinement of a design-oriented stress-strain model for FRP-confined concrete", in preparation.

Wei, S., Mau, S. T., Vipulanandan, C. and Mantrala, S. K. (1995). "Performance of new sandwich tube under axial loading: experiment", *Journal of structural Engineering*, ASCE, 121 (12), 1806-1814.

Wu, Z. S., Iwashita, K., Hayashi, K., Higuchi, T., Murakami, S. and Koseki, Y. (2003). "Strengthening prestressed-concrete girders with externally prestressed PBO fiber reinforced polymer sheets", *Journal of Reinforced Plastics and Composites*, 22 (14), 1269-1286.

Xia, S. H. and Teng, J. G. (2005). "Behaviour of FRP-to-steel bonded joints", *Proceedings, International Symposium on Bond Behaviour of FRP in Structures*, 7-9 December, Hong Kong, China, 419-426.

Xiao, Y. (2004). "Application of FRP composites in concrete columns", *Advances in Structural Engineering*, 7 (4), 335-341.

Xiao, Y., He, W. H. and Choi, K. K. (2005). "Confined concrete-filled tubular columns", *Journal of Structural Engineering*, ASCE, 131 (3), 488-497.

Yao, J. (2005). *Debonding Failures in RC Beams and Slabs Strengthened with FRP Plates*. PhD Thesis, The Hong Kong Polytechnic University.

Yao, J. and Teng, J. G. (2006). "Plate end debonding in FRP-plated RC beams-I: Experiments", submitted for publication.

Yao, J., Teng, J. G. and Lam, L. (2005a). "Experimental study on intermediate crack debonding in FRP-strengthened RC flexural members", *Advances in Structural Engineering*, 8 (4), 365-396.

Yao, J., Teng, J. G. and Chen, J. F. (2005b). "Experimental study on FRP-to-concrete bonded joints", *Composites-Part B: Engineering*, 36 (2), 99-113.

Ye, L. P., Lu, X. Z. and Chen, J. F. (2005). "Design proposals for the debonding strengths of FRP strengthened RC beams in the Chinese design code", *Proceedings, International Symposium on Bond Behaviour of FRP in Structures*, Hong Kong, China, 45-54.

Yu, T. (2006). *Structural Behavior of Hybrid FRP-Concrete-Steel Double-Skin Tubular Columns*, PhD Thesis, The Hong Kong Polytechnic University.

Yu, T. and Teng, J. G. (2006). "Structural design of concrete-filled FRP tubular members", in preparation.

Yu, T., Wong, Y. L., Teng, J. G., Dong, S. L. and Lam, E. S. S. (2006). "Flexural behavior of hybrid FRP-concrete-steel double skin tubular members", *Journal of Composites for Construction*, ASCE, 10 (5), 443-452.

Yuan, H., Teng, J. G., Seracino, R., Wu, Z. S. and Yao, J. (2004). "Full-range behavior of FRP-to-concrete bonded joints", *Engineering Structures*, 26 (5), 553-565.

Zhang, L., Hollaway, L. C., Teng, J. G. and Zhang, S. S. (2006). "Strengthening of steel bridges under low frequency vibrations", *Proceedings, Third International Conference on FRP Composites in Civil Engineering*, 13-15 December, Miami, Florida, USA.

Zhang, L. and Teng, J. G. (2006). "Elastic flexural-torsional buckling of beams strengthened with bonded plates" in preparation.

Zhao, X. L. and Han, L. H. (2006). "Double skin composite construction", *Progress in Structural Engineer-*

ing and Materials, 8 (3), 93-102.

Zhao, X. L., Fernando, D. and Al-Mahaidi, R. (2006). "CFRP Strengthened RHS subjected to transverse end bearing force", *Engineering Structures*, 28 (11), 555-1565.

Zhao, X. L., Grzebieta, R. and Elchalakani, M. (2002). "Tests of Concrete-Filled Double Skin CHS Composite Stub Columns", *Steel and Composite Structures*, 2 (2), 129-146.

International Forum on Advances in Structural Engineering

IFASE 2006

首届"结构工程新进展国际论坛"简介

论坛主题——新型结构材料与体系

会议地点：北京
会议时间：2006年11月12日至13日
支持单位：中华人民共和国建设部
中华人民共和国建设部科学技术司

主办单位： 中国建筑工业出版社

 同济大学《建筑钢结构进展》编辑部

 香港理工大学《结构工程进展》编委会

承办单位： 清华大学土木工程系

主 持 人： 清 华 大 学　韩林海 教授
香港理工大学　滕锦光 教授

关 于 论 坛

"结构工程新进展国际论坛"旨在促进我国结构工程界对学术成果和工程经验的总结及交流,汇集国内外结构工程各方面的最新科研信息,提高专业学术水平,推动我国建筑行业科技发展。论坛划分为四个领域,以每年一个主题的形式轮流出现。分别为:

- 新型结构材料与体系(包括组合结构、膜结构等)
- 钢结构(包括空间结构、大跨度结构)
- 混凝土结构
- 结构防灾、监测与控制(包括抗震、抗风、抗火等)

论坛主题将每四年一个轮回,循环往复,充分而及时地体现研究成果的更新与交替。论坛有两大特点。其一,论坛会议采取特邀报告人专题演讲的形式,其人选在81位包括院士、长江学者、国家杰出青年基金获得者及业内专家学者在内所组成的论坛学术委员会推荐的基础上产生,从而保证演讲内容的学术性、代表性及领先水平。其二,"论坛文集"的学术定位:文集由特邀报告人对演讲论文进行深度展开或延伸,充分表达其学术成果,按主题每年一册,于论坛召开之前由中国建筑工业出版社正式出版。论坛文集不仅是论坛的重要标志,更重要的是,它将成为结构工程研究领域的风向标,为教学科研人员、在校研究生提供一个既有前瞻性又有可行性的、能引导研究方向的重要文献资料。同时,该书的连续出版,也将成为结构工程科研历程的全记录,对整个研究工作的沿革起着承上启下的作用。

关于首届论坛

首届论坛的主题为"新型结构材料与体系"。特邀报告人针对索穹顶结构、膜结构、钢—混凝土组合结构、钢管混凝土结构、FRP在结构工程中的应用、新型结构材料等研究领域,阐述其变革及发展的最新信息。给与会者提供一个与专家学者互动并且获取宝贵经验的直接机会。

组织委员会

主　任：仇保兴（建设部 副部长）
　　　　袁　驷（全国人大常委，清华大学土木水利学院院长，长江学者特聘教授）
副主任：赖　明（建设部科学技术司　司长）
　　　　赵　晨（中国建筑工业出版社　社长）
　　　　李国强（同济大学副校长，《建筑钢结构进展》主编）
　　　　滕锦光（香港理工大学协理副校长，《结构工程进展》主编）
　　　　韩林海（清华大学土木工程系　教授）
委　员：胡永旭（中国建筑工业出版社　副总编辑兼社长助理）
　　　　王静峰（清华大学土木工程系　博士后）
　　　　陆　烨　何亚楣　刘玉姝　（同济大学）
　　　　赵梦梅　刘婷婷　　　　　（中国建筑工业出版社）

学术委员会

丁大钧	仇保兴	牛荻涛	王永昌（英国）	王光远
王国周	叶可明	叶列平	石永久	龙驭球
刘锡良	吕西林	吕志涛	孙利民	江亿
江欢成	吴波	吴智深（日本）	张佑启	张慕胜（中国香港）
李杰	李惠	李宏男	李国强	李宗津（中国香港）
李忠献	李秋胜（中国香港）	李爱群	杨永斌（中国台湾）	汪大绥
沈世钊	沈祖炎	肖岩（美国）	陈建飞（英国）	陈绍蕃
陈惠发（美国）	陈锦顺（中国香港）	陈肇元	周绪红	周福霖
周锡元	岳清瑞	欧进萍	范立础	金伟良
柯长华	胡庆昌	赵国藩	赵晓林（澳大利亚）	郝洪（澳大利亚）
钟善桐	项海帆	容柏生	徐建	徐磊（加拿大）
徐幼麟（中国香港）	徐有邻	徐培福	柴昶	聂建国
袁驷	钱七虎	顾明	顾强	高赞明（中国香港）
崔鸿超	梁以德（中国香港）	傅学怡	曾庆元	程懋堃
葛汉彬（日本）	董石麟	谢礼立	韩林海	蓝天
赖明	鲍广鑑	蔡克铨（中国台湾）	蔡春声（美国）	蔡益燕
滕锦光（中国香港）				

首届论坛特邀报告人简介

（按论坛报告顺序）

袁　驷　清华大学土木水利学院院长、教授、博士生导师。教育部长江学者特聘教授。袁教授于1989年创立了有限元线法，并对其作了系统的开发与发展；1993年出版了独著的国内外首部有关该法的英文专著《The Finite Element Method of Lines》。已发表学术论文150余篇，多篇被SCI、EI收录。先后获国家教委科技进步一、二、三等奖（第二完成人）各一项。1991年获"做出突出贡献的中国博士"，1995年获宝钢教育奖优秀教师特等奖，1996年获国家杰出青年科学基金，1996年首批入选"国家百、千、万人才工程"第一、二层人选，1997年获北京市优秀教师等奖励，1998年获中国土木工程学会第三届优秀论文二等奖第一名。2000年被聘为国家教育部特聘教授。2001年分别获北京市和国家级教学成果一等奖，2002年分别获全国高等学校优秀教材奖一等奖和二等奖各一项。

董石麟　中国工程院院士，空间结构工程专家，浙江大学建筑工程学院院长，教授、博士生导师。董院士长期从事薄壳结构、网架结构、网壳结构、塔桅结构等空间结构及升板结构的科研与教学工作，同时主持设计了大量重要空间结构工程。代表性工程有北京325米大气污染监测塔、新乡百货大楼组合网架楼层结构、南海观音大佛多层多跨高耸网架、首都体育馆等。其科研成果十八次荣获国家级科技进步三等奖及省部级科技进步一、二等奖，主要成果有：新型空间结构的强度、稳定性和动力性能的研究（国家科技进步三等奖，排名第1）、多层大跨建筑组合网架楼层结构应用技术（国家科技进步三等奖，排名第1）、空间网格结构CAD系统的研制及其工程应用（国家科技进步三等奖，排名第2）。发表专著两部、学术论文150余篇，还共同主持了数本国家规范规程的编制和修订工作。于1994年创办了《空间结构》杂志并担任主编至今。

张其林　同济大学土木工程学院副院长，教授、博士生导师。张教授主要从事钢结构、索结构、膜结构、幕墙结构等专业的研究和教学工作。主要著作有《索和膜结构》、《轻型门式刚架结构》，译著《钢管截面的结构应用》，并在国内外期刊和学术会议发表论文100余篇。编制和参编了中国建设标准化协会规程《点支式玻璃幕墙技术规程》、《索膜结构设计规程》，以及上海市标准《膜结构技术标准》、《建筑结构用索技术标准》等。组织研制了大型钢结构CAD软件3D3S。获教育部、上海市科技进步二等奖各一项、三等奖各一项。讲授《钢结构》、《结构稳定理论》、《索膜结构》等课程。代表性设计工程"三亚美丽之冠会展中心"、"郑州杂技馆"等。

孙　伟　中国工程院院士，土木工程材料专家，现任东南大学纤维与纤维混凝土技术研究所所长，教授、博士生导师。孙院士长期从事土木工程材料领域的教学、科研和人才培养工作。先后承担了国家级、省部级、国家自然科学基金重点项目、重大工程项目和国际或境外合作项目40多项。获国家发明三等奖"钢纤维混凝土路面性能设计与施工技术"一项、江苏省科技进步一等奖"高耐久混凝土的评价与失效机理及寿命预测"一项、教育部科技进步一等奖"钢纤维混凝土结构设计与施工规程"一项、省部级自然科学和科技进步二等奖六项。发表论文300多篇，其中SCI、EI收录126篇次，SCI引用100多篇次。

李宗津　香港科技大学土木系教授及工学院副院长，香港工程师协会注册工程师。国际标准化组织TC71技术委员会委员，美国混凝土协会中国分会创始主席。李教授在水泥基绿色建材及非破损检验领域做了大量研究工作。研究范围包括早龄期混凝土性能的测定，纤维与水泥及骨料与水泥界面的性能研究，素混凝土及纤维混凝土的断裂性能研究，高性能混凝土，以挤压技术开发建筑型材，功能材料的研发，以及开发适用于土木工程的非破损检验技术。共得到超过2800万港币的科研经费，已发表180余篇论文及著作，其中被SCI、EI收录的有100余篇。获得5项专利，其中3项已被开发成产品。

叶列平　清华大学土木工程系教授、博士生导师。叶教授主要从事混凝土结构基本理论、钢骨混凝土结构、高强混凝土结构、预应力混凝土结构、高层建筑结构抗震与减震分析理论和设计方法、纤维增强复合材料及其工程应用、混凝土结构加固技术和计算理论等领域的研究工作。负责国家自然科学基金、国家自然科学基金重点项目、国家杰出青年海外青年学者合作基金（国内合作者）、国家高技术研究发展计划（863计划）项目子项各1项和30余项科研项目。发表论文120余篇，SCI收录9篇，ISTP收录11篇，EI收录29篇，独立编著和参加编著5本专著和教材。获1995年北京市优秀青年教师，2000年宝钢教育优秀教师奖，教育部自然科学一等奖（排名第1）等奖励。

聂建国　清华大学土木工程系教授，结构工程研究所所长，结构工程与振动教育部重点实验室主任。教育部长江学者特聘教授。聂教授的主要研究领域为钢—混凝土组合结构。主要学术成就包括作为第一完成人获国家科技进步二等奖1项，作为第一、二完成人分别获省（部）级科技进步一、二等奖6项。2005年获全国优秀博士后称号。独著专著1本，主编教材1本。作为第一、二作者在《Journal of Structural Engineering – ASCE》、《土木工程学报》等期刊发表学术论文132篇，发表会议论文38篇。研究成果中有多项已经被国家相关标准、规程所采纳，在工程中获得了成功的应用并取得了显著的技术经济效益和社会效益。

韩林海 清华大学土木工程系教授、博士生导师。主要研究领域为组合结构与混合结构、工程防火、钢结构。韩教授为首批"国家百千万人才工程"第一、二层次人选（1996年），建设部"有突出贡献的中青年专家"（1997年），国务院特贴专家（1997年）；曾获霍英东教育基金（1996年）和霍英东教育基金会高校青年教师奖（2004年）；2004年获国家杰出青年科学基金资助。出版学术专著3部，在国内外重要学术期刊上发表论文100多篇（其中SCI收录26篇）。论著近年被他引900多次（其中SCI期刊他引31次）。部分成果为10余部工程建设标准所采纳，并在一些典型工程中应用。曾以第一、二获奖人获国家教委科技进步一等奖等科技奖励五次。

Brian Uy 澳大利亚伍伦贡大学（University of Wollongong）土木、采矿与环境工程学院院长，钢结构教授。从事钢结构和钢—混凝土组合结构的研究已超过15年，发表论文200多篇。许多研究得到澳大利亚研究委员会和工业界的资助。现任澳大利亚组合结构标准委员会、美国钢结构协会（AISC）组合结构委员会、美国土木工程师协会（ASCE）组合结构技术委员会，以及国际桥梁及结构工程师协会（IABSE）第二工作委员会（钢、木及组合结构）委员。同时担任钢和组合结构领域内7本主要国际期刊的编委。曾任2005年颁布的香港钢结构规范的国际顾问。

Aftab A. Mufti 加拿大曼尼托巴大学（University of Manitoba）土木工程教授，加拿大新型结构及智能监测研究网络（ISIS）主席及项目负责人，国际智能结构健康监测协会（ISHMII）主席。研究领域包括：FRP、光纤传感器、有限元、桥梁工程，以及结构健康监测和诊断。已出版9本专著，发表了200多篇学术论文。Mufti博士曾获得16次奖励，其中包括卓有声望的国际Mirko–Rǒs奖（杰出个人终身成就奖）及2005年的NSERC团队合作奖。作为主要开发者提出了无钢筋混凝土桥面板的概念，并且在该方面拥有多项专利。此外，还担任加拿大公路桥梁设计规范技术委员会的技术分会主席，该分会最近完成了有关FRP材料应用设计条文的编写工作（包括新建桥梁及桥梁加固的设计）。

滕锦光 香港理工大学协理副校长，结构工程讲座教授。国际土木工程FRP学会（IIFC）主席，国际学术期刊《Advances in Structural Engineering》主编，美国土木工程师协会《Journal of Structural Engineering》副主编。滕教授的主要研究领域包括FRP复合材料在结构工程中的应用、钢结构、薄壳结构及结构力学。发表学术论文约300篇，包括SCI国际学术期刊论文约110篇，并为英文专著《FRP加固混凝土结构》的第一作者。研究成果被其他学者以及相关规范/规程大量引用，仅2006年首7个月SCI他引次数已达60余次。所做研究工作曾在国内外多次获奖，包括美国土木工程师协会（ASCE）优秀论文奖，英国土木工程师学会Howard奖章，（国际）实验力学学会Harting奖，国家杰出青年科学基金等。